14th International Workshop on Slow Positron Beam Techniques and Applications 2016 (SLOPOS14)

Journal of Physics: Conference Series Volume 791

Matsue, Japan
22 – 27 May 2016

Editors:

Masanori Fujinami
Yasuyuki Nagashima
Brian E. O'Rourke

ISBN: 978-1-5108-3693-8
ISSN: 1742-6588

Printed from e-media with permission by:

Curran Associates, Inc.
57 Morehouse Lane
Red Hook, NY 12571

CURRAN ASSOCIATES INC. proceedings.com

Some format issues inherent in the e-media version may also appear in this print version.

Printed by Curran Associates, Inc. (2017)

For permission requests, please contact the Institute of Physics
at the address below.

Institute of Physics
Dirac House, Temple Back
Bristol BS1 6BE UK

Phone: 44 1 17 929 7481
Fax: 44 1 17 920 0979

techtracking@iop.org

Additional copies of this publication are available from:

Curran Associates, Inc.
57 Morehouse Lane
Red Hook, NY 12571 USA
Phone: 845-758-0400
Fax: 845-758-2633
Email: curran@proceedings.com
Web: www.proceedings.com

Table of contents

Volume 791

14th International Workshop on Slow Positron Beam Techniques & Applications
22–27 May 2016, Matsue, Japan

Accepted papers received: 8 December 2016
Published online: 22 February 2017

Preface

Papers

IOP Conf. Series: Journal of Physics: Conf. Series **791** (2017) 011001 doi:10.1088/1742-6596/791/1/011001

Preface

These proceedings arose from the 14th International Workshop on Slow Positron Beam Techniques (SLOPOS14), which was held at Kunibiki Messe, Matsue, Shimane prefecture, Japan, from the 22nd–27th May 2016. Meetings in the SLOPOS series are held every three years.

The SLOPOS workshop series has traditionally been devoted to investigations on the production of positron and positronium beams, their fundamental physics and chemistry, and their applications to materials such as metals, semiconductors and soft matter. During the workshop numerous applications using positron and positronium beams were also presented, clearly demonstrating the usefulness of such beams to the determination of surface structure, defect characterization as well as to fundamental scientific studies.

For SLOPOS14 the main subjects of the workshop included the following:

- Positron transport and beam technology
- Pulsed positron beams and positron traps
- Defect profiling in bulk and layered structures
- Nano structures, porous materials and thin films
- Surface and interface analysis
- Positronium formation, emission and beams
- Positron and positronium interactions with atoms and molecules
- Many positrons and anti-hydrogen
- Improvement of experimental techniques

106 delegates from 14 countries participated in the SLOPOS14, including 31 student delegates, which was a most encouraging sign for the future. The scientific program comprised 5 plenary talks, 22 invited talks, 32 contributed talks and 46 posters presented during two poster sessions. Student prizes were awarded for the best presented scientific contributions by 4 students from University of College London, Universität der Bundeswehr München and The University of Tokyo.

On a sad note, delegates paid tribute to the contributions of our recently deceased colleagues, Prof. A. Seeger, Prof. R.N. West, Prof. T.C. Griffith, and Prof. Z. Tang. Memorial talks were given and a one minute silence was observed before the opening session of workshop in their memory.

The workshop would not have been possible without the hard work of the Local Organizing Committee and the International Organizing Committee. We would also like to express our gratitude to the students from Chiba University for their hospitality and efficient support during the workshop.

Finally we would like to thank all of the attendees of SLOPOS14 for their scientific contributions to the workshop, and for fruitful discussions in an informal atmosphere.

IOP Conf. Series: Journal of Physics: Conf. Series **791** (2017) 011001 doi:10.1088/1742-6596/791/1/011001

SLOPOS15 will be held in Romania in 2019 and we are all greatly looking forward to the occasion.

Masanori Fujinami, Yasuyuki Nagashima, and Brian E. O'Rourke
Guest Editors

December 2016

Local Organizing Committee of SLOPOS14

- Masanori Fujinami (Chair) *Chiba University*
- Kenji Ito *AIST*
- Yasuyuki Nagashima *Tokyo University of Science*
- Brian O'Rourke *AIST*
- Nagayasu Oshima *AIST*
- Ryoichi Suzuki *AIST*
- Ken Wada *QST*
- Chiba University team

International Organizing Committee

- G. Amarendra *Indira Gandhi Centre for Atomic Research* (India)
- M.-F. Barthe *CEMHTI-CNRS* (France)
- R.S. Brusa *University of Trento*(Italy)
- S. Buckman *Australian National University* (Australia)
- D. Cassidy *University College London* (UK)
- C. Corbel *CEMHTI-CNRS* (France)
- M. Fujinami *Chiba University* (Japan)
- C. Hugenschmidt *Technische Universität München* (Germany)
- R. Krause-Rehberg *Universität Halle* (Germany)
- K. Lynn *Washington State University* (USA)
- H. Schut *Delft University of Technology* (Netherland)
- R. Suzuki *AIST* (Japan)
- F. Tuomisto *Aalto University* (Finland)
- A.H. Weiss *The University of Texas at Arlington* (USA)

IOP Conf. Series: Journal of Physics: Conf. Series **791** (2017) 011001 doi:10.1088/1742-6596/791/1/011001

Delegates of SLOPOS14, 22-27 May 2016, Matsue, Japan

Peer review statement

All papers published in this volume of *Journal of Physics: Conference Series* have been peer reviewed through processes administered by the proceedings Editors. Reviews were conducted by expert referees to the professional and scientific standards expected of a proceedings journal published by IOP Publishing.

IOP Conf. Series: Journal of Physics: Conf. Series **791** (2017) 012001 doi:10.1088/1742-6596/791/1/012001

Symphonic use of quantum beams for materials science -now and future-

K Yamada[1] and H Hiraka[2]

[1] Institute of Materials Structure Science, KEK, Oho1-1, Tsukuba, Ibaraki 305-0801, Japan
[2] Graduate School of Science and Engineering, Ibaraki University, 2-1-1 Bunkyo, Mito, Ibaraki 310-8512, Japan

E-mail: yamadak@post.kek.jp

Abstract.

Symphonic use of multiple probes for materials science is discussed through two recent studies on high-T_c superconductors, and a future prospect of the symphonic use is given. For a hydrogen-doped iron-based superconductor, which shows a double-peaked superconducting dome as a function of hydrogen concentration, symphonic use of neutron, muon and synchrotron *x*-ray probes discovered a second antiferromagnetic (AF) phase and its contrasting feature compared to the first AF phase. A drastic change of the AF order by Ni impurities in slightly hole-doped La_2CuO_4, a Mott insulator, has been studied by using neutron and synchrotron *x*-ray probes. The symphonic use of these probes clarified the microscopic dual electronic states for both Cu and Ni in this system.

1. Introduction

Recent research in materials science has widely utilized quantum beams such as synchrotron *x*-ray, neutron and other beams to understand the origin of novel properties. In particular, many researches on novel materials such as multiferroic materials and high-T_c superconductors require "symphonic use" of quantum beam probes or multi-probe use (MPU). There are two main reasons for such MPU. One is the recent development of quantum beam techniques. For example, an energy tunable *x*-ray beam with high brilliance and intensity is produced in advanced synchrotron facilities. Also, intense pulsed neutron beams are produced in spallation neutron facilities. Then these quantum beams can provide detailed information on crystal, magnetic and electronic structures of a wide variety of materials. The other reason is complicate coexistence of hierarchical structures with different length, energy and temporal scales in novel materials. Such coexistence is possibly related with inter-coupled electronic, vibrational and magnetic properties in novel materials. Therefore, only single probe use is not enough to obtain invaluable information on these multi-scale materials. In general, use of different types of beam is done one by one. On the other hand, future MPU of different types of probe or symphonic use will be done simultaneously and the information by each probe can be exchanged and can affect running experiments. MPU then enhances the effectiveness of each experiment, helps to understand unknown phenomenon, and is suitable for heterogeneous multi-scale systems. In the following session, we present several examples of MPU.

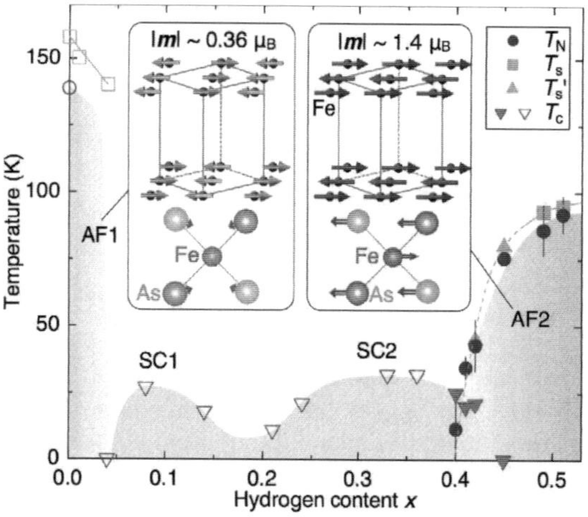

Figure 1. Magnetic and superconducting phase of hydrogen-doped iron-based superconductor (reprinted with permission from Macmillan Publishers Ltd: Nature Physics [1], copyright 2014). AF1 and AF2 denote antiferromagnetic phases determined by muon and neutron beams. SC1 and SC2 denote superconducting phases. Insets show arrangements of Fe spins in each magnetic phase (above) by neutron beam and atomic displacements of Fe and As (below) at structural phase transitions above the magnetic transition determined by x-ray beam.

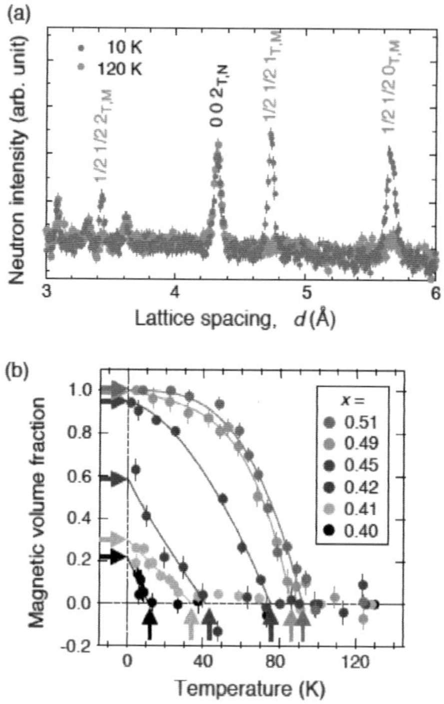

Figure 2. (a) Pulsed neutron powder diffraction pattern of $LaFeAsO_{1-x}H_x$ with $x = 0.51$ (reprinted with permission from Macmillan Publishers Ltd: Nature Physics [1], copyright 2014). The magnetic reflections of $(1/2, 1/2, n)_{T,M}$ ($n = 0, 1, 2$) were observed clearly at 10 K, which disappears at ~120 K. (b) Temperature dependence of magnetic volume fraction for samples in the range of hydrogen concentration $x = 0.40 \sim 0.51$.

2. Examples of symphonic use for superconducting materials

2.1 Discovery of the second magnetic phase of an iron-based superconductor

Role of magnetism is one of the most important and longstanding issues for the novel superconductors. In these systems, the superconductivity appears by destroying a magnetic phase in a parent compound by carrier doping or band filling control. In that sense, a magnetic phase transforms to a superconducting phase. A recent study reported an advanced doping method using a hydrogen anion instead of fluorine in the La1111 type iron-based superconductor that has surpassed the doping-limit of fluorine, and uncovered the concealed second superconducting dome (SC2), in addition to the first superconducting dome (SC1) [1]. If the origin of two superconducting domes is different or two distinct superconducting phases exist, the existence of a second magnetic phase is highly suggested. Then to search for a hidden magnetic phase started beyond the SC2 doping region, MPU has been performed using neutron, muon, and x-ray beams.

Shortly after the start of the MPU project, a second anti-ferromagnetic phase (AF2) was discovered and the overall phase diagram of AF2 and the detailed properties were determined (Figure 1). In the project, muon spins injected in the hydrogen-doped sample first detected a local magnetic field developing at low temperature, which strongly suggests the existence of magnetic order for $x > 0.4$ in Figure 1. Then neutron scattering confirmed a long range magnetic order and determined the spin structure as well as the size of the ordered moment, both of them were found to be different between AF1 and AF2 as shown in Figure. 1. Just above the magnetic phase transition in AF2 synchrotron *x*-ray diffraction observed a structural phase transition and determined the anomalously large atomic shifts at the transition.

Furthermore, the muon experiment strongly suggests a heterogeneous feature near the boundary of SC2 and AF2 due to the imperfect antiferromagnetic volume fraction at the base temperature (Figure. 2 (b)). On the other hand, the peak-width of the magnetic Bragg reflection by neutron diffraction experiment is instrumental resolution-limited (Figure. 2(a)) at $x = 0.51$, which means the length scale of the magnetically ordered region is larger than 10~100 nm. A neutron diffraction experiment at $x \sim 0.4$ will clarify if the coexistence between magnetic and superconducting phases near the boundary is microscopic or mesoscopic in nature.

As shown in this topic, MPU made it possible to discover a new magnetic phase (AF2) and to extract the contrasting feature compared to AF1. Scientifically, the discovery of AF2 reconfirmed the appearance of magnetic and superconducting phases in pairs. Also, the contrasting feature between AF1 and AF2 strongly suggests contrasting superconducting property and origin between SC1 and SC2.

In general, each probe in a MPU study not only contributes to solve a target issue but also sometimes provides supplemental information by applying the unique feature of each probe. For this example, high sensitivity to hydrogen atoms of neutron beam discovered a tunneling motion of the doped hydrogen atoms in the iron-based superconductor. Interestingly, the tunneling motion was found to be drastically affected by the opening of superconducting gap. Therefore, the detail analysis of the tunneling motion can provide valuable information on the superconducting gap [2].

2.2 Ni impurity effect in a doped cuprate superconductor

MPU plays important roles to understand unusual phenomena. It has been well known that a tiny amount (~1%) of hole-carrier doped into La_2CuO_4, an antiferromagnetic Mott insulator, drastically destroys the antiferromagnetic order. Such effect has been believed to occur by forming a so-called

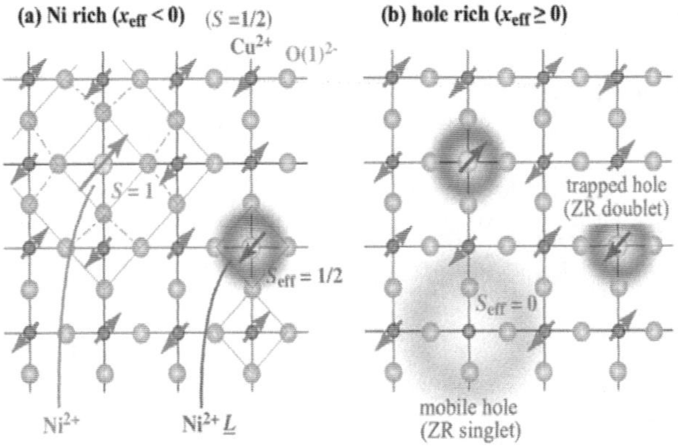

Figure 3. Schematics of the Ni-doped CuO_2 plane in the (a) Ni-rich and (b) hole-rich groups (reprinted with permission from [5], copyright 2009, the American Physical Society). The red (blue) clouds around Ni (Cu) represent spatially trapped (mobile) holes. Spin direction of Ni^{2+} ($S = 1$) in (a) is not confirmed experimentally. With increasing the number of Ni^{2+} the spin direction should rotate 90 degrees and form a spin structure of La_2NiO_4.

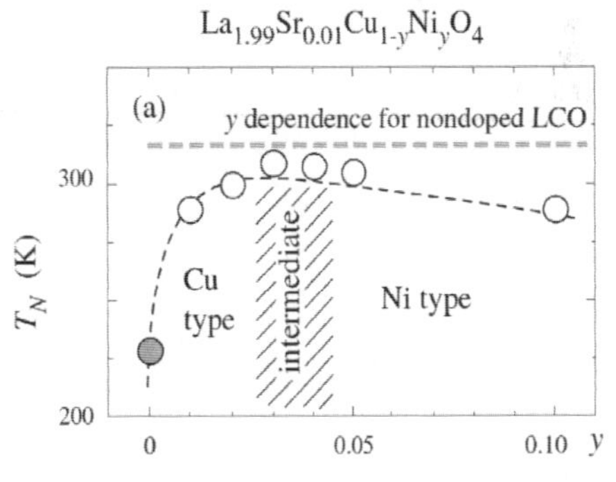

Figure 4. Magnetic phase-diagram of $La_2Cu_{1-y}Ni_yO_4$ determined by neutron scattering (reprinted with permission from [4], copyright 2005, the Physical Society of Japan). The data point at $y = 0$ (closed circle) was taken from a magnetization measurement. The curved broken line presents results from susceptibility measurements [3] and the thick broken line represents T_N for $La_2Cu_{1-y}Ni_yO_4$.

Zhang-Rice (ZR) singlet state which rapidly moving on the antiferromagnetic ordered CuO_2 plane (Figure 3(b)). However, a fascinating phenomenon was discovered when a tiny amount of Ni-impurity is compensated for the doped hole in the magnetically diminished samples. In contrast to common sense Ni-impurities recover the magnetic order! This recovery only occurs in hole-doped samples. Ni-impurities in non-doped La_2CuO_4 weakly affect the magnetic order due to the dilution effect of magnetic moment (as shown by a broken line in Figure. 4). The first observation of this effect was made by a magnetization measurement [3]. But there was no interpretation for the reason. Then a neutron diffraction measurement has been performed using single crystals of $La_{1.99}Sr_{0.01}Cu_{1-y}Ni_yO_4$ [4]. The neutron scattering not only reconfirmed the recovery of magnetic order by Ni but also discovered the change in the magnetic structure with increasing Ni concentration. A switch of spin structure from the La_2CuO_4 type to the La_2NiO_4 type occurs by a spin rotation in CuO_2 planes by 90 degrees. This effect strongly suggests that for $y > 0.05$ the Ni spin direction should be equal to that of La_2NiO_4 and each Ni spin has $S = 1$ (Ni^{2+}). Then the question is the Ni spin state for the dilute region $y < 0.03$ where the magnetic order recovers with Ni impurity. A hypothesis to explain the effect is the "blotting paper effect", when the substituted Ni absorbs or traps a hole, the effective Ni spin state changes to $S_{eff} = 1/2$ (Ni^{3+}, more exactly $Ni^{2+}L$, where L represents a ligand hole bound on oxygen atoms around Ni), the same as Cu^{2+}. The local spin direction at the Ni-site is also expected to be identical to that of Cu. Thereby, the hole-trapped Ni decreases the number of ZR singlets and recovers the magnetic order (Figure 3). In order to experimentally confirm this hypothesis local information on the electronic state of Ni is indispensable. Then x-ray-absorption fine structure (XAFS) experiment to study the element-selective electronic as well as crystal structure around a Ni dopant has been performed using single crystals over a wide hole-doping range. Then the near-edge structure is found to be consistent with that of Ni^{2+} for the Ni rich and $Ni^{2+}L$ for hole-rich samples. Furthermore, as shown in Figure 5, the two types of extended XAFS (EXAFS) pattern were obtained only for the x-ray polarization parallel to CuO_2 plane. The two types of EXAFS pattern correspond to two types of in-plane Ni-O bond length as shown in Figure 6 (a).

Combined with previous results on hole-localization effects ("blotting paper effect") of Ni, two types of charge states are strongly indicated for Ni. This duality disqualifies a magnetic-impurity picture for Ni dopant in the superconducting phase of cuprates. In order to understand the microscopic reason why the "blotting paper effect" occurs by Ni in hole-doped La_2CuO_4 it is important to distinguish the four types local charge state, Cu^{2+}, ZR-singlet, Ni^{2+} and $Ni^{2+}L$. One of the best probes is a resonant inelastic x-ray scattering (RIXS). In fact, a preliminary RIXS experiment has been performed for both Ni-substituted La_2CuO_4 and Cu-substituted La_2NiO_4 [6]. The capability of RIXS to investigate the local excitations around the substituted

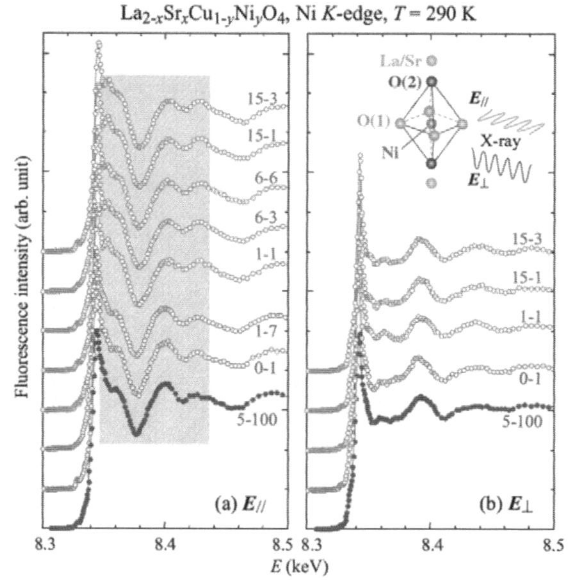

Figure 5. Raw data of polarized XAFS at Ni K-edge of $La_{2-x}Sr_xCu_{1-y}Ni_yO_4$ for polarization of x-ray beam parallel (a) and perpendicular (b) to CuO_2 plane (reprinted with permission from [5], copyright 2009, the American Physical Society). The Ni^{2+}standard sample of (5–100) is included (closed symbol). Here, the hole and Ni concentrations are expressed as ($100x$ -$100y$).

element was demonstrated; that is, electronic excitations in $La_2Cu_{0.95}Ni_{0.05}O_4$ and $La_2Cu_{0.05}Ni_{0.95}O_4$ were successfully observed at the *K*-edge of the substituted element. When the incident photon energy is tuned to the well-screened core hole state, charge transfer excitations from O 2*p* to substituted Cu or Ni 3*d* states appear in the spectra. The edge position of the charge transfer excitations is found to be different between the bulk state and the substitute state even if the same element is targeted by RIXS. These observations strongly suggest that RIXS on the substituted atoms can be useful for exploring their electronic states affected by their surroundings.

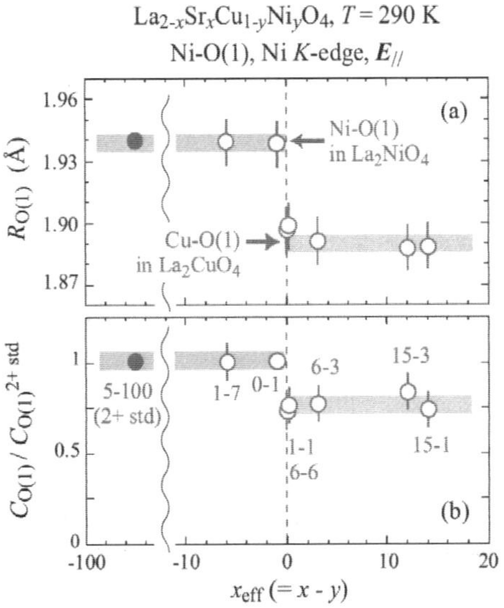

Figure 6. (a) The bond length $R_{O(1)}$ and (b) $C_{O(1)}$ which approximately corresponds to the in-plane oxygen coordination number around Ni (reprinted with permission from [5], copyright 2009, the American Physical Society). $C_{O(1)}{}^{2+std}$ presents the $C_{O(1)}$ for standard sample of (5-100). The blue and red arrows in (a) correspond to the Ni-O(1) distance in La_2NiO_4 and the Cu-O(1) distance in La_2CuO_4, respectively.

3. Future of symphonic use of quantum beams

In the near future, thanks to further development of quantum beam probes, symphonic use of multi-probe will open a new paradigm for materials science. In one way, such development will expand the accessible region in a length-frequency (***Q***, ω) space of each probe. For example, both *x*-ray and neutron inelastic scattering will observe the same elementary excitations in materials such as phonons and magnons with the same instrumental resolution. Then one can discuss an "inner structure" of elementary excitations. For example, since neutrons are sensitive to nuclei while *x*-rays are sensitive to the surrounding electrons, phonons by neutron dominantly reflect the motions of nuclei, while those by *x*-ray reflect the motion of surrounding electrons. Therefore, by combining information from each probe one can extract the inner structure of phonons or, in other words, valuable information on electron-phonon interaction in materials. In the other direction, symphonic use will be carried out to elucidate a phenomenon from multi-direction in (***R***, *t*)-(***Q***, ω) space, for example, elucidation of phonons in heterogeneous multi-scale system requires both information on atomic motion in the local(***R***)-real time (*t*) space and wave-like nature in length-frequency (***Q***, ω) space.

The ultimate symphonic use of multiple probes does not simply mean a combination of various probes but means a new type of research complex where inter-correlated and inter-affected experiments (or sometimes including theories) are performed using different types of probes.

References

[1] Hiraishi M *et al.* 2014 *Nature Physics* **10** 300
[2] Hiraka H *et al.* in preparation
[3] Watanabe N, Masui T, Itoh Y, Machi T, Kato I, Koshizuka N and Murakami M 2003 *Physica C* **388–389** 241
[4] Hiraka H, Machi T, Watanabe N, Itoh Y, Matsuda M and Yamada K 2005 *J. Phys. Soc. Jpn.* **74** 2197
[5] Hiraka H, Matsumura D, Nishihata Y, Mizuki J and Yamada K 2009 *Phys. Rev. Lett.* **102** 037002
[6] Ishii K, Tsutsui K, Ikeuchi K, Jarrige I, Mizuki J, Hiraka H, Yamada K, Tohyama T, Maekawa S, Endoh Y, Ishii H and Cai Y Q 2012 *Phys. Rev. B* **85** 104509

IOP Conf. Series: Journal of Physics: Conf. Series **791** (2017) 012002 doi:10.1088/1742-6596/791/1/012002

Present and future experiments using bright low-energy positron beams

Christoph Hugenschmidt

Heinz Maier-Leibnitz Zentrum (MLZ) and Physik Department E21, Technische Universität München, Lichtenbergstraße 1, 85748 Garching, Germany

E-mail: hugen@frm2.tum.de

Abstract. Bright slow positron beams enable not only experiments with drastically reduced measurement time and improved signal-to-noise ratio but also the realization of novel experimental techniques. In solid state physics and materials science positron beams are usually applied for the depth dependent analysis of vacancy-like defects and their chemical surrounding using positron lifetime and (coincident) Doppler broadening spectroscopy. For surface studies, annihilation induced Auger-electron spectroscopy allows the analysis of the elemental composition in the topmost atomic layer, and the atomic positions at the surface can be determined by positron diffraction with outstanding accuracy. In fundamental research low-energy positron beams are used for the production of e.g. cold positronium or positronium negative ions. All the aforementioned experiments benefit from the high intensity of present positron beam facilities. In this paper, we scrutinize the technical constraints limiting the achievable positron intensity and the available kinetic energy at the sample position. Current efforts and future developments towards the generation of high intensity spin-polarized slow positron beams paving the way for new positron experiments are discussed.

1. Introduction

Mono-energetic positron beams are widely applied in various experiments for fundamental research as well as in materials science, condensed matter and surface physics. As the SLOPOS conference series is devoted to slow positron beam techniques and applications the applied experimental techniques can be subdivided into two categories.

The first category comprises established "bulk" methods of positron annihilation spectroscopy (PAS) which are carried out with low-energy positron beams in order to gain near surface or depth and/or laterally resolved information on lattice defects. Doppler-broadening spectroscopy (DBS) of the positron electron annihilation line as function of positron implantation energy can simply be applied by using a single high-resolution Ge detector. Additional scanning of the beam allows for imaging defect distributions in order to analyze e.g. the spatial variation of the oxygen deficiency in single crystalline $YBa_2Cu_3O_{7-\delta}$ thin films [1]. Coincident DBS (CDBS) provides additional information on the chemical surrounding of open volume defects or the presence of precipitates in thin layers or near the surface. Although it is challenging to achieve a sufficient high timing resolution several facilities are in operation world-wide for depth dependent positron annihilation lifetime spectroscopy (PALS) in order to e.g. characterize the free volume in thin polymers [2] or to identify the species of vacancies in thin oxide films [3]. Only a few measurements of the age-momentum correlation (AMOC), where the positron lifetime and the Doppler-shift are detected simultaneously for each annihilation event, have been performed so far, since a high beam intensity is crucial for this technique. The same holds for the determination of the angular correlation of annihilation radiation (ACAR) in order to study the electronic structure of matter with low energy positron beams [4].

The second group of experiments includes specifically surface and fundamental studies, where positron beams with narrow band width are absolutely required. In surface science positron annihilation induced Auger-electron spectroscopy (PAES) is applied for element selective surface studies of the topmost atomic layer [5], and (total) reflection high-energy positron diffraction ((T)RHEPD) allows surface structure determination with outstanding accuracy [6]. Bright beams are used in atomic physics for positron and positronium (Ps) scattering experiments, and in fundamental research for the production of a neutral positron electron pair plasma [7] and bound leptonic systems such as cold Ps, the positronium negative ion Ps^- [8] or the Ps_2 molecule [9]. An overview of the slow positron beam techniques and recent applications are presented in detail in a comprehensive review on positrons in surface physics [10].

In the past all these techniques have benefited from the development of evermore intense mono-energetic positron beams. On the one hand most experiments profit from a high beam intensity which leads to a drastically reduced measurement time and to a considerably enhanced signal-to-background ratio. In particular, a high positron intensity is beneficial for all kinds of coincidence techniques applied in solid state physics. On the other hand high intensity positron beams with high brightness have enabled the development of novel techniques in fundamental research and new positron applications such as *time-dependent* PAES, e.g. for the observation of surface segregation *in situ* [11]. In addition, intense positron beams are desired for the realization of micro-beams using multiple remoderation for brightness enhancement.

The development of future high-intensity positron beams, however, is challenging in many respects: a spin-polarization to a high degree is desired, the positron energy at the sample should be variable within 0-30 keV, the strength of the positron source should be improved and the beam brightness should be as high as possible, preferably simultaneously. The improvement of positron moderators, in fact, would lead to beams with enhanced brightness as well [12] but is not subject of this paper. In the following the limitations of present positron beam setups will be briefly reviewed before the focus will be put on recent progress at current beam facilities and on developments of future positron sources enabling new positron beam experiments.

2. Coping limitations of beam intensity and brightness

2.1. Limits of positron source strengths

Positrons are basically generated either via β^+ decay or by pair production. In conventional (bulk) PAS and in tabletop beam setups the achievable positron yield is limited by the positron self-absorption in the β^+ source. For this reason, positron beams of *high* intensity, i.e. $> 10^7$ moderated positrons per second [13], are produced by pair production at large scale facilities. In the last decades great efforts have been made to generate positrons by pair production using intense γ sources in order to develop low-energy positron beams with high intensity. Such positron sources use either bremsstrahlung emitted from decelerating electrons or γ radiation released from nuclear processes.

At research reactors, positrons can be generated by the absorption of high energy γ radiation released either from nuclear fission processes or from the de-excitation of excited nuclear states after neutron absorption. The first process is applied at the Delft reactor, where an assembly of thin W foils inside a beamtube is located close to the reactor core in order to absorb the γ radiation from nuclear fission for pair production [14]. The other approach was pursued at the NEutron induced POsitron source MUniCh (NEPOMUC) where the nuclear reaction $^{113}Cd(n,\gamma)^{114}Cd$ generates high-energy prompt γ rays, which are absorbed in Pt foils to provide 10^9 moderated positrons per second (see [15, 16] and references therein). The brightness of the primary 1 keV positron beam is enhanced by a positron remoderator using a W single crystal (Φ^+=-3.0 eV [12]) in back reflection geometry yielding a total efficiency of about 5% and a beam diameter of less than 2 mm (FWHM) in a 6 mT guiding field [17]. The variable kinetic energy of the remoderated beam is usually set to 20 eV for most experiments. It is noteworthy that the brightness of the primary beam at NEPOMUC is similar to that of a realizable ^{22}Na based beam (using a high activity of 50mCi and efficient moderation) but the remoderation leads to a brightness enhancement of at least two orders of magnitude.

For positron production at electron linear accelerators (linacs), high energy bremsstrahlung is released by decelerating electrons in heavy targets (typically Ta or W) in order to convert the γ radiation into positron-electron pairs [18, 19, 20, 21]. Since a linac is intrinsically pulsed the produced positron beam shows a pulsed structure too. At linacs releasing bremsstrahlung up to around 20 MeV the largest fraction of the bremsstrahlung spectrum is absorbed due to the photo and Compton effects. Therefore, elaborate cooling devices have to be applied to dissipate the heat at the converter. For example, at the positron source at the ELBE facility a stack of W plates has to be cooled by a high-pressure water cycle [22].

It has to be emphasized that at both, linacs and reactors, the maximum positron yield is mainly limited by the maximum allowed temperature of the conversion target and the supporting structure caused by the so-called γ-heating. In positron sources based on pair production most of the energy of the γ radiation is dissipated by the photo effect and Compton scattering due to their high cross sections, particularly at lower energy. According to the behavior of the pair production cross section, $\sigma_{pp} \propto Z^2 lnE$, the crossover from energy loss by Compton scattering to pair production is at a photon energy of E=5 MeV for W with atomic number Z=74. As an example, figure 1 shows the product of the relative γ intensity [23] and the respective energy of the γ cascade emitted by the de-excitation of ^{114}Cd. Even assuming complete absorption in W, the maximum fraction of γ's contributing to pair production is only of secondary importance, i.e. > 75% of the deposited energy does not contribute to pair production.

2.2. Gamma-beam based positron generation

A new approach is the application of a brilliant high-energy γ *beam* for the creation of positron-electron pairs by pair production in a suitable target. Currently large efforts are devoted world-wide to the development of γ beams created by inverse Compton scattering of photons from an intense laser with electrons (see [24] and references therein). For this purpose, relativistic electron beams will be applied since the inverse Compton effects basically boosts the photon energy by the Lorentz factor squared. For example, using a green high-power laser with 2.5 eV photons hitting a relativistic electron bunch of 500 MeV would yield in 9.6 MeV γ's.

Such a γ beam based positron source exhibits several advantageous features. In principle, the energy of the γ beam can be varied in the range of several MeV that allows the maximization of the positron production rate and hence the yield of moderated positrons. Due to the narrow band width of the γ beam, which can be well below 1%, no γ's hit the converter target with an energy below the pair production threshold. Therefore, compared to linac or reactor based positron sources the heat load is intrinsically considerably lower. The low emittance of the γ beam allows the adaptation of a converter and positron moderator in an efficient positron source geometry. Hence, the intrinsic small diameter of the γ beam can be used to create a high-brightness moderated positron beam. It has to be mentioned that the source area of the γ beam is easily accessible which greatly facilitates e.g. the change of the source setup. The time structure of the positron beam is basically provided by the pulsed γ beam which might be beneficial for positron beam experiments using coincidence techniques. Moreover, a spin-polarized positron beam can be created using a polarized γ beam (see below). A detailed overview of the novel concept of a positron source based on γ beams produced via Compton back-scattering can be found in [24]. At present, a γ beam based positron source is being

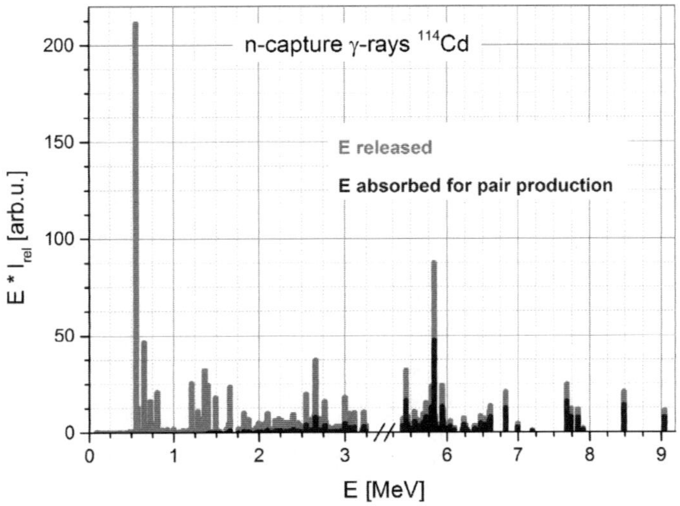

Figure 1. γ radiation as emitted by de-excitation of ^{114}Cd (data from [23]). Red: The product of the γ energy E and its relative intensity I_{rel}; black: the fraction of the deposited energy converted into positron-electron pairs.

developed at the so-called ELI-NP facility, which is under construction in Bucarest, Romania [25]. As demonstrated most recently the polarization of electrons can be transferred to positrons produced via pair production of polarized bremsstrahlung in a high-Z target [26] that might open up another possibility to generate spin-polarized slow positrons in the future.

3. Variable beam energy

3.1. Present status of low-energy positron beam setups

Mono-energetic positron beams with variable energy are applied for all kinds of depth dependent measurements. Usually, the positron implantation energy is adjusted by biasing the sample or the source-moderator assembly. Hence, the maximal possible potential difference between positron emitter and sample constrains the implantation energy to typically 30-40 keV. The potential of the positron source, however, cannot be raised very high at the positron emitting area at most laboratory setups or in reactor-based sources as well as at positron remoderators. Consequently, in such cases the acceleration voltage has to be applied to the sample stage.

This constraint can be adverse for using remoderation devices or in surface studies. It is especially crucial for positron re-emission measurements where the sample has to be fixed to ground or even to a positive potential to allow for low-energy positrons to reach the grounded detector. For this reason, only an increase of the *total energy* of the beam *between* source and sample can bypass this constraint enabling any desired implantation energy. An increase of the potential energy of a positron pulse, without altering its velocity, can be achieved by a so-called positron energy elevator using time dependent fields [27].

3.2. New positron beam elevator

Recently, two novel pulsed setups for positron acceleration based on elevation of the total beam energy have been successfully put into operation within two different projects at the NEPOMUC

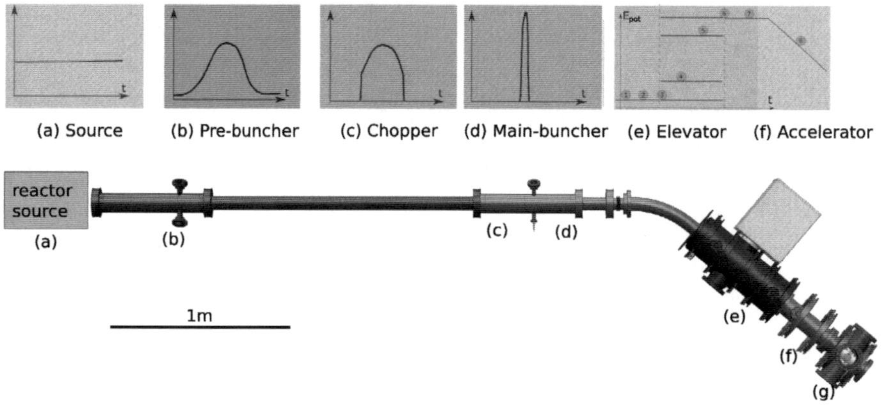

Figure 2. Novel acceleration facility including a beam energy elevator at NEPOMUC: (a) continuous positron beam, (b) pre-buncher for pulse generation with low background, (c) chopper for further background reduction, (d) main-buncher for pulse compression to below 2 ns (FWHM), (e) positron elevator for raising the potential energy of the pulse, (f) electrostatic accelerator for conversion of potential into kinetic energy, and (g) beam monitor at entrance of surface spectrometer. Figure from ref. [28].

positron beam facility in order to overcome the restriction of the applied target potential. The first one basically consists of a two-stage beam bunching system, a positron energy elevator, and an electrostatic accelerator as presented in [27]. This system is devoted to positron re-emission experiments at the surface spectrometer at NEPOMUC where the hemispherical state-of-the-art electron energy analyzer is grounded [28]. The second device was installed at an interface at the NEPOMUC beamline to enable the connection of the positron microscope with additional remoderation unit [29].

Based on the elevator principle proposed in [27] it was also shown that positrons stored in a buffer gas trap can be extracted with a frequency of 10 Hz [30]. It is noteworthy, that a harmonic high-voltage buncher [31] attached to a positron accumulator achieved a minimum beam energy of 0.75 keV with an energy spread of about 1 keV [32]. Such trap-based devices, however, are not suitable due to their more than five orders lower repetition rate compared to pulsed positron beams, which are typically applied in solid state and surface physics for depth dependent PAS.

Figure 2 exemplarily shows the setup for pulsing the positron beam with subsequent elevation of the total beam energy. It is noteworthy, that this new device is versatile with regard to the species of charged particles to be accelerated, and in principle it can be used for efficient deceleration. The viability of the new concept has been demonstrated by elevating the energy of a 20 eV mono-energetic positron beam to 3.5 keV. After passage through the elevator the time structure is preserved with a pulse width of < 1.6 ns (FWHM), and it was shown that the whole device operates with a high efficiency of 88% [28]. Since the new elevator technique is demonstrated to compensate the constraints of source and samples being near ground potential positron re-emission experiments are envisaged within the next beamtimes.

4. Spin-polarized beams

4.1. Spin-polarized positron experiments

Positrons emitted from β^+ sources are intrinsically right-handed longitudinally spin-polarized due to parity violation in the weak interaction. Polarized positrons can also be generated by pair production either using polarized electron induced bremsstrahlung [26] or via the inverse Compton effect by using a circularly polarized laser [33]. The helicity H of the emitted positrons, which is basically given by the ratio of the positron velocity v and the speed of light c, can be calculated from the average energy E_{av} of the β^+ spectrum [34]: $H \equiv v_{av}/c = \sqrt{1-(1+E_{\mathrm{av}}/mc^2)^{-2}}$. For this reason, ^{68}Ge is a good candidate for polarization dependent positron experiments due to its high endpoint energy of 1.9 MeV. For bulk studies, however, ^{22}Na has been used as well to exploit the positron spin-polarization for the investigation of e.g. the spin channels of the electron density of states in ferromagnetic materials [35, 36].

It is important to note that despite the high number of inelastic scattering events during positron moderation the depolarization is negligible. Hence, positron beams generated with a β^+ emitter retain their axial polarization to a high degree [37] and therefore are ideally suited for non-destructive polarization dependent surface studies. So far, two research groups have been working with polarized slow positron beams. The setup of the Michigan group provided a beam intensity of $1.4 \cdot 10^4$ positrons per second with an energy between 0.3 and 1.5 keV and a polarization of P=0.50(3) [37, 38]. At JAEA a polarized slow positron beam was set up in close vicinity to a cyclotron used for the production of ^{68}Ga yielding an intensity of typically $5 \cdot 10^3$ moderated positrons per second with a polarization of P=0.47(8) [39, 40].

4.2. Development of new low-energy spin-polarized positron beams

A circularly polarized laser beam interacting with a GeV electron beam would produce positrons with a high degree of polarization [33]. Therefore, it is intriguing to apply the aforementioned technique to produce positrons with a polarized γ beam in order to generate a spin-polarized positron beam with high brightness [24]. Since the positron polarization is almost entirely maintained during moderation, spin-polarized positron beam experiments will become feasible at such a γ based positron source.

At ELI-NP in Bucarest it is envisaged to generate spin-polarized positrons based on the interaction of a circularly polarized γ beam with an optimized W converter. Simulations of the low energy γ beam (< 3.5 MeV) with an intensity of $2.4 \cdot 10^{10}\gamma$/s predict a positron beam intensity of $> 10^6$ slow positrons per second. In case of total circular polarization of the γ beam the degree of spin polarization of the slow positron beam is expected to be in the range of 33-45% dependent on the thickness of the W foils used in the γ-positron converter [25]. It has to be mentioned that in the lab-scale spin-polarized positron beams with intensities of $> 10^6$ slow positrons per second and a degree of polarization of about 30% are realizable. In the long term, however, γ beam facilities as a matter of principle offer the great potential to develop intense positron beams with significant spin-polarization.

5. Conclusion and future positron beam experiments

The availability of new positron sources providing (spin-polarized) positron beams with enhanced brightness would greatly improve all kinds of positron beam applications in material science, solid-state, surface, and atomic physics. In general, a beam with increased intensity and same emittance would immediately lead to a higher quality of the experiments in terms of higher signal-to-noise ratio, shorter measurement time and/or improved statistics. In addition, a higher beam intensity is also advantageous for the generation of remoderated positron micro-beams and for scanning beam applications.

For fundamental experiments, a high intensity of moderated positrons is crucial e.g. for the creation of an energy variable mono-energetic Ps beam by photodetaching an electron from Ps$^-$, spectroscopy of Ps and Ps$^-$, generation of a neutral leptonic plasma, and the future creation of a Ps Bose-Einstein condensate. In materials science and solid-state physics, such *scanning* micro-beams are highly in demand for spatially resolved defect spectroscopy using PALS and CDBS. For surface studies the maximum available positron intensity is crucial for positron diffraction experiments TRHEPD and for future *spatially resolved* surface analysis using PAES.

With the new positron beam elevator the kinetic energy of the particle becomes independent on potentials of both the source and the sample. This allows for the spectroscopy of re-emitted positrons and hence the investigation of the material dependent positron work function as well as the determination of depth dependent moderation efficiency. In the future it would be most intriguing to analyze the electronic structure in thin layers and to observe the evolution of the Fermi surface from the bulk to the surface using *depth-dependent* 2D-ACAR. Moreover, in ferromagnetic materials this could even be done for each spin channel of the electron density of states by using spin-polarized low-energy positrons.

Acknowledgment
Funding of several projects is greatfully acknowledged: DFG-projects "Determination of the electronic structure using positrons" within the Transregional Collaborative Research Center TRR 80 "From electronic correlations to functionality", and BMBF-projects 05K10WOB and 05K13WO1 for the development of various positron beam techniques in surface science and solid state physics. Furthermore, the author would like to thank JSPS for awarding a research fellowship in Japan (2014) and DAAD for financial support of the participation at the SLOPOS 14 in Matsue, Japan.

References

[1] Reiner M, Gigl T, Jany R, Hammerl G and Hugenschmidt C 2015 *Appl. Phys. Lett.* **106** 111910
[2] Rätzke K, Wiegemann M, Shaikh M Q, Harms S, Adelung R, Egger W and Sperr P 2010 *Acta Biomater.* **6** 2690–2694
[3] Keeble D J, Mackie R A, Egger W, Löwe B, Pikart P, Hugenschmidt C and Jackson T J 2010 *Phys. Rev. B* **81** 064102
[4] Falub C V, Eijt S W H, Mijnarends P E, Schut H and van Veen A 2002 *Nucl. Instr. Methods Phys. Res. A* **488** 478–192
[5] Weiss A, Mayer R, Jibaly M, Lei C, Mehl D and Lynn K G 1988 *Phys. Rev. Lett.* **61** 2245–2248
[6] Kawasuso A and Okada S 1998 *Phys. Rev. Lett.* **81** 2695–2698
[7] Saitoh H, Stanja J, Stenson E V, Hergenhahn U, Niemann H, Pedersen T S, Stoneking M R, Piochacz C and Hugenschmidt C 2015 *New J. Phys.* **17** 103038
[8] Nagashima Y 2014 *Phys. Rep.* **545** 95 – 123
[9] Cassidy D B and Mills A P 2007 *Nature* **449** 195
[10] Hugenschmidt C 2016 *Surf. Sci. Rep.* **71** 547
[11] Mayer J, Hugenschmidt C and Schreckenbach K 2010 *Phys. Rev. Lett.* **105** 207401
[12] Coleman P 2000 *Positron beams and their applications* (Singapore: World Scientific)
[13] Hugenschmidt C 2010 Positron Sources and Positron Beams *Physics with many positrons (Proceedings of the International School of Physics "E. Fermi"* vol 174) ed A Dupasquier, A P Mills Jr, R S Brusa (Amsterdam: IOS Press)pp 399-417
[14] van Veen A, Schut H, de Roode J, Labohm F, Falub C, Eijt S and Mijnarends P 2001 *Mat. Sci. For.* **363-365** 415–419
[15] Hugenschmidt C, Löwe B, Mayer J, Piochacz C, Pikart P, Repper R, Stadlbauer M and Schreckenbach K 2008 *Nucl. Instr. Methods Phys. Res. A* **593** 616–618
[16] Hugenschmidt C, Ceeh H, Gigl T, Lippert F, Piochacz C, Pikart P, Reiner M, Weber J and Zimnik S 2013 *J. Phys.: Conf. Ser.* **443** 012079
[17] Piochacz C, Kögel G, Egger W, Hugenschmidt C, Mayer J, Schreckenbach K, Sperr P, Stadlbauer M and Dollinger G 2008 *Appl. Surf. Sci.* **255** 98–100
[18] Suzuki R, Ohdaira T, Mikado T, Uedono A, Ohgaki H, Yamazaki T and Tanigawa S 1997 *Mat. Sci. For.* **255-257** 114
[19] Akahane T, Chiba T, Shiotani N, Tanigawa S, Mikado T, Suzuki R, Chiwaki M, Yamazaki T and Tomimasu T 1990 *Appl. Phys. A* **51** 146–150
[20] Krause-Rehberg R, Brauer G, Jungmann M, Krille A, Rogov A and Noack K 2008 *Appl. Surf. Sci.* **255** 22–24
[21] Wada K, Hyodo T, Yagishita A, Ikeda M, Ohsawa S, Shidara T, Michishio K, Tachibana T, Nagashima Y, Fukaya Y, Maekawa M and Kawasuso A 2012 *Euro. Phys. J. D* **66** 37
[22] Werner M, Altstadt E, Jungmann M, Brauer G, Noack K, Rogov A and Krause-Rehberg R 2011 *J. Phys.: Conf. Ser.* **265** 012028
[23] https://www.orau.org/ptp/ptp%20library/library/doe/bnl/tnc/byn/page102.html

[24] Hugenschmidt C, Schreckenbach K, Habs D and Thirolf P 2012 *Appl. Phys. B* **106**(1) 241–249
[25] Djourelov N, Oprisa A and Leca V 2016 *Nucl. Instr. Methods Phys. Res. A* **806** 146 – 153
[26] Abbott D, et al. 2016 *Phys. Rev. Lett.* **116**(21) 214801
[27] Piochacz C and Hugenschmidt C 2013 *J. Phys.: Conf. Ser.* **443** 012093
[28] Grill N, Piochacz C, Zimnik S and Hugenschmidt C 2016 *Rev. Sci. Instr.* **87** 053304
[29] Dickmann M, Mitteneder J, Kögel G, Egger W, Sperr P, Ackermann U, CPiochacz and GDollinger 2016 *Nucl. Instr. Methods Phys. Res. A* **821** 40 – 43
[30] Cooke D A, Barandun G, Vergani S, Brown B, Rubbia A and Crivelli P 2016 *J. Phys. B* **49** 014001
[31] Chu S and Mills A P 1982 *Phys. Rev. Lett.* **48** 1333–1337
[32] Cassidy D B, Hisakado T H, Tom H W K and Mills A P 2011 *Phys. Rev. B* **84**(19) 195312
[33] Omori T, Fukuda M, Hirose T, Kurihara Y, Kuroda R, Nomura M, Ohashi A, Okugi T, Sakaue K, Saito T, Urakawa J, Washio M and Yamazaki I 2006 *Phys. Rev. Lett.* **96** 114801
[34] Major J 2000 Spin Polarized Positron Beams in Condensed-Matter Studies *Positron beams and their applications* ed P Coleman (Singapore: World Scientific) chapter 9 pp 259 - 306
[35] Weber J A, Bauer A, Böni P, Ceeh H, Dugdale S B, Ernsting D, Kreuzpaintner W, Leitner M, Pfleiderer C and Hugenschmidt C 2015 *Phys. Rev. Lett.* **115**(20) 206404
[36] Ceeh H, Weber J A, Böni P, Leitner M, Benea D, Chioncel L, Ebert H, Minar J, Vollhardt D and Hugenschmidt C 2016 *Sci. Rep.* **6** 20898
[37] Zitzewitz P W, Van House J C, Rich A and Gidley D W 1979 *Phys. Rev. Lett.* **43** 1281–1284
[38] Gidley D W, Köymen A R and Capehart T W 1982 *Phys. Rev. Lett.* **49** 1779–1783
[39] Kawasuso A, Fukaya Y, Maekawa M, Zhang H, Seki T, Yoshino T, Saitoh E and Takanashi K 2013 *J. Mag. Mag. Mater.* **342** 139 – 143
[40] Maekawa M, Fukaya Y, Yabuuchi A, Mochizuki I and Kawasuso A 2013 *Nucl. Instr. Methods Phys. Res. B* **308** 9 – 14

Research progress at the Slow Positron Facility in the Institute of Materials Structure Science, KEK

T Hyodo[1], K Wada[1]*, I Mochizuki[1], M Kimura[1], N Toge[2], T Shidara[2], Y Fukaya[3], M Maekawa[4], A Kawasuso[4], S Iida[5], K Michishio[5] and Y Nagashima[5]

[1]Institute of Materials Structure Science, KEK, Tsukuba, 305-0801, Japan.
[2]Accelerator Laboratory, KEK, Tsukuba, 305-0801, Japan.
[3]Advanced Science Research Center, JAEA, Tokai, 319-1195, Japan.
[4]Quantum Beam Science Center, QST, Takasaki, 370-1292, Japan.
[5]Department of Physics, Tokyo University of Science, Tokyo 162-8601, Japan.
*Present address: Quantum Beam Science Center, QST, Takasaki, 370-1292, Japan.

Email: toshio.hyodo@kek.jp

Abstract. Recent results at the Slow Positron Facility (SPF), Institute of Materials Structure Science (IMSS), KEK are reported. Studies using the total-reflection high-energy positron diffraction (TRHEPD) station revealed the structures of rutile-TiO_2(110) (1×2), graphene on Cu (111) and Co (0001), and germanene on Al (111). First observations of the shape resonance in the Ps^- photodetachment process were made using the positronium negative ion (Ps^-) station. Experiments using the positronium time-of-flight (Ps-TOF) station showed significant enhancement of the Ps formation efficiency and the energy loss in the Ps formation-emission process. A pulse-stretching section has been implemented, which stretches the positron pulse width from 1.2 μs up to almost 20 ms.

1. Introduction

At the Slow Positron Facility (SPF), Institute of Materials Structure Science (IMSS), High Energy Accelerator Research Organization (KEK), pulsed positrons generated through bremsstrahlung pair production at the beam dump of an electron linac are moderated using W foils which have a negative positron work function (-3 eV). Prior to transport, they are accelerated to the desired energy to form an energy-tunable monenergetic positron beam, i.e. a slow positron beam [1].

The intensity of the beam is 5×10^7 e^+/s in the accelerator long pulse mode (electron pulse width 1.2 μs) and 5×10^6 e^+/s in the variable short pulse mode (1-12 ns). Figure 1 shows the present layout of the beam line branches in the facility. The connections of the stations to the beam line branches have been exchanged from the original arrangement a few years ago [1]. The currently used stations are for total-reflection high-energy positron diffraction (TRHEPD) on branch SPF-A3, for the positronium negative ion (Ps^-) related experiments on SPF-B1 and for the positronium time-of-flight (Ps-TOF) measurements on SPF-B2. A station for low energy positron diffraction (LEPD) is under construction on SPF-A4.

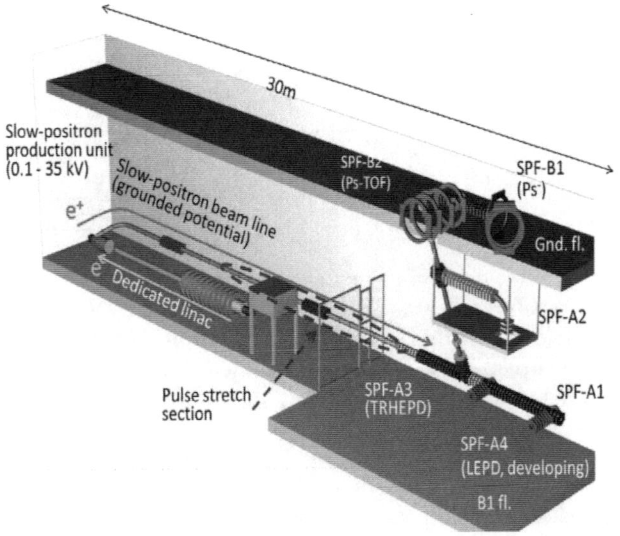

Figure 1. Layout of the stations at SPF as of March, 2016. SPF-A3, SPF-B1, SPF-B2 are for TRHEPD, Ps^- spectroscopy and Ps-TOF, respectively. A LEPD stations is under construction on SPF-A4. A pulse stretching section has been added to the beam line.

2. TRHEPD and its applications

The properties of materials are determined by the kinds of atoms composing them and their precise positions. Thus knowledge of the structure of materials is critical in discerning their properties. In bulk materials research, X-ray diffraction is acknowledged as the standard technique used to determine the structure of the crystals of new materials, proteins, *etc*.

In surface research, however, a technique for the definitive determination of the topmost- and near-surface atomic configurations is yet to be established. This has resulted in a number of studies where possible atomic geometries consistent with the observed characteristics are proposed; as a consequence, sometimes many different models are proposed by various authors.

Recently, the TRHEPD technique, formerly called RHEPD [2-4], is developing to be a standard method for surface structure determination. Its surface sensitivity originates from the inelastic scattering, as is the case for electron diffractions (LEED and RHEED), i.e., once a particle undergoes inelastic scattering, it loses coherence to contribute to the diffraction pattern. In addition, TRHEPD has another feature which gives its unique in its surface sensitivity, i.e., the electrostatic potential inside every solid is positive and thus the surface is repulsive to a positron. As shown in figure 2, a positron directed at a surface with a glancing angle smaller than a certain critical angle, θ_c, whose value depends on the material and the energy of the incident positron, is totally reflected. In this condition, the diffraction pattern depends only on the atoms on the topmost surface [5]. It is also possible to get information on the immediate subsurface by increasing the glancing angle across θ_c. In this condition, the positron is refracted into the crystal in the direction toward the surface, in contrast to the electron which is refracted away from the surface and penetrates deeper. TRHEPD is the only technique in which the Bragg diffraction condition and the total reflection condition happen to be simultaneously satisfied in a considerable angular range of measurement. Furthermore, since TRHEPD is based on the self-interference of an elastically scattered positron, just as the other diffraction techniques, it is very sensitive not only to the positions of the atoms but also to the kinds of atoms.

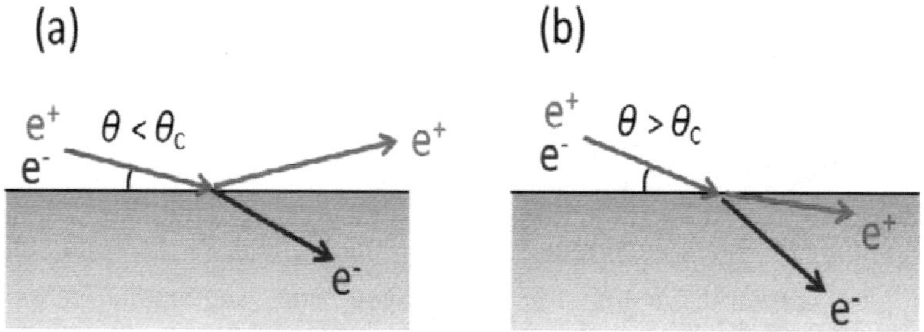

Figure 2, When the incident glancing angle, θ, is smaller than a critical angle θ_c, positrons are totally reflected (a). When $\theta > \theta_c$, positrons are refracted into the crystal in the direction towards the surface (b). The electrons are always refracted into the crystal in the direction away from the surface.

To achieve useful data in diffraction experiments, brightness enhancement of the beam is important. The positron beam for TRHEPD is transported with an energy of 15 keV, guided by a magnetic field. It is released into a magnetic-field free region and then focussed onto a spot of diameter smaller than 1 mm on a 100 nm-thick W foil remoderator, kept at an electrostatic potential of 10 kV[6]. The positrons thermalize in the foil and are emitted from the back of the foil as a quasi-point source with a kinetic energy of 3 eV. The electrostatic potential of the foil then accelerates the positrons to 10 keV as required for the TRHEPD experiments. This process reduces the normalized emittance by two orders of magnitude, while the intensity is reduced by one order of magnitude owing to annihilation inside the foil. The brightness, *i.e.* intensity/ (normalized emittance)^2, is thus enhanced by three orders of magnitude. The spot positron source is analogous to a W filament point electron source.

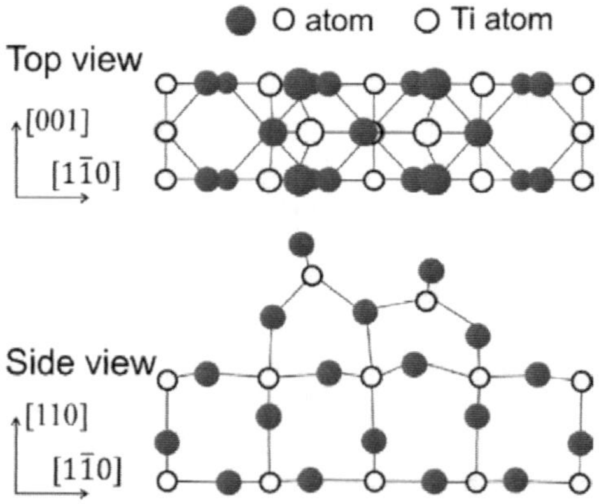

Figure 3. The top view and the side view of the structure of rutile-TiO_2 (110) (1×2) surface determined by TRHEPD(experiment) [7] and USPEX (density functional theory) [11]

Recently, TRHEPD has been used in the determination of the structure of the rutile-TiO_2(110) (1×2) surface [7], which had been under debate over the past 35 years [8]. TiO_2 is used in a variety of applications including photo-catalysts, metal-nanoparticle catalyst supports, gas sensors and corrosion-protective coating materials [9]. Knowledge of the surface structure, where the catalytic processes occur, is crucial for studying the fundamental basis of the reactivity and reaction mechanisms of solid catalysts. The rocking curves of the 00-spot obtained from the TRHEPD patterns were compared to the curves for various models calculated based on a dynamical diffraction theory, followed by further adjustment of the atomic positions to get a better agreement. The experimental curve matched the curve for the model consisting of Ti_2O_3 composition, originally suggested by Onishi and Iwasawa [10], but with a further modification of atomic positions as shown in figure 3. This was found to be close to the model proposed by Wang et al. [11] after a global search made by changing both the chemical composition of the surface and the atomic positions with the USPEX code.

Graphene is attracting increasing attention with focus on its features such as high carrier mobility, high thermal conductivity and robust mechanical properties [12]. Theoretical calculations demonstrated that the nature of the spacing between graphene and a metal substrate can be classified into two groups, depending on the interaction between them: weak interaction with simple and noble metal substrates; and strong interaction with transition metal substrates [13-15]. Thus, the experimental determination of the spacing between graphene and the substrate is crucial to elucidate the origin of the electronic properties of graphene. Although the spacing had been extensively investigated theoretically, reports on its experimental determination were limited. For example, the spacing between graphene and the Cu(111) substrate was yet to be confirmed experimentally, whereas a scanning tunneling microscopy study showed that the spacing between graphene and Co(0001) was in the range 1.5 to 2.2 Å [16].

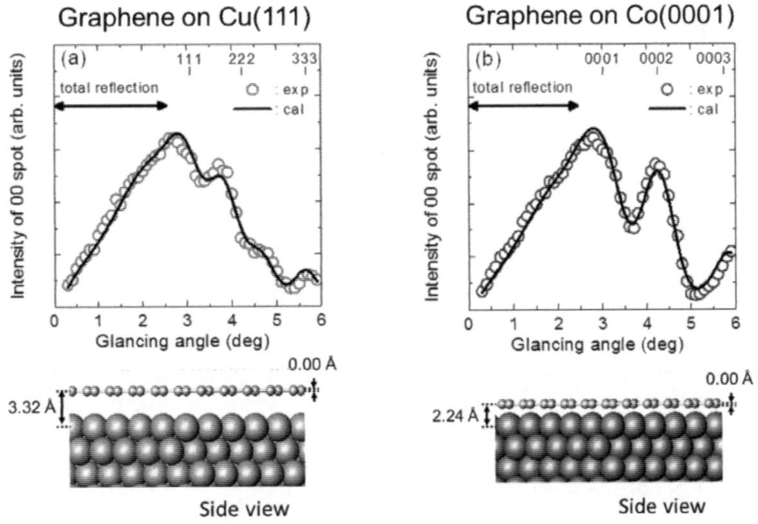

Figure 4. The TRHEPD rocking curve analysis revealed that the spacing between graphene and the Cu surface (left) is much larger than that between graphene and the Co surface (right), as predicted by theory.

As demonstrated in the study of silicene on a Ag(111) substrate [17], TRHEPD is most suitable for studying the possible buckling of a single atomic layer and its distance from the substrate surface. Rocking curve analysis based on the dynamical diffraction theory shows that the spacing between graphene and a Cu(111) substrate is 3.34 Å and that between graphene and a Co(0001) substrate is 2.06 Å (figure 4) [18]. The former is close to the interlayer spacing in graphite, experimentally verifying a weak interaction with a noble metal substrate, whereas the latter small value verifies a strong interaction with a transition metal substrate. The study also confirmed that graphene on these substrate metals do not have buckling, in agreement with what had been expected.

Recently the structure of germanene on Al(111) was measured and a structure which was not expected from the previous theoretical and experimental studies was revealed [19].

3. Observation of a resonance in the photodetachment process of Ps negative ions (Ps^-)

Positronium (Ps) negative ions, Ps^- ($e^-e^+e^-$), are efficiently produced from positrons which are injected into a Na-coated W surface and come back to the surface after thermalization [20]. The ions are accelerated to a desired energy by an electrostatic field, followed by the removal of one of the electrons to make an energy-tunable, neutral Ps beam [21]. Since the constituents of the ion are solely point-like leptons with the same mass, Ps^- provides an excellent testing ground for the three-body problem in quantum mechanics. While theoretical works on its energy levels and dynamics *etc.*, have been extensively performed, experimental investigations of its characteristics had been hampered by the small ion yield and short annihilation lifetime.

Theoretical studies suggest that Ps^- does not have an excited state but only a ground state where the two electrons have opposite spins [22, 23]. However, quasi-bound states (resonances) are predicted in the vicinities of the formation thresholds of the excited states of Ps (of principal quantum number $n \geq 2$) [24]. Photodetachment via this state is possible and the interaction of this process with the direct detachment process gives characteristic structures on the cross section.

Recently, a laser spectroscopic study of Ps^- was performed and a shape resonance near the Ps (n=2) formation threshold was observed [25]. The width of the positron pulse using the short pulse mode at KEK, 12 ns, is suitable for investigating the interaction with a high-intensity pulsed laser beam of similar width. A positron beam with a kinetic energy of 4.2 keV and a repetition of 50 Hz was transported to the measurement chamber on branch SPF-B1. It was deflected by 45° along a curved magnetic field, and impacted onto a W target coated with 0.3 monolayer of Na. Some positrons diffusing back to the surface after thermalization in the bulk form Ps^- ions, which are then emitted spontaneously with a small kinetic energy. The proportion of Ps^- ions to the total incident positron flux is reported to be as high as 2% [20].

The Ps^- ions, accelerated by an electrostatic potential, were intersected by a laser beam from a tunable dye laser in the electric field-free region between two grids kept at the same electrostatic potential. Neutral Ps atoms were formed both by the direct photodetachment process and via the shape resonance. *Para*-Ps and *ortho*-Ps are formed but the *para*-Ps self-annihilates quickly (125 ps). Only *ortho*-Ps travels to the MCP, which was placed at a distance of 0.88 m from the Ps^- formation target, and thus detected. The charged particles were removed by the curved magnetic field.

Figure 5 shows the Ps count rate measured as a function of the laser wavelength for the acceleration voltages of V=3400 V and 1500 V. Peaks corresponding to the resonance were clearly observed in both cases. The solid lines are fitted model profiles in the laboratory frame, with the Doppler-broadening and the longitudinal Doppler-effect taken into account. From this data, the resonance energy in the rest frame of the ions was deduced to be 5.437(1) eV.

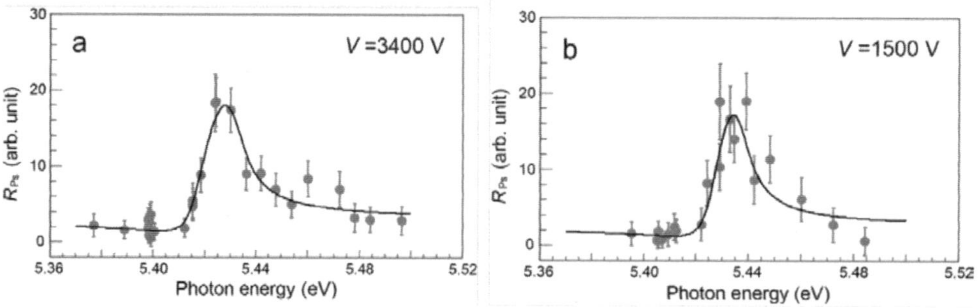

Figure 5. Shape resonance of the Ps^- photodetachment at different acceleration voltage *V*[25].

4. Ps time-of-flight (Ps-TOF) spectroscopy from alkali-metal-coated W surfaces

The mechanism of large enhancement of the Ps^- production from alkali-metal-coated W surfaces is not yet fully understood. Changes in the work functions of the electron and the positron upon coating may be relevant. If this is the main cause of the enhancement, it may be possible that the coating does not have an effect on the Ps formation probability since Ps consists of one electron and one positron, whereas Ps^- consists of one positron and two electrons.

In order to check this idea, Ps-TOF measurements were performed. Positrons of an energy of 4.5 keV and a width of 12 ns (electron linac short pulse mode) were transported to the Ps-TOF measurement chamber on SPF-B2. The beam was collimated by an aluminum plate with an aperture of 6 mm in diameter before entering the chamber.

One of the γ-rays emitted when *ortho*-Ps self-annihilates in flight was detected by a scintillation counter behind a lead slit set at a fixed distance from the sample. The annihilation γ-rays from the target were also recorded and used to determined time 0 of the TOF spectra.

As shown in figure 6, Ps emission was remarkably enhanced by the addition of sub-monolayers of various alkali metals [26, 27]. It was also revealed that the low energy tail of the spectra has a component with a specific energy loss which varied with the alkali metal [27]

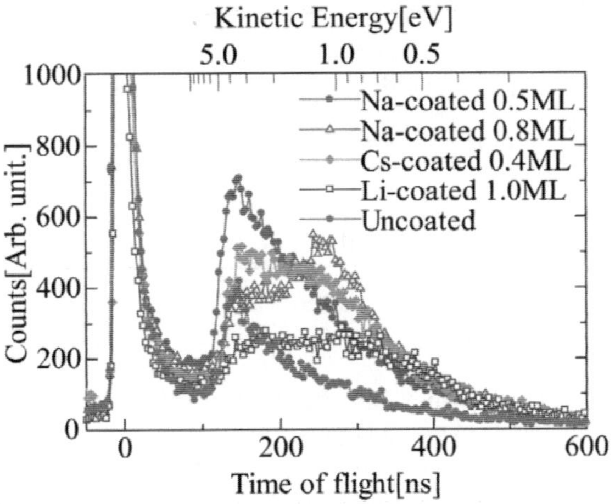

Figure 6. Ps-TOF spectra of W surfaces coated with submonolayers of alkali metal as indicated.

5. Positron beam pulse-stretching section

A pulse-stretching section has been installed in the beam line, which stretches the slow positron pulse of a width of 1.2 μs (in the long pulse mode) to a variable width of up to almost 20 ms at a beam energy of 5 keV. The purpose of this system is twofold. Firstly, the stretched beam will be used for the position-sensitive delay-line detector of the LEPD system, for future spectroscopic studies of the 511-keV annihilation γ-rays such as coincidence Doppler broadening (CDB) spectroscopy and angular correlation of annihilation radiation (ACAR), and for the construction of a short pulse producing system for positron lifetime spectroscopy (PALS). Secondly, it provides the know-how to improve a similar system for PALS experiments at the National Institute of Advanced Industrial Science and Technology (AIST). There, after stretching the original pulse of width 1-4 μs [28], the beam is chopped and pulsed to have a ~100-ps width, which is suitable for PALS [29, 30]. The efficiency of the system will be improved if we have a pulse-stretching section with low intensity loss and with almost constant outlet beam energy.

The system at KEK-SPF is essentially a Penning-Malmberg trap as is the one at AIST [28]. The entrance electrode voltage, normally kept at 5.5 kV, is temporarily lowered to 4.5 kV to accept a 1.2

μs wide positron pulse with an energy of 4.8±0.05 keV into a 6 m long trapping electrode. The voltage of the entrance electrode is then raised back up to 5.5 keV before the positrons reflected at the exit electrode, kept at 5.0 kV, return. The voltage of the trapping electrode is then increased gradually, letting the positrons spill over the exit electrode, which is kept at a constant voltage of 5.0 kV. By adjusting the sweeping speed of the trapping electrode voltage, we obtain a beam of controllable pulse width of up to almost 20 ms with a fixed energy of 5.0 keV, operated at 50 Hz synchronized with the linac operation.

The performance of the pulse-stretching section was examined by letting the positrons annihilate on a closed gate valve downstream. The annihilation γ-rays were detected by a plastic scintillator mounted on a photomultiplier tube. Its anode signals were recorded by a digital oscilloscope as shown in figure 7 and counted. It was confirmed that the 50 Hz positron pulses were stretched up to ~20 ms at the same frequency with an efficiency of 70%. This technique will be transferred to the AIST PALS system.

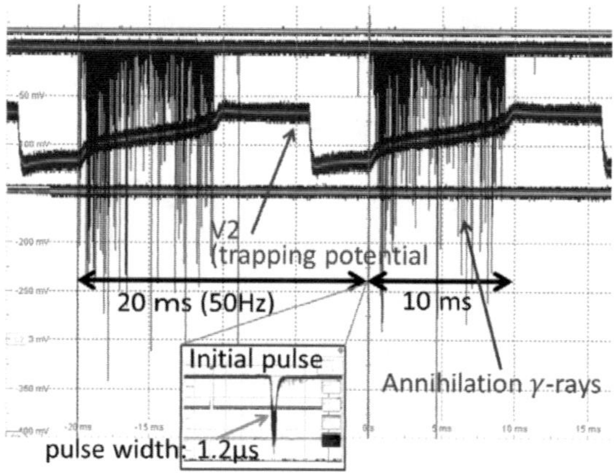

Figure 7. Performance of the pulse-stretching section. The original positron pulse of 1.2 μs width (inset at the bottom) is stretched to a width of 10 ms. Stretching up to ~20 ms is possible.

6. Summary

Description of the layout of the measurement stations at the Slow Positron Facility in the Institute of Materials Structure Science, KEK is updated. Recent results obtained by using the active stations have been reported. Total-reflection high-energy positron diffraction (TRHEPD) determined the structures of rutile-TiO_2(110) (1×2), graphene on Cu (111) and Co (0001), and germanene on Al (111). The unique surface sensitivity of TRHEPD is attributed to the positive electrostatic potential common to all the materials. A shape resonance in the Ps^- photodetachment process was observed for the first time. Coating of a W surface with submonolayers of alkali metals, which enhances Ps^- formation, has been found to enhance the formation of neutral Ps as well. A pulse-stretching section which stretches the positron pulse width from 1.2 μs up to almost 20 ms has been added to the beam line.

Acknowledgements: We thank the staff of the Photon Factory and the Accelerator Laboratory at KEK for their support in the operation of the Slow Positron Facility. This work was supported by JSPS KAKENHI Grant Numbers 24221006, 24221007, 25800182 and 26800170, Toray Science and Technology Grant and the Cross-Ministerial Strategic Innovation Promotion Program (SIP, unit D66). Experiments were conducted under the approval of the PF PAC (proposal Nos. 2012G653, 2013U002, 2013S2-005, 2014S2-004) and partly under the auspices of the JAEA- KEK Joint Development Research and KEK-Hokkaido University Joint Research Program.

References

[1] Wada K, Hyodo T, Yagishita A, Ikeda M, Ohsawa S, Shidara T, Michishio K, Tachibana T, Nagashima Y, Fukaya Y, Maekawa M, Kawasuso A 2012 *Eur. Phys. J.* **66** 37
[2] Ichimiya A 1992 *Solid State Phenom.* **28-29** 143
[3] Kawasuso A and Okada S 1998 *Phys. Rev. Lett.* **81** 2695
[4] Fukaya Y, Maekawa M, Mochizuki I, Wada K, Hyodo T and Kawasuso A 2013 *J. Phys.: Conf. Series* **443** 012068
[5] Fukaya Y, Maekawa M, Kawasuso A, Mochizuki I, Wada K, Shidara T, Ichimiya A and Hyodo T 2014 *Appl. Phys. Express* **7** 056601
[6] Maekawa M, Wada K, Fukaya Y, Kawasuso A, Mochizuki I, Shidara T and Hyodo T 2014 *Eur. Phys. J. D* **68** 165
[7] Mochizuki I, Ariga H, Y. Fukaya Y, Wada K, Maekawa M, Kawasuso A, Shidara T, Asakura K and Hyodo T 2016 *Phys. Chem. Chem. Phys.* **18** 7085
[8] Diebold U 2003 *Surf. Sci. Rep.* **48** 53
[9] Fujishima A, Nakata K, Ochiai T, Manivannan A and Tryk D 2013 *Electrochem. Soc. Interface* **22** 51
[10] Onishi H and Iwasawa Y 1994 *Surf. Sci.* **313** L783
[11] Wang Q, Oganov A R, Zhu Q and Zhou X F 2014 *Phys. Rev. Lett.* **113**, 266101
[12] Geim A K 2009 *Science* **324** 1530
[13] Giovannetti G, Khomyakov P A, Brocks G, Karpan V M, van den Brink J and Kelly P J 2008 *Phys. Rev. Lett.* **101** 026803
[14] Batzill M 2012 *Surf. Sci. Rep.* **67** 83
[15] Silvestrelli P L and Ambrosetti A 2015 *Phys. Rev. B* **91** 195405
[16] Eom D, Prezzi D, Rim K T, Zhou H, Lefenfeld M, Xiao S, Nuckolls C. Hybertsen M S, Heinz T F and Flynn G W 2009 *Nano Lett.* **9** 2844
[17] Fukaya Y, Mochizuki I, Maekawa M, Wada K, Hyodo T, Matsuda I and Kawasuso A 2013 *Phys. Rev. B* **88** 205413
[18] Fukaya Y, Entani S, Sakai S, Mochizuki I, Wada K, Hyodo T and Shamoto S 2016 *Carbon* **103** 1
[19] Fukaya Y, Matsuda I, Feng B, Mochizuki I, Hyodo T and Shamoto S 2016 *2D Materials* **3** 035019
[20] Nagashima Y 2014 *Phys. Rep.* **545** 95
[21] Michishio K, Tachibana T, Suzuki R H, Wada K, Yagishita A, Hyodo T and Nagashima Y 2012 *Appl. Phys. Lett.* **100** 254102
[22] Mills A P Jr 1981 *Phys. Rev. A* **24** 3242
[23] Bhatia A K and Drachman R J 1983 *Phys. Rev. A* **28** 2523
[24] Ho Y K 1979 *Phys. Rev. A* **19** 2347
[25] Michishio K, Kanai T, Kuma S, Azuma T, Wada K, Mochizuki I, Hyodo T, Yagishita A and Nagashima Y 2016 *Nature Communications* **7** 11060
[26] Terabe H, Iida S, Yamashita T, Tachibana T, Barbiellini B, Wada K, Mochizuki I, Yagishita A, Hyodo T and Nagashima Y 2015 *Surface Science* **641** 68
[27] Iida S, Wada K, Mochizuki I, Tachibana T, Yamashita T, Hyodo T and Nagashima Y 2016 *J. Phys.: Condens. Matter* **28** 475002
[28] Akahane T, Chiba T, Shiotani N, Tanigawa S, Mikado T, Suzuki R, Chiwaki M, Yamazaki T, Tomimasu T 1990 *Appl. Phys. A* **51** 146
[29] Suzuki R, Kobayashi Y, Mikado T, Ohgaki H, Chiwaki M, Yamazaki T, Tomimasu T 1991 *Jpn. J. Appl. Phys.* **30** L532
[30] Suzuki R, Mikado T, Chiwaki M, Ohgaki H, Yamazaki T 1995 *Appl. Surf. Sci.* **85** 87

Positron annihilation lifetime spectroscopy at a superconducting electron accelerator

A Wagner[1], W Anwand[1], A G Attallah[2,3], G Dornberg[4], M Elsayed[2], D Enke[4], A E M Hussein[2], R Krause-Rehberg[2], M O Liedke[1], K Potzger[5] and T T Trinh[1,6]

[1]Helmholtz-Zentrum Dresden-Rossendorf, Institute of Radiation Physics Bautzner Landstraße 400, 01328 Dresden, Germany
[2]Martin-Luther-Universität Halle, Institut für Physik, 06099 Halle, Germany
[3]Physics Department, Faculty of Science, Minia University, 61519 El-Minia, Egypt
[4]Institute of Chemical Technology, Universität Leipzig, 04103 Leipzig, Germany
[5]Helmholtz-Zentrum Dresden-Rossendorf, Institute of Ion Beam Physics and Materials Research, Bautzner Landstraße 400, 01328 Dresden, Germany
[6]Technische Universität Dresden, 01062 Dresden, Germany

E-mail: a.wagner@hzdr.de

Abstract. The Helmholtz-Zentrum Dresden-Rossendorf operates a superconducting linear accelerator for electrons with energies up to 35 MeV and average beam currents up to 1.6 mA. The electron beam is employed for production of several secondary beams including X-rays from bremsstrahlung production, neutrons, and positrons. The secondary positron beam after moderation feeds the Monoenergetic Positron Source (MePS) where positron annihilation lifetime (PALS) and positron annihilation Doppler-broadening experiments in materials science are performed in parallel. The adjustable repetition rate of the continuous-wave electron beams allows matching of the pulse separation to the positron lifetime in the sample under study. The energy of the positron beam can be set between 0.5 keV and 20 keV to perform depth resolved defect spectroscopy and porosity studies especially for thin films.

1. Introduction

Positron annihilation lifetime spectroscopy (PALS) serves as a unique tool for the characterization of lattice defects in materials science. While the annihilation lifetime yields characteristic information about the size of open-volume defects ranging from single atomic vacancies up to even porous microstructures, the kinematical Doppler-broadening of annihilation radiation tells about the local electron momentum distribution at the annihilation site. Long annihilation lifetimes in the order of ns result from production of positronium, especially in porous media and in polymers. The well-established technique of employing radionuclides simultaneously emitting gamma-rays and positrons of typically up to several hundreds of keV is widely applied. Annihilation lifetimes are monotonously related to the void size with a surprisingly low dependence on the material surrounding the void [1]. The quantitative determination of positron annihilation lifetimes nevertheless is hampered when thin

IOP Conf. Series: Journal of Physics: Conf. Series **791** (2017) 012004 doi:10.1088/1742-6596/791/1/012004

films or layered structures with sub-μm thicknesses which are of high technological relevance are considered. The isotropic emission and the high energy of common positron emitters cause penetration depths which are on the mm-scale. The installation described here overcomes this limitation by using mono-energetic positron beams with kinetic energies between 0.5 keV and 20 keV thus enabling depth-dependent thin film studies on the nm to μm scale [2].

2. The Mono-energetic Positron Source MePS

The Mono-energetic Positron Source MePS has been set up at the superconducting electron linear accelerator ELBE (Electron LINAC with high Brilliance and low Emittance) [3] at Helmholtz-Zentrum Dresden-Rossendorf in a collaborative effort between Martin-Luther University Halle-Wittenberg and the Helmholtz-Zentrum Dresden-Rossendorf. Figure 1 shows the layout of the system schematically. Some of the details of the installation are described below.

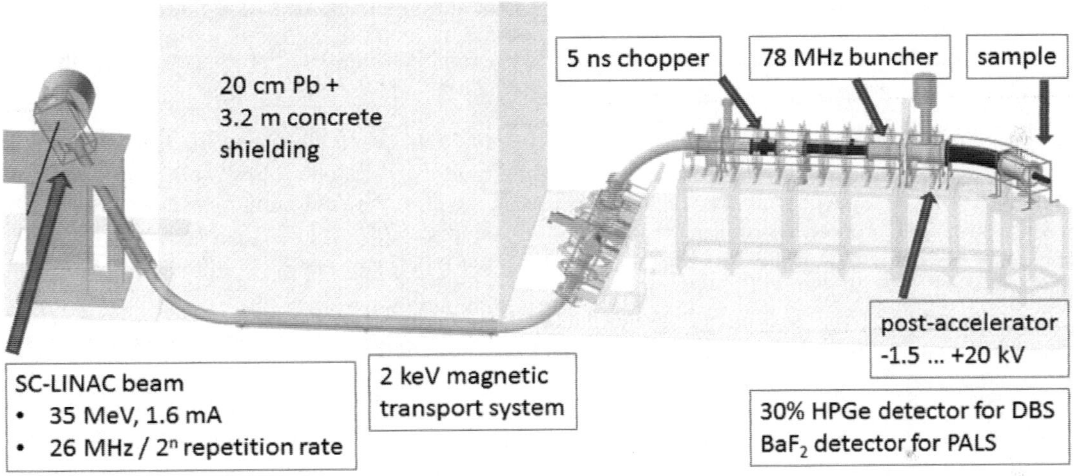

Figure 1. Layout of the monoenergetic positron source.

The incident electron beam is delivered by the superconducting (SC) electron linear accelerator ELBE at energies up to 35 MeV. In contrast to conventional normal conductive accelerators superconducting technology makes possible the application of continuous radio-frequency accelerating electromagnetic fields with electric field gradients up to 20 MV/m. The injection of electrons from an injector can be done in principle with any divisor of the base frequency which is 1.3 GHz for ELBE. However, pulse-preforming by the electron beam bunching system prior to injection restricts the maximum applicable pulse repetition rate to 260 MHz. For positron annihilation studies repetition rates of 1.625 MHz, 6.5 MHz, 13 MHz, and 26 MHz have been employed up to now. The repetition rates have been selected in order to match to the annihilation lifetimes in the samples. Minimization of the effect of pulse-overlap (lower repetition rate favored) has been done while maximizing the detection efficiency (higher repetition rate favored) at given electron beam bunch charges. If we request a pulse-overlapping fraction of not more than one part in a thousand, maximal annihilation lifetimes should be 90 ns, 22 ns, 11 ns, and 6 ns for the *canonical* repetition rates given above. The electron beam is directed towards a water-cooled tungsten bremsstrahlung converter consisting of 50 foils of 100 μm thickness each with a separation of 100 μm each as shown in Figure 2, left side. The converter is enclosed in a stainless steel casing with attached stainless steel cooling pipes as shown in Figure 2, right side. The converter has been designed to cope with average beam powers of up to 40 kW out of which only 4 kW have been used by now. Electron bremsstrahlung with a continuous spectrum up to the electron beam energy is then converted in turn by pair production into electrons and positrons inside the converter and also inside the following tungsten moderator. Thermalized positrons from the moderator are being accelerated with a bias potential of +2 kV towards a conductive mesh at a few mm distance. Further positron transport is accomplished by a longitudinal magnetic field of 8 mT with variations of less than 10%.

Figure 2. The bremsstrahlung converter. Partially assembled stack of 50 Tungsten foils with 100 μm thickness each (left). Assembled converter as viewed by the electron beam (right).

More than 3 m of iron-enriched concrete and 20 cm of lead shield the accessible laboratory from the positron generating converter. A water-cooled aluminium beam catcher of 600 mm length absorbs most of the radiation penetrating the converter and the moderator. Pure aluminium has been selected in order to minimize photo-neutron production owing the neutron separation energy of 13 MeV for ^{27}Al. The design of the beam catcher has been derived from the liquid lead photo-neutron source which is operated at ELBE, as well [4].

After transporting the beam through the radiation shielding the beam is sent through a double beam chopper which imposes a transversal electric field of up to 500 V_{peak}/cm, see Figure 3 (left). The pulsed electric field has a Gaussian shape with a FWHM of 5 ns while the electrodes are separated by 4 mm. Out-of-phase positrons are deflected towards an adjustable aperture after half a gyration length of 126.3 mm. After removal of those non-phase matched positrons the beam is further longitudinally compressed with a double-slit buncher operating at a frequency of 78 MHz. A resonant circuit drives the central isolated drift tube while generating an electric field strength of up to 2.5 kV/cm at the two gaps, see Figure 3 (right). The two buncher gaps are separated by 509 mm which matches to 3/2 times the buncher period at 2 keV positron transport energy.

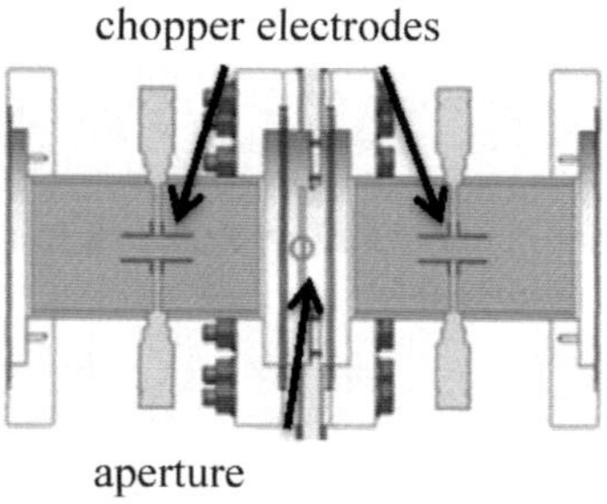

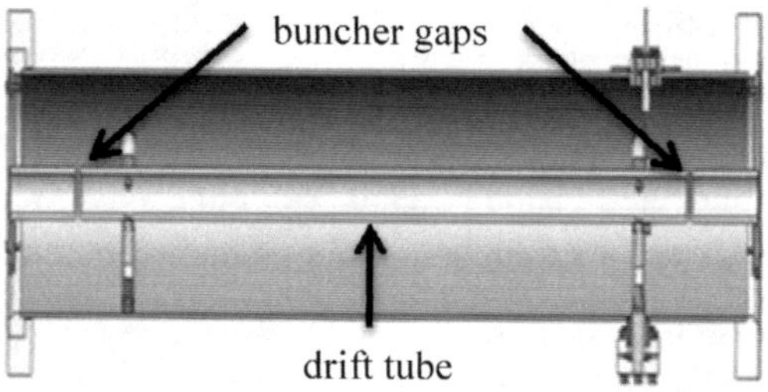

Figure 3. Beam chopper (left) and double-slit buncher (right). The beam enters from the left and additional beamline sections are omitted.

After longitudinal bunch compression by the chopper a 6-stage electrostatic acceleration structure follows which allows the positron kinetic energy and thus the penetration depth inside the sample to be varied. Prior to the target the magnetic guidance system is bent by 45° thus suppressing positrons reflected from the sample and bounced by the accelerator field to impact again onto the sample.

The sample station is kept at a high potential of up to 20 kV. About 3 cm behind the sample a μ-metal shielded BaF_2 scintillation detector is employed for annihilation lifetime measurements. The timing reference is derived from the precision master oscillator of the superconducting accelerator and is phase-matched to the electron bunches of the beam.

A magnetic beam transport system guides positrons to the samples under investigation. A dedicated chopper/buncher system is used to maintain a high timing resolution for depth-dependent annihilation lifetime studies in thin films. The signal-to-noise ratio is above 10^4 while lifetime resolutions of around 230 ps FWHM have been obtained. Figure 4 shows the positron annihilation lifetime of two reference samples with well-known annihilation lifetimes for an incident positron kinetic energy of 10 keV. On the left side, single crystalline Y_2O_3-stabilized ZrO_2 (YSZ) is shown together with a Gaussian distribution with a derived time resolution of 230 ps FWHM. On the right hand side, mono-crystalline silicon is shown. In both cases, the annihilation lifetimes of 181 ps and 218 ps, respectively, which were determined in conventional lifetime studies are well reproduced. Besides the main contribution an additional stable background lifetime of 720 ps of unknown origin is visible which is subtracted in all subsequent analysis. For both samples a signal to noise ratio of $5x10^5$ is obtained.

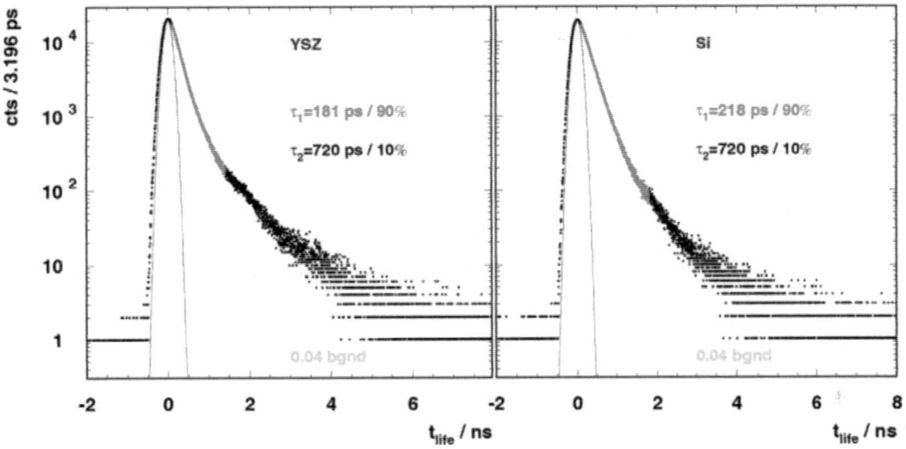

Figure 4. Positron annihilation lifetime distributions for Y_2O_3-stabilized ZrO_2 (left) and wafer-grade silicon (right). A Gaussian distribution with 230 ps FWHM (blue) is included in order to visualize the timing resolution.

An example for longer positron annihilation lifetimes is shown in Figure 5. It shows the variation of ortho-positronium lifetimes with the incident positron energies for nano-porous glasses. The μm-thin glass films have been produced by stimulated phase separation in sodium borosilicate glass into silica and an alkali borate phase thus generating sponge-like porous structures with unknown porosity [5]. Nano-porous glasses feature a tunable pore width and adjustable surface properties which makes them ideal candidates for separation membranes, chemo-sensors, drug delivery, optical coatings, and many other applications. Determination of porosity is hampered when employing conventional porosimetry methods. Both, Hg intrusion or LN_2 sorption methods fail in the case of thin films due to weak signals from small surface areas and the small pore volume. Furthermore, systems with closed porosity cannot be investigated by means of intrusion techniques. In contrast, PALS serves as an ideal tool in such cases. Experimentally, the lifetime distributions are almost free from distortions and the peak to background ratios are between $5x10^4$ and $8x10^4$ due to a slight pulse-to-pulse overlap with the selected beam repetition rate of 1.625 MHz which corresponds to 615 ns pulse-to-pulse spacing. Due to imperfect bunching the timing resolution decreases for positron beam energies below 2 keV. The sample had been covered with a thin carbon layer to prevent positronium from escaping the surface.

Figure 5 shows a scanning electron microscope picture of a sample cut at the left. The nano-porous film has a thickness of about 4 μm.

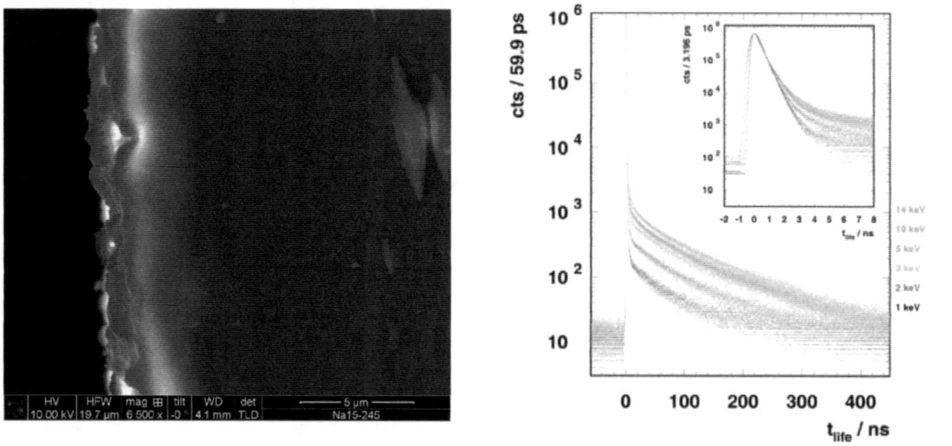

Figure 5. Left panel: SEM picture of the nano-porous film with the surface on the left side and the non-porous structure extending from the centre to the right side. Right panel: Positron annihilation lifetime distributions for nano-porous glasses (inset shown with increased time dispersion).

MePS has also been employed successfully for porosity studies of ultra low-k dielectric thin films as promising materials for advanced inter-connects scheme of ultra-large scale integrated devices. The aim is the reduction of resistance-capacitance delay and cross-talk noise by the introduction of nano-scale voids inside the insulating SiO_2 [6,7]. In the future, MePS will be complemented by an ultra-high vacuum system for in-situ defect studies called Apparatus for In-Situ Defect Analysis (AIDA). Here, the introduction and the annealing of defects during ion-irradiation, temperature treatment and during thin film growth will be studied.

3. The Gamma-induced Positron Source GiPS

A second accelerator-driven positron source is realized in the so-called Gamma-induced Positron Source (GiPS) [8]. A pioneering technology called "accelerator-based γ-ray-induced PAS" or AGAPS at a normal conducting electron LINAC had been developed at Idaho State University and it was used for Doppler-broadening spectroscopy experiments [9]. The setup employs high-energy bremsstrahlung for pair production right inside the sample under study. The GiPS setup additionally allows annihilation lifetime experiments, as well, and it is especially suited for extended bulk samples, or samples which cannot be exposed to external positrons sources like positron beams or radioactive sources or because they are imposing hazardous conditions (high pressure, high temperature, intrinsic radioactivity), or if the sample handling imposes difficulties (fluids, gases, organic samples). Derived from a setup for nuclear resonance fluorescence studies [10] special emphasis has been put on background reduction and shielding. Recently, positron annihilation lifetime studies on point defects in the skyrmion-lattice compound MnSi [11] and on the origin of luminescence and scintillation in ZnO [12] have been published. As for the MePS source the beam repetition rate can be matched to the positron annihilation lifetime thus optimizing for high average intensity and low detector pile-up distortions.

4. The Slow Positron System of Rossendorf

While the accelerator-driven positron sources require sharing beam time with users of beam lines, complementary Doppler-broadening spectroscopy studies are being performed at the ^{22}Na-based mono-energetic positron beam SPONSOR (Slow POsitroN System Of Rossendorf) [13]. Here, Doppler-broadening studies are performed with higher statistics and more energy steps. Additionally, with both Germanium detectors arranged face-to-face and perpendicular to the beam line coincidence Doppler-broadening experiments are performed which allow for a two order of magnitude improvement in signal to noise ratio. ^{22}Na-based lifetime experiments for bulk materials can be performed at a conventional lifetime setup. As a precursor for the MePS installation a pre-stage of AIDA has been implemented at the SPONSOR beam and in-situ experiments on Fe-Al alloys have been performed [14].

5. Conclusion

The mono-energetic positron source MePS and the Gamm-induced Positron Source GiPS represent the first accelerator-based setups for positron annihilation lifetime, Doppler-broadening and age-momentum correlation measurements operated in a continuous wave mode. Adjustable beam repetition rates are selected to match annihilation lifetimes in order to obtain efficient measurements and low pile-up distortions. Conventional positron sources complement the accelerator-based setups for in-depth and precursor studies.

Acknowledgements

The pre-stage system of AIDA was funded by the Initiative and Networking Fund of the Helmholtz-Association. (FKZ VH-VI-442 Memriox). Thanks go to the ELBE accelerator crew for providing stable beams, to M. Görler, A. Müller, D. Stach, and G. Staats for their contributions to the beam transport, buncher and chopper system, and especially to M. Jungmann for his contributions to the MePS setup.

References

[1] Wada K and Hyodo T 2013 *J. Phys: Conf. Ser.* **443** 012003

[2] Krause-Rehberg R, Brauer G, Jungmann M, Krille A, Rogov A and Noack K 2008 *Appl. Surf. Sci.* **255** 22

[3] Gabriel F, Gippner P, Grosse E, Janssen D, Michel P, Prade H, Schamlott A, Seidel W, Wolf A and Wünsch R 2000 *Nucl. Instrum. Methods Phys. Res. B* **161** 1143

[4] Altstadt E, Beckert C, Freiesleben H, Galindo V, Grosse E, Junghans A R, Klug J, Naumann B, Schneider S, Schlenk R, Wagner A and Weiss F-P 2007 *Ann. Nucl. Energy* **34** 36

[5] Uhlig H, Adouane G, Bluhm C, Zieger S, Krause-Rehberg R and Enke D 2016 *J. Porous Mat.* **23** 139

[6] Jungmann M, Haeberle J, Krause-Rehberg R, Anwand W, Butterling M, Wagner A, Johnson J M and Cowan T E 2013 *J. Phys: Conf. Ser.* **443** 012088

[7] Uedono A, Armini S, Zhang Yu, Kakizaki T, Krause-Rehberg R, Anwand W and Wagner A 2016 *Appl. Surf. Sci.* **368** 272

[8] Butterling M, Anwand W, Cowan T E, Hartmann A, Jungmann M, Krause-Rehberg R, Krille A and Wagner A 2011 *Nucl. Instrum. Methods Phys. Res. B* **269** 2623

[9] Selim F A, Wells D P, Harmon J F, Williams J 2005 *J. Appl. Phys.* **97** 113539

[10] Schwengner R, Beyer R, Dönau F, Grosse E, Hartmann A, Junghans A R, Mallion S, Rusev G, Schilling K D, Schulze W and Wagner A 2005 *Nucl. Instrum. Methods Phys. Res. A* **555** 211

[11] Reiner M, Bauer A, Leitner M, Gigl T, Anwand W, Butterling M, Wagner A, Kudejova P, Pfleiderer C and Hugenschmidt C 2016 *Sci. Rep.* **6** 29109

[12] Ji J, Colosimo A M, Anwand W, Boatner L A, Wagner A, Stepanov P S, Trinh T T, Liedke M O, Krause-Rehberg R, Cowan T E and Selim F A 2016 *Sci. Rep.* **6** 31238

[13] Anwand W, Brauer G, Butterling M, Kissener H R and Wagner A 2012 *Defect and Diffusion Forum* **331** 25

[14] Liedke M O *et al.*, 2015 *J. Appl. Phys.* **117** 163908

IOP Conf. Series: Journal of Physics: Conf. Series **791** (2017) 012005 doi:10.1088/1742-6596/791/1/012005

Efficient positron moderation with a commercial 4H-SiC epitaxial layer

A M M Leite, P Debu, P Pérez, J-M Reymond, Y Sacquin, B Vallage and **L Liszkay**
Irfu, CEA, University Paris-Saclay F-91191 Gif-sur-Yvette Cedex, France
E-mail: amelia.leite@cea.fr

Abstract. We have studied the properties of a commercially available 4H-SiC epitaxial layer and evaluated its potential application as an efficient positron remoderator. A remoderation efficiency of more than 65% has been measured for incident positrons with 1 keV energy. We have determined the work function and the energy distribution of the emitted slow positrons, a property which is essential for practical applications. Comparison of the positron moderation properties of the epitaxial layer with results from a n-type 4H-SiC single crystal, indicate that the epitaxially grown layer is a superior secondary moderator than its substrate counterpart.

1. Introduction

Slow monoenergetic positron beams rely on radioactive or pair creation sources. Positrons are moderated by a dense material with negative work function for positrons, or a solid rare gas.

It has been long established that n type silicon carbide is an efficient moderator material [1], although its low density makes it unsuitable to be used as a primary moderator. Previous work by Suzuki et al. [2] suggests that an epitaxially grown 6H-SiC film with a low carrier density is a good candidate for a secondary moderator. Other studies on as grown SiC wafers have shown the potential of this material, but revealed a small remoderation efficiency in comparison to the epitaxially grown wafers [3][4]. Today, large (10 cm) good quality epitaxial wafers are readily available with a very low carrier concentration and defect density, due to improvements in layer growth technology and defect control.

In this paper we present measurements on the remoderation properties of a commercial 4H-SiC epitaxial layer, and show that it can be efficiently used as a positron secondary moderator.

2. Experimental setup

We used the positron beam facility at Irfu/CEA Saclay, France [5]. It consists of a small linear electron accelerator (linac) that generates 4.3 MeV electrons in 2.5 μs long pulses with 200 Hz repetition frequency. The electrons hit a tungsten target, producing gamma radiation by Bremsstrahlung, which induces positron-electron pairs. Directly behind the target there is a stack of well annealed thin tungsten meshes that is used as a moderator. The positron moderator slows down the high energy positrons to a few eV kinetic energy range. A pulsed beam of slow positrons ($\sim$ 50 eV) leaves the moderator structure and is adiabatically transported in a 8 mT magnetic field, generated by solenoids wound around the vacuum tube. We routinely obtain a positron flux of $2\times10^6 e^+/s$.

A positron pulse stretcher forms a quasi-continuous beam from the 2.5 μs pulses generated by the linac. It is a low-field Penning-Malmberg trap with three electrodes. The second electrode is 4 m long, in order to accommodate the entire slow positron pulse. The other two electrodes open and close the trap, synchronized with the linac pulses. For this setup the voltage pulse applied to the exit electrode of the stretcher was modified to produce short positron pulses ($\sim$ 100 ns FWHM).

Near the target the positron beam is accelerated at the entry of a 30 cm tube, kept at a negative bias V_{tube}, and hits a 4H-SiC epitaxial layer held at a bias V_{sample}.

The epitaxial layer was grown on a n-type 4H-SiC (0001) substrate, with 16-28 mΩ cm resistivity, and is nitrogen doped with a carrier density smaller than 10^{15} cm^{-3}, and 5.5 μm thick. It was purchased from Norstel. The epitaxial layer was not subjected to any form of surface treatment, it was used as received. The detection of the photons resulting from the annihilation of the positrons was done by means of a plastic scintillator.

All measurements reported were performed on the epitaxial layer and on the SiC substrate (backside of the 4H-SiC wafer).

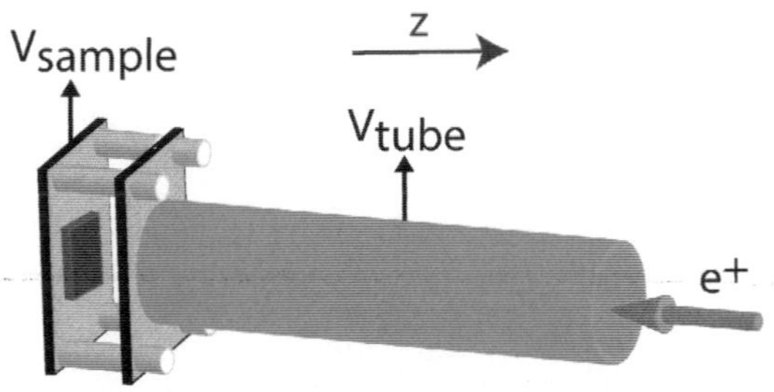

Figure 1. Scheme of the experimental setup. A 30 cm length tube held at V_{tube} accelerates the positrons onto the 4H-SiC target with a bias V_{sample}.

3. Results and discussion

Since 4H-SiC has a negative work function $\Phi^+ \sim$ -2 eV [3][4], incident positrons are re-emitted from the surface. If their energy is high enough to overcome the potential difference between the tube and the sample (ΔV), they can execute several round trips between the sample and the end of the tube, while a fraction of positrons is annihilated at the target in each trip. This process can be observed in Figure 2 for both the epitaxial layer and substrate. In here V_{sample} was set at -2000 V and V_{tube} was varied. One observes that the time it takes the positrons to do a round trip increases with increasing ΔV, due to the decreasing energy of the re-emitted positrons.

A remoderation efficiency, ratio of the number of re-emitted positrons to the number of incident positrons, higher than 65% has been obtained for 1 keV positron energy while for the substrate the maximum efficiency is 60%. The positron re-emission yield (Y) is defined as the fraction of the incident positron beam that was remoderated $Y = 1 - \frac{A_1}{A}$, where A_1 is the integrated area of the first positron annihilation peak (Figure 3), when remoderation occurred, and A the integrated area of the total annihilation peak (i.e the tube was set at -1995 V, hence all the positrons were annihilated).

The work function Φ^+ was measured by setting the sample at -2000 V, and varying the tube voltage from -1990.5 V to 2010.6 V. Figure 4 shows the re-emission yield as a function

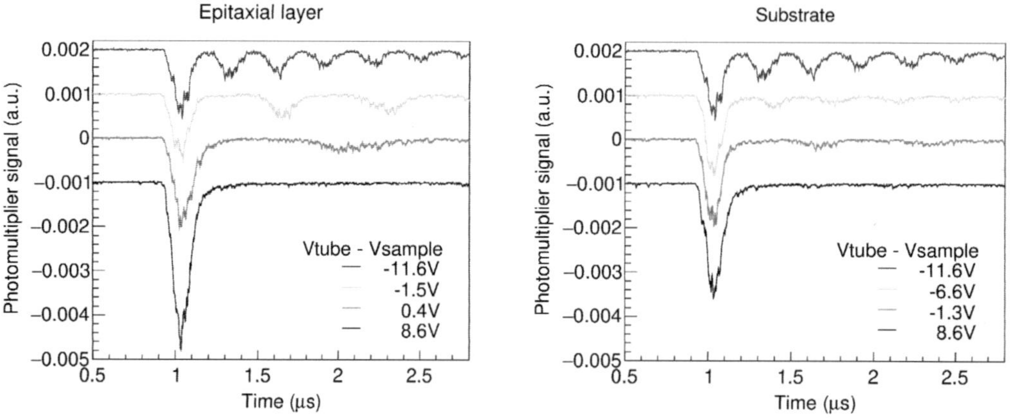

Figure 2. Signal obtained by the photomultiplier for the epitaxial layer (right) and substrate (left) for different voltage differences between the tube and the SiC sample. The energy of the incident positrons is 2 keV. The curves were shifted vertically for clarity.

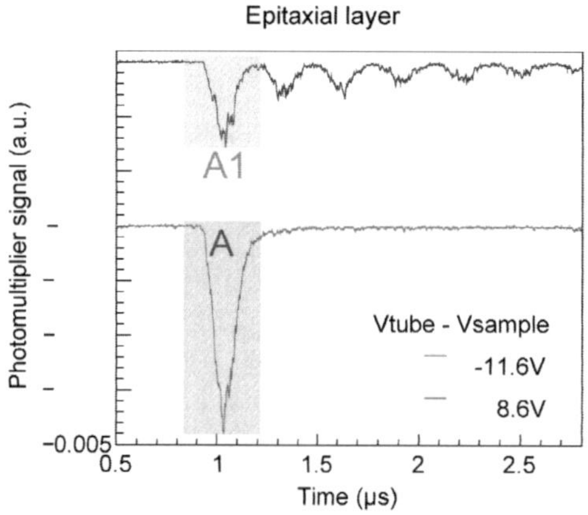

Figure 3. The positron re-emission yield (Y) is defined as the fraction of the incident positron beam that was remoderated. A_1 is the integrated area of the first positron annihilation peak, when remoderation occurred, and A the integrated area of the total annihilation peak.

of the voltage difference between the tube and the sample (ΔV). The data was fitted to the complementary error function. The positron work function Φ^+ can be found by subtracting the bias for which all the positrons emerge from the sample and the bias corresponding to the point of steepest descent of Figure 4. We have measured Φ^+ = -2.1 eV for the epitaxial layer and Φ^+ = -2.7 eV for the substrate, in agreement with the results of [3] and [4].

We can obtain the energy distribution of the surfacing positrons by differentiating the complementary error function. These had an energy spread of σ = 0.91 eV $\pm$ 0.05 eV for the epitaxial layer and σ = 1.13 eV $\pm$ 0.09 eV for the substrate.

The positron re-emission yield was also measured as a function of the incident positron beam energy. Our results were fitted using the program VEPFIT [6] to calculate the probability distribution, $P(E_{e^+}, z)$, as a function of the depth z, implantation profile and effective diffusion length L. We obtained an effective diffusion length of L = 267 nm for the epitaxial layer and

IOP Conf. Series: Journal of Physics: Conf. Series **791** (2017) 012005 doi:10.1088/1742-6596/791/1/012005

L = 51 nm for the substrate. The smaller diffusion length of the substrate is consistent with a higher amount of defects that trap the positrons. Furthermore, we observed a higher yield for a larger positron energy range for the epitaxial layer, which can be an advantage for large energy spread beams.

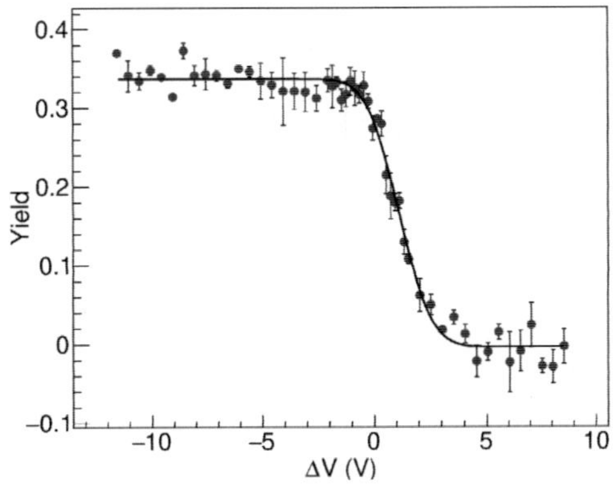

Figure 4. Positron re-emission yield as a function of the voltage difference between the tube and the SiC sample (ΔV) for the substrate.

Although the kinetic energy distribution of the remoderated positrons is significantly broader than that of the best single crystal metal moderators, high efficiency and easy handling makes SiC an attractive alternative in many cases where significant improvement is needed in brightness but narrow energy distribution is not essential. Such cases are when brightness enhancement is needed after a positron buncher with large amplitude or positron ejection from an accumulator with a deep potential well.

4. Conclusion

We demonstrate that a commercially available n-type 4H-SiC epitaxial layer grown on a SiC substrate is an efficient positron remoderator. A remoderation efficiency higher than 65% has been measured for 1 keV implanted positrons. The measured work function $\Phi^+ = -2.1$ eV is in good agreement with [3] and [4].

Our results indicate that the epitaxially grown layer is a superior secondary moderator than its substrate counterpart. Being easily available and requiring minimal maintenance makes the epitaxial 4H-SiC an attractive solution for positron moderation, facilitating the construction of laboratory-based positron beams with higher beam intensity and brightness.

References

[1] Stormer J, Goodyear A, Anwand W, Brauer G, Coleman P G and Triftshauser W 1996 *Journal of Physics: Condensed Matter* **8** 89
[2] Suzuki R, Ohdaira T, Uedono A, Cho Y K, Yoshida S, Ishida Y, Ohshima T, Ito H, Chiwaki M, Mikado T, Yamazak T and Tanigawa S 1998 *Japanese Journal of Applied Physics, Part 1: Regular Papers and Short Notes and Review Papers* **37** 4636
[3] Nangia A, Kim J H, Weiss A H and Brauer G 2002 *Journal of Applied Physics* **91** 2818
[4] Ling C C, Weng H M, Beling C D and Fung S *Journal of Physics: Condensed Matter* **14** 6373
[5] Liszkay L, Comini P, Corbel C, Debu P, Grandemange P, Pérez P, Rey J M, Reymond J M, Ruiz N, Sacquin Y and Vallage B 2014 *Journal of Physics: Conference Series* **505** 012036
[6] Veen A V, Schut H, Vries J D, Hakvoort R A and Ijpma M R 1991 *Positron Beams for Solids and Surfaces* (*American Institute of Physics Conference Series* vol 218) (New York: AIP) p 171

IOP Conf. Series: Journal of Physics: Conf. Series **791** (2017) 012006 doi:10.1088/1742-6596/791/1/012006

Micrometer positron beam characterization at the Scanning Positron Microscope Interface

J Mitteneder[1], M Dickmann[2], G Kögel[1], W Egger[1], P Sperr[1] and G Dollinger[1]

[1] Universität der Bundeswehr München, Inst. f. Angewandte Physik und Messtechnik, Werner-Heisenberg-Weg 39, 85577 Neubiberg, Germany

[2] FRM II, Technische Universität München, Lichtenbergstr. 1, 85748 Garching, Germany

E-mail: johannes.mitteneder@unibw.de

Abstract. For the investigation of inhomogeneous defect distributions the Scanning Positron Microscope (SPM) of the Universität der Bundeswehr München provides a pulsed positron beam with a diameter of about 1 μm and a time resolution of 250 ps (FWHM). To increase the count-rate the SPM is currently transferred to the intense positron source NEPOMUC.

To connect the SPM to the NEPOMUC source a special interface was build, which transforms the NEPOMUC beam to the requirements of the SPM. In this contribution we will give an overview of the SPM interface, and its performance. The beam is characterized at the finale stage of the interface, the positron elevator, where the potential energy of the beam is increased, without altering other beam parameters.

From our measurements we are able to predict the performance of the SPM at NEPOMUC. In future position resolved measurements will be possible with an improved spatial resolution of about 0.3 μm and an event rate of about 3.7 kHz.

1. Introduction

To characterize inhomogeneous defect distributions on the micrometer scale, e.g. precipitation hardened alloys, defect distributions close to fatigue cracks or grain boundaries with positron annihilation lifetime spectroscopy (PALS) a pulsed beam of variable energy and a spot size in the micrometer range is required.

The Scanning Positron Microscope (SPM) [1, 2] complies with these requirements. To achieve positron pulses of 200 ps duration and a spot size below 2 μm a narrow intense positron beam of about 5 mm^2 cross section and 30 meV transverse energy spread is required. Therefore, even with a primary source of 0.5 GBq ^{22}Na, only event rates of 100 to 500 Hz are achievable. This leads to exceedingly long measurement times. To overcome this limitation, the SPM is currently being connected to the intense positron source NEPOMUC [3, 4].

The primary beam of NEPOMUC delivers up to $5{\cdot}10^8\,{}^{e^+}\!/_s$ with a transverse phase space of about 3 000 $mm^2{\cdot}1$ eV. A first re-moderation stage reduces this phase space to about 5 $mm^2{\cdot}1.3$ eV and a intensity of $6{\cdot}10^7\,{}^{e^+}\!/_s$ for the re-moderated beam at the experiment entrance [5].

For further reduction of the phase space and pulsing the continuous NEPOMUC beam a dedicated interface was constructed for the SPM [6]. The newly develop final component of the interface, the positron elevator, increases the potential energy of the beam to the required potential at the entrance of the SPM. A brief description of the interface is given in section 2.

The knowledge of the actual beam parameters is important for further developments of the SPM interface and the SPM. Therefore we performed and report measurements of the beam characteristics and the influence of the elevator on the time structure. Finally, we predict the expected performance of the SPM at NEPOMUC.

2. The SPM interface at NEPOMUC

Figure 1 shows an overview of the SPM interface and the SPM. For the desired spatial resolution a narrow, monoenergetic beam with small energy spread is crucial. Therefore, the re-moderated positron beam of NEPOMUC is directed onto the SPM interface, which includes an additional remoderator (2nd remoderator). The last re-moderation stage (3rd remoderator) is a part of the SPM. In both re-moderation stages, magnetic lenses and W(100) single crystals in reflection geometry are used.

For positron lifetime measurements the continuous beam of NEPOMUC has to be pulsed. A narrow pulse width is essential for a good time resolution. To use as many positrons as possible a sawtooth prebuncher compresses the beam. In addition, two sine wave bunchers reduce the pulse width. A chopper cuts out the positrons between the pulses, thus reducing the background.

In each re-moderation stage several keV of kinetic energy are lost. The elevator compensates these losses between the re-moderation steps. Therefore, the 3rd remoderator can be set close to ground potential, as in the laboratory set up of the SPM. The elevator keeps the required high-voltage low and allows implantation energies up to 20 keV (see figure 1).

With a properly adjusted SPM interface, the 3rd remoderator provides a pulsed mono-energetic positron source of a few micrometer in diameter. Up to now, the SPM interface works as expected.

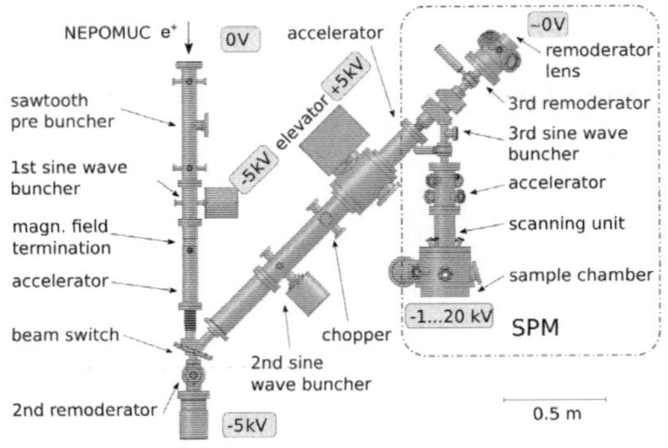

Figure 1. Schematic of the SPM (dashed box) and the SPM interface at NEPOMUC

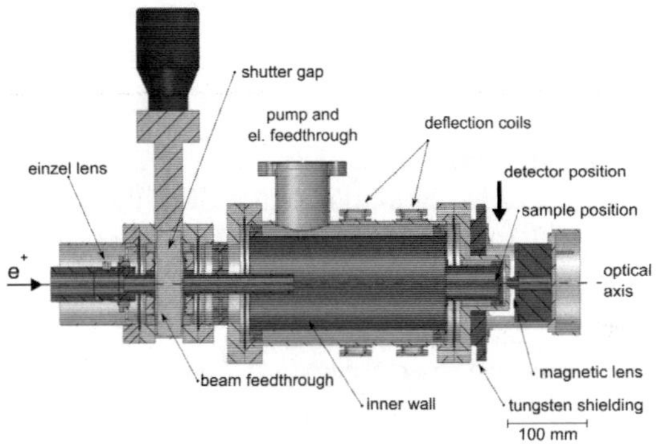

Figure 2. Schematic of the test chamber for beam characterization

3. Test chamber and beam characterization

For proper adjustment the beam characteristics have to be determined in front of the elevator. Therefore, a special test chamber was build for the SPM interface and installed at the position of the elevator (see figure 1).

3.1. Test chamber

The test chamber is shown in figure 2. The positron beam enters from the left and is electro-statically guided via the shutter gap into the chamber. The beam is deflected by scanning coils and magnetically focussed at the sample position. The detector is placed behind a tungsten shield. Due to the large inner volume of the chamber, the homogeneous electrical field inside and the shielding of the detector, no backscattered positrons are detected. For a better analysis of the beam characteristics the magnetic lens of the 3rd remoderator stage is used at the test chamber [7].

3.2. Experimental results

The sample consists of a conductive carbon tape and metallic glass, as shown in figure 3(a). These materials differ in positron lifetime due to the metallic quality of the glass and positronium production in the polymer.

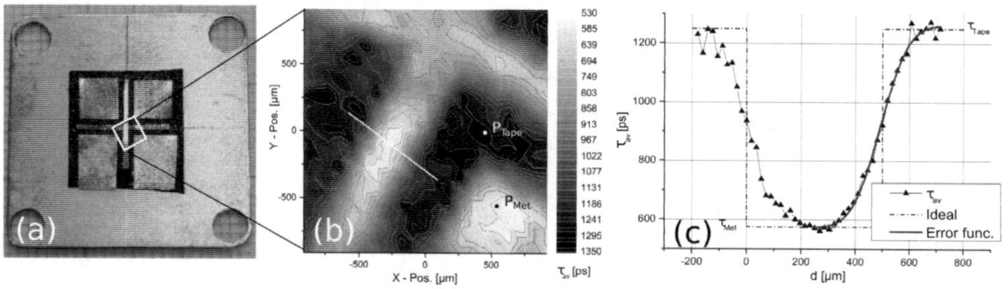

Figure 3. (a) Illustration of the sample and (b) mean positron lifetime map of the scanned area. In (c) the measured line scan is shown as indicated with the white line in (b).

All measurements were recorded with an implantation energy of 1.5 keV. In figure 3(b) the spatial distribution of the mean lifetime is shown. The scanning area is about 1.8 x 1.8 mm^2 and free of image distortion. The beam diameter of (180 $\pm$ 10) µm is estimated from the full width at half maximum (FWHM) of the slope of the line scan in figure 3(c).

Two positron lifetime spectra for the different sample materials are also obtained with high statistics (P_{Tape} and P_{Met} in figure 3(b)). The spectra are mainly free of backscatter peaks with a peak to background ratio of 1000. From these spectra a pulse duration of (257 $\pm$ 3) ps (FWHM) is derived.

4. The positron elevator

The elevator is the final stage of the SPM interface. It elevates the potential energy as indicated in figure 1 from -5 keV to +5 keV. In this way the positrons can be accelerated to the required energies for the implantation into the remoderator or the sample. Here we follow closely the detailed description in [8].

4.1. Setup of the positron elevator

The positron elevator consists of three electrodes, the entrance-, central- and exit-electrode. The entrance-electrode is on a negative potential of $U_{ent} = -5\,\text{kV}$, the exit electrode is at $U_{exit} = +5\,\text{kV}$. The central-electrode is on a static voltage of 0 V superimposed by a 10 kV (peak to peak) sine wave signal of $f = 50$ MHz; $U_{cent}(t) = 0\,\text{V} + 5\text{kV}\sin(2\pi f\,t)$.

The incoming pulsed positron beam is focused into the middle of the central electrode. Adjusting the beam velocity and the phase of the elevator to the positron pulses, they pass the gaps when they are field free. Thus the only effect of the elevator on the positron beam is the raised potential energy. The setup includes also a monitoring system with an Einzel-lens,

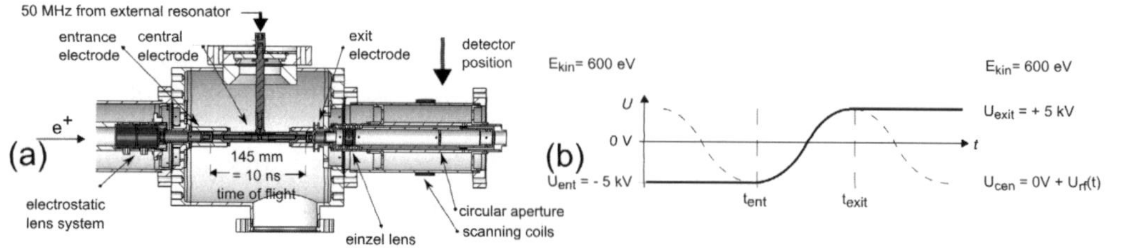

Figure 4. (a) Schematic of the positron elevator and the beam monitoring system. (b) Working principle and desired potential setup of the elevator.

scanning coils and a circular aperture. The beam can be scanned over the aperture in order to determine the beam diameter and the transverse momentum of the beam. These measurements are possible for both, the elevated and the non-elevated beam.

4.2. Experimental results

By adjusting the elevator correctly, almost all positrons with an energy gain of 1keV are transmitted through the gaps which are 4 mm in diameter.

We investigated the influence of the beam elevation on the time structure as shown in figure 5(a). The two spectra are normalized and shifted in time. There is no broadening of the peak, which means that the influence of the elevator on the time structure is negligible.

Figure 5(b) shows a scanning image of the aperture. The dark region (high count rate) is attributed to be the aperture surface where the count rate becomes higher due to the positron annihilation. From the rectangular areas in figure 5(b) the beam diameter can be calculated as shown in figure 5(c). We obtain beam diameters of 0.6 mm and 1.2 mm (FWHM) in x- and y-direction respectively.

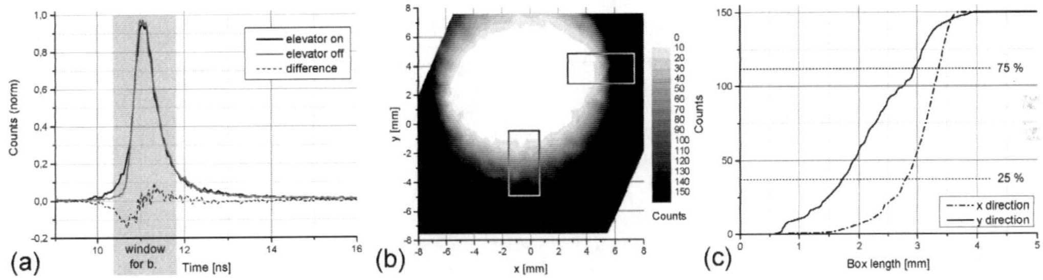

Figure 5. Experimental results of the positron elevator commissioning. (a) influence on the time structure, (b) count rate map of the scanned aperture, (c) transition plotted as profiles with respect to the indicated boxes in (b).

5. Conclusion and outlook

The newly implemented SPM interface with the positron elevator has been used to adapt the NEPOMUC beam properly to the required input characteristics of the SPM: the continuous beam is pulsed, enhanced in brightness and elevated to 5 keV potential energy with respect to ground.

Thus we expect for the SPM at NEPOMUC a time resolution of 250 ps and an improved spatial resolution of about 300 nm (FWHM) at a drastically improved count rate of 3700 events per second in comparison to the laboratory setup [8].

Acknowledgments

Funding from the German BMBF (Projects 05K10WNA-Posimethod and 05K13WN1-Posianalyse) and financial support by the Munich research reactor FRM II is gratefully acknowledged.

References

[1] Kögel G 1997 *Appl. Surf. Sci.* **116** 108 – 113
[2] David A, Kögel G, Sperr P and Triftshäuser W 2001 *Phys. Rev. Lett.* **87** 067402
[3] Kögel G and Dollinger G 2006 *Appl. Surf. Sci.* **252** 3111 – 3120
[4] Hugenschmidt C and Piochacz C 2015 *Journal of large-scale research facilities* **1** A22
[5] Stanja J, Hergenhahn U, Niemann H, Paschkowski N, Pedersen T S, Saitoh H, Stenson E, Stoneking M, Hugenschmidt C and Piochacz C 2016 *Nucl. Instrum. Methods Phys. Res., Sect. A* **827** 52 – 62
[6] Piochacz C 2009 *Generation of a high-brightness pulsed positron beam for the Munich scanning positron microscope* Ph.D. thesis TUM
[7] Britton D, Uhlmann K and Kögel G 1995 *Appl. Surf. Sci.* **85** 158 – 164
[8] Dickmann M, Mitteneder J, Kögel G, Egger W, Sperr P, Ackermann U, Piochacz C and Dollinger G 2016 *Nucl. Instr. Methods in Phys. Res. A* **821** 40 – 43

IOP Conf. Series: Journal of Physics: Conf. Series **791** (2017) 012007 doi:10.1088/1742-6596/791/1/012007

Study on Bose-Einstein condensation of positronium

K Shu[1], T Murayoshi[1], X Fan[1], A Ishida[1], T Yamazaki[1], T Namba[1], S Asai[1], K Yoshioka[2], M Kuwata-Gonokami[1], N Oshima[3], B E O'Rourke[3] and R Suzuki[3]

[1]Department of Physics, Graduate School of Science, and International Center for Elementary Particle Physics (ICEPP), The University of Tokyo, 7-3-1 Hongo, Bunkyo-ku, Tokyo 113-0033, Japan

[2]Photon Science Center (PSC), Graduate School of Engineering, The University of Tokyo, 7-3-1 Hongo, Bunkyo-ku, Tokyo 113-0033, Japan

[3]National Institute of Advanced Industrial Science and Technology (AIST), 1-1-1 Umezono, Tsukuba, Ibaraki 305-8568, Japan

E-mail: kshu@icepp.s.u-tokyo.ac.jp

Abstract. A new method to realize Bose-Einstein condensation (BEC) of positronium (Ps) is proposed and a Monte-Carlo simulation was performed to evaluate the cooling efficiency of Ps atoms. Ps-BEC was found to be possible by a method including laser cooling of Ps. Laser specifications for Ps cooling are challenging and development is ongoing. A new temperature measurement method for cold Ps and development of an intense, pulsed, slow-positron microbeam are also necessary. They are under study as outlined in this contribution.

1. Introduction

Positronium (Ps), a bound state of an electron and a positron, is the best candidate to form Bose-Einstein condensation (BEC) with anti-matter for the first time because of the following two advantages: (1) the number of atoms which can be created is much larger for Ps by accumulating slow positrons than any other bosons with anti-matter, (2) its mass is so light that the critical temperature of BEC transition is quite high. Many applications with Ps-BEC are expected, such as the realization of a 511 keV gamma-ray laser via coherent decays of Ps-BEC[1, 2] and a precision measurement of gravity on a positron using an atomic interferometer[3].

Two key technologies are essential to realize Ps-BEC: (1) creation of a dense ensemble of Ps atoms, (2) cooling them down to low enough temperature in a few hundreds of nanoseconds because the lifetime of long-lived *ortho*-positronium is as short as 142 ns. The state-of-the-art techniques achieved 150 K for the temperature[4] and $10^{15}\,\mathrm{cm}^{-3}$ for the density[3]. The target phase-space density to realize Ps-BEC is $10^{18}\,\mathrm{cm}^{-3}$ at 14 K. In order to create such dense and cold Ps atoms, we will develop a new slow-positron microbeam and a Ps cooling system which combines a cold cavity and laser cooling.

2. Cooling of positronium

A new experimental method to realize Ps-BEC was recently proposed[5]. In this method, 10^7 spin-polarized slow positrons are focused into a 100 cube nm silica cavity, resulting in around 4000 spin-polarized Ps atoms being created and trapped inside the cavity. The Ps atoms will then be cooled down to 10 K via the following two processes:

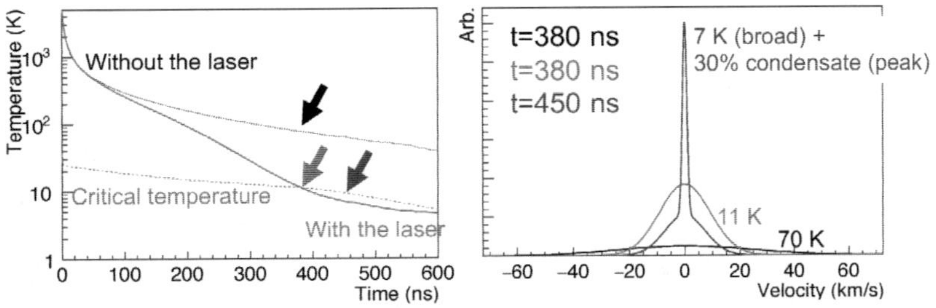

Figure 1. (Left) Temperature evolution of Ps. (Right) Ps velocity distributions. Black: without the laser at $t = 380\,\text{ns}$. Red and blue: with the laser at $t = 380\,\text{ns}$, and $t = 450\,\text{ns}$, respectively. Each arrow in the left panel shows the point at which the velocity distribution with the same color is drawn.

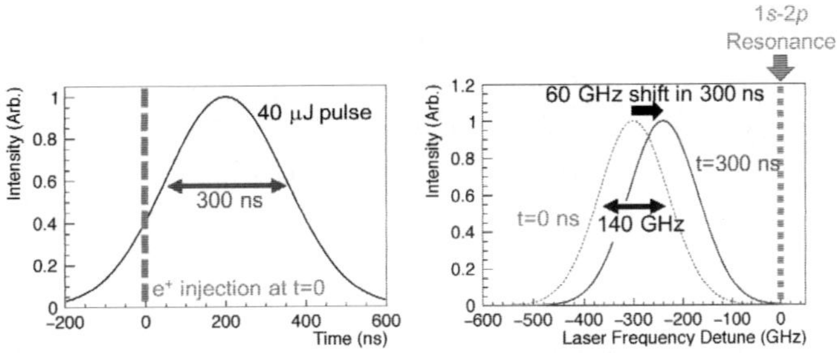

Figure 2. (Left) Time profile of the laser. (Right) Frequency profile of the laser.

(i) Exchange of kinetic energy between the cold (1 K) silica and Ps (a thermalization process). This process is efficient for Ps from its creation (around 6000 K) down to 300 K.

(ii) Laser cooling (Doppler cooling) using the $1s - 2p$ transitions. This process is efficient from 300 K down to 10 K.

The cooling efficiency of Ps in this scheme was evaluated by a Monte-Carlo simulation, which was explained in detail elsewhere[5]. The simulation included (1) the thermalization process, (2) Ps-Ps two-body scatterings, (3) interactions between Ps and the laser photons. Figure 1 shows calculated temperature evolutions and selected Ps velocity distributions with and without laser cooling. As the temperature evolution graph shows, cooling with the laser is efficient enough to cool Ps down below the critical temperature with a phase transition to the BEC phase at $t = 380\,\text{ns}$. Ps-BEC can be observed by a sharp peak in the velocity distribution as shown in the right panel of figure 1. Around 30% of the remaining Ps atoms are estimated to be in the BEC phase at $t = 450\,\text{ns}$.

3. Development of the cooling laser

The wavelength of the cooling laser for Ps is 243 nm (1.23 PHz) ultraviolet light for the $1s - 2p$ transition. The laser needs some challenging specifications. The first requirement is a special time profile, which is shown in the left panel of figure 2. The laser must operate in pulsed mode in order to achieve enough power to saturate the transitions. The pulse must be long enough to continuously cool Ps down for more than 300 ns. The second requirement is a special frequency profile as shown in the right panel of figure 2. The Doppler effect for a Ps atom is quite large (e.g. 240 GHz at 300 K) due to its light mass. In order to cool Ps with various

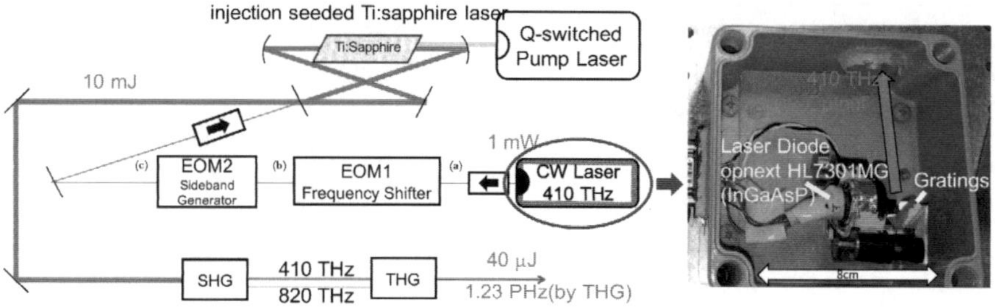

Figure 3. A schematic diagram of the laser system for Ps cooling, and an external cavity diode laser for seeding. Details are in [5].

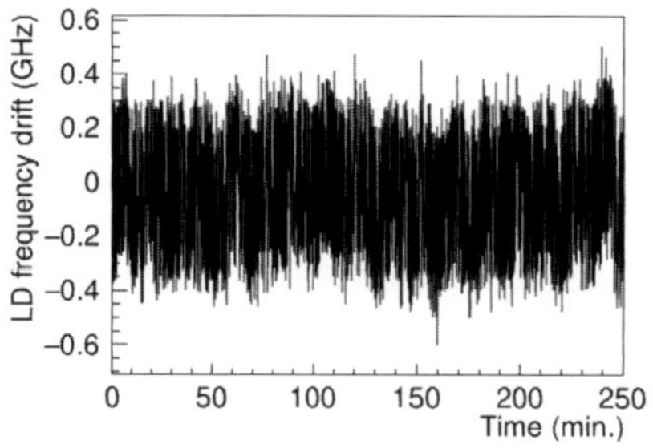

Figure 4. Frequency drift of the ECDL. The frequency should be stabilized within around 1 GHz for the later frequency shift and the sideband generation.

velocities, the frequency distribution must be as wide as 140 GHz. The third requirement is the frequency shift which is also demonstrated in the right panel of figure 2. The center frequency should be shifted in accordance with the temperature of Ps because more Ps need photons whose frequencies are close to the $1s-2p$ resonance for Ps at rest. The shift should be 60 GHz in 300 ns. This frequency shift during the pulse is a new challenge for optics.

In order to realize the desired laser specifications, a new laser system was designed[5]. Figure 3 show a schematic diagram of the system. A Ti:sapphire seed injection laser generates 729 nm pulsed light of 300 ns. The frequency is tripled via harmonic generations (SHG and THG). Fine tuning of the frequency is performed for a continuous-wave (CW) laser, which is seeded into the Ti:sapphire afterwards. These controls will be done by electro-optic modulators (EOMs).

Development of the Ps cooling laser is ongoing. The CW laser for seeding is now ready as shown in figure 3. An external cavity diode laser (ECDL) was constructed for this seeding light. Performance of the ECDL was examined to confirm that it was powerful and stable (see figure 4) enough for the later stages. The injection seeding into Ti:sapphire is the next step.

4. Temperature measurement method for cold Ps

A method to measure the temperature of Ps as cold as 10 K is necessary for Ps-BEC experiment. We are planning to perform Doppler spectroscopy using the $1s-3s$ transition (410 nm visible wavelength) because the laser system can be easily constructed. Figure 5 shows required laser specifications for 10 K Ps and an expected resonance curve with 10^7 Ps atoms by a Monte-Carlo simulation. A sufficiently high precision can be obtained during a several-day run with a modern intense slow-positron beamline. Development of this laser is currently under study. Spectroscopy for Ps in cold silica is planned to confirm the ability of the method and also to

Wavelength	410 nm
Pulse energy	1 mJ
Pulse width	A few ns
Bandwidth (FWHM)	25 GHz
Freqency tunable range	150 GHz
beam waist (2σ)	850 μm
Repetition rate	100 Hz

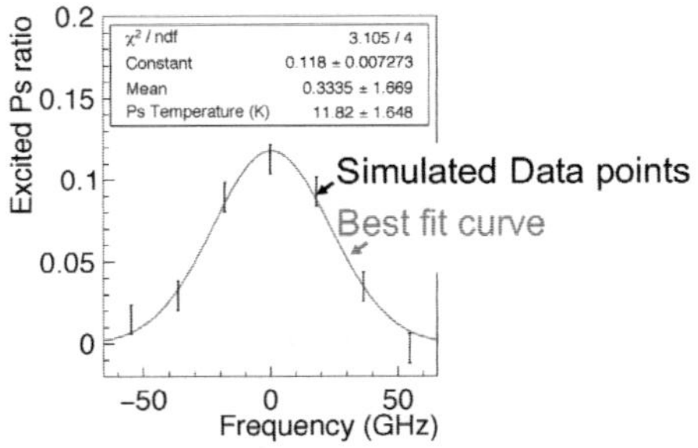

Figure 5. (Left) Required laser specifications for the $1s - 3s$ Doppler spectroscopy. (Right) An expected resonance curve for 10^7 Ps in total at $t = 300\,\text{ns}$, 10 K. A gamma ray detector is assumed to be $LaBr_3(Ce)$ scintillator (20 ns fall time and 5% FWHM energy resolution at 511 keV) with 5% detection efficiency for 2 gamma 511 keV decay.

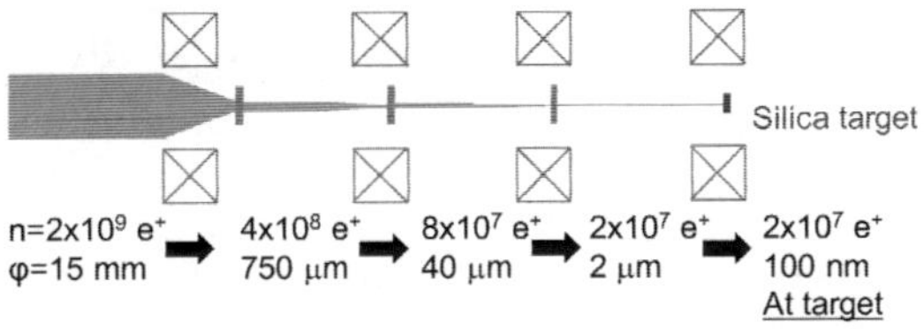

Figure 6. A proposed 3-stage brightness enhancement system to achieve 10^7 slow positrons with a 100 nm beam waist. The desired beam compression is 1/20 at each stage.

perform a precise measurement of thermalization process between Ps and silica.

5. Overview of the slow-positron microbeam

Development of a slow-positron microbeam with a 100 nm beam waist is essential to obtain a high enough density of Ps to realize Ps-BEC. A brightness enhancement technique[6] will be implemented and improved for this purpose. Considering a reduction of beam diameter is by a factor of around 20 at each stage of brightness enhancement, with 2×10^9 bunched slow positrons, and transmission type remoderators with 20% efficiency each, 10^7 positrons in 100 nm is possible using a 3-stage brightness enhancement system as shown in figure 6. Design of the buncher and the beam optics are currently under study.

6. Roadmap for Ps-BEC

The first step is the temperature measurement by Doppler spectroscopy in two years. After establishing the temperature measurement method, laser cooling of Ps will be performed in four years. The slow positron system will be developed at the same time and combined for Ps-BEC.

References

[1] Vanyashin V 1994 *Letters in Mathematical Physics* **31** 143–149
[2] Avetissian H K, Avetissian A K and Mkrtchian G F 2014 *Phys. Rev. Lett.* **113** 023904
[3] Cassidy D B and Mills A P 2007 *physica status solidi (c)* **4** 3419
[4] Mariazzi S, Bettotti P and Brusa R S 2010 *Phys. Rev. Lett.* **104** 243401
[5] Shu K, Fan X, Yamazaki T, Namba T, Asai S, Yoshioka K and Kuwata-Gonokami M 2016 *Journal of Physics B: Atomic, Molecular and Optical Physics* **49** 104001
[6] Oshima N, Suzuki R, Ohdaira T, Kinomura A, Narumi T, Uedono A and Fujinami M 2008 *Materials Science Forum* **607** 238

Low energy positron beam system for the investigation of 2D and porous materials

M D Chrysler[1], V A Chirayath[1], A D Mcdonald[1], R W Gladen[1], A J Fairchild[1], A R Koymen[1] and A H Weiss[1]

[1]The University of Texas at Arlington, Arlington Texas, USA

Email: matthew.chrysler@mavs.uta.edu

Abstract. An advanced variable energy positron beam (~2 eV to 20 keV) has been designed, tested and utilized for coincidence Doppler broadening (CDB) measurements at the University of Texas at Arlington (UTA). A high efficiency solidified rare gas (Neon) moderator was used for the generation of a slow positron beam. The gamma rays produced as a result of the annihilation of positrons with the sample electrons are measured using a high purity Germanium (HPGe) detector in coincidence with a NaI(Tl) detector. Modifications to the system, currently underway, permits simultaneous measurements utilizing Positron annihilation induced Auger Electron Spectroscopy (PAES) and CDB. The tendency of positrons to become trapped in an image potential well at the surface will allow the new system to be used in measurements of the chemical structure of surfaces, internal or external and interfaces. The system will utilize a time of flight (TOF) technique for electron energy measurements. A 3m flight path from the sample to a micro-channel plate (MCP) in the new system will give it superior energy resolution at higher electron energies as compared to previous TOF systems utilizing shorter flight paths.

1. Introduction

Substantial interest has developed in the field of two-dimensional (2D) materials since the isolation of atomic layer graphene [1]. The rich electronic properties observed in 2D materials, such as the superconducting transitions [2] and semiconducting properties observed in 2D transition metal dichalcogenides (TMDs), provide materials and electronics engineers the prospect of designing and controlling 2D material properties and devices based on them [3,4]. Not only is there interest in 2D materials, but also in the properties specific to the surfaces of materials [5]. In this regard, it is desirable to study these systems with techniques that can probe material properties with atomic layer precision without probe modification of the surface properties and with minimal sample preparation.

Positron beams have been used to study depth dependent properties of overlayers with positron energies typically larger than 100 eV [6]. By reducing the positron beam energy to a few eV it has been shown that positrons can be efficiently deposited into a surface localized state in a process known as Auger mediated positron sticking (AMPS) [7]. Due to this localization, the signals resulting from the annihilation of surface trapped positrons are highly surface specific. Consequently, low energy

IOP Conf. Series: Journal of Physics: Conf. Series **791** (2017) 012008 doi:10.1088/1742-6596/791/1/012008

positrons are well suited to the investigation of two dimensional materials.

Nanoporous metals are another class of materials which are developed for applications in catalysis, energy storage, bio sensing, etc [8]. An understanding of the chemical and structural evolution of the internal surfaces of these high aspect ratio materials are important in tuning their properties to increase their efficiency in photo catalysis [9] or to understand their stability in service conditions [10]. Conventional methods used in characterizing external surfaces are unsuitable for internal surfaces due to the inability of the electron or optical signals to escape from the internal surfaces. Investigations of internal surfaces have relied on destructive or indirect measurements such as EXAFS [11].

Positrons implanted at a few keV energies into porous materials will predominantly reach the internal surfaces of pores and voids after thermalization and will get trapped in a surface state on the internal surfaces resulting in annihilation with electrons at the top atomic layer. The ability of the annihilation gammas to reach the detector from buried interfaces or from the internal surfaces of pores without the loss of the information that can be derived from the gamma spectra permits annihilation gamma spectroscopies to be used in the characterization of internal surfaces and interfaces [12].

In this paper we describe a new positron beam system that has been developed at the University of Texas at Arlington with the aim of studying 2D materials and internal surfaces of porous metals. The system will allow simultaneous measurement of the Doppler broadening (DB) of the annihilation gamma and Positron annihilation induced Auger electron spectra (PAES). The system is capable of depositing positrons of kinetic energies as low as 2 eV to as high as 20 keV on 2D or porous samples. This will enable us to investigate chemical information (PAES and DB) and open volume defect (DB) distribution on 2D materials when low energy (a few eV) positrons are used. With positrons at a few kilo electron volt energies, structural defects and chemical information of internal surfaces of porous materials may be obtained. DB spectra and PAES spectra collected in coincidence will enable us to correlate the Doppler broadened gamma-ray line shape to positron annihilations with electrons belonging to specific core levels of an element as demonstrated by Kim *et al* and Eshed *et al* [13, 14]. This correlation will be utilized to quantify the Doppler broadened gamma line shape so as to deduce chemical information from the inner surfaces of porous materials through the analysis of the gamma spectrum.

2. Experimental Apparatus

The new positron beam system shown in Figure 1 consists of three main components: 1) A rare gas moderator (RGM), 2) A 3m TOF region, and 3) Sample preparation and analysis chamber.

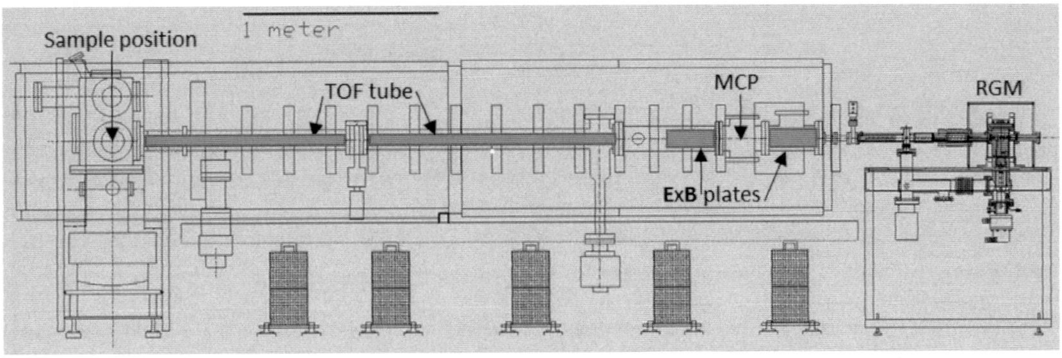

Figure 1. The rare gas moderator (RGM) on the far right is used to produce a slow positron beam. The positrons will be bent around a Microchannel Plate (MCP) electron detector using **E**×**B** deflection plates. A 3 m electric field-free region is used for TOF measurements. The sample chamber at the end houses the sample manipulator and is the site of gamma ray detection.

2.1. Rare Gas Moderator

A solidified rare gas moderator (RGM), produced by condensing Neon gas onto a cold head kept at 6.8 K, is used to generate slow positrons from a 0.5 mCi ^{22}Na radioactive source. A moderator efficiency of ~ 0.2 % has been achieved initially. Further optimization is required in the solid neon growth parameters and in the extraction optics of the slow positron beam to achieve the 1% efficiency reported for solid neon moderator [15]. A typical moderator will degrade to ~ 75% of its initial count rate after 6 days. Control of the positron beam kinetic energy is accomplished by applying a positive bias to the moderator. Once the slow positrons are emitted from the Neon moderator, they are accelerated by the electric field created by the positive bias on the moderator and guided by an axial magnetic field. The positrons are deflected by a magnetic field through an aperture in a tungsten barrier offset from the beam axis. The tungsten barrier serves to filter out fast positrons which are not deflected enough to make it through the offset hole. The positrons which make it through the filter are introduced into the TOF region using a magnetic field of 0.03 T. A beam diameter of ~ 6 mm at the exit of the RGM is decided by the aperture at the exit.

2.2. Time-of-Flight (TOF) Spectrometer

The spectrometer consists of a TOF tube which can be independently biased with respect to the grounded beam tube, a set of **E**×**B** plates, and a MCP electron detector. Positrons entering the TOF region are magnetically guided to the sample using a series of coils which produce an axial magnetic field of ~ 0.006 T. A set of rectangular correction coils are used to counteract the Earth's magnetic field or any other stray magnetic fields along the beam line and to help align the beam on the center of the sample. The **E**×**B** plates will be used to bend the positron beam around the MCP and to bend electrons emitted from the sample into the MCP. The 3 m long flight path of the new spectrometer is expected to lead to a 3-fold improvement in resolution compared to previous TOF systems [16]. Equation 1 shows that the flight time, t, of an electron traveling from the sample to the MCP is proportional to the travel distance, l. What is measured as a flight time is converted to channel number using a time-to-amplitude converter (TAC) and a multichannel analyzer and thus a difference in flight time will correspond to a proportional difference in channel numbers. As the flight path is increased to 3 m from 1 m, the difference in flight times of two electrons of nearby energies increases by three times thereby increasing the separation between them in the TOF spectra to three times their previous separation. This means that the energy per channel is reduced to one third of its previous value. High energy Auger peaks which were irresolvable because of being too close (< 2 ns) in time of arrival will now be resolved as they show up as separate peaks.

$$\# \propto t = \sqrt{\frac{m}{2E}}\, l \qquad (1)$$

2.3. Sample Preparation and Analysis Chamber

The sample chamber, capable of achieving a vacuum of 1.33×10^{-9} Pa, is equipped with a sputtering system to clean the sample, a LEED system for surface structure determination, and an electron Auger Electron Spectrometer (EAES) for additional characterization of surfaces. A permanent magnet producing a magnetic field of 0.07 T at the sample position is used to parallelize the electrons leaving the sample. The magnet also reduces the positron beam to ~4 mm on the sample allowing analysis of small samples. The sample is mounted on an x-y-z stage manipulator which helps in moving the sample from the beam axis to the sputtering position and to the analysis position of LEED and EAS. The sample is electrically isolated from the manipulator and is biased through the copper rod of a HV feed through. The system is equipped with a high resolution HPGe detector and a fast NaI(Tl) detector for coincidence doppler broadening (CDB) measurements.

3. System Testing

Having two oppositely placed gamma detectors in coincidence mode (Figure 2) reduces the background by an order of magnitude as shown in figure 3 which is comparable to the improvement obtained with similar coincidence systems [17]. In coincidence mode the output of the NaI(Tl) detector is used to gate the amplified signals from the HPGe detector at the multichannel analyzer (MCA). The histogram of the gamma counts generated by the MCA is converted to energy spectra by calibrating the detector system using four known gamma peaks of ^{133}Ba. The energy resolution of the HPGe detector was found to be 0.98 keV at the 356 keV peak of ^{133}Ba.

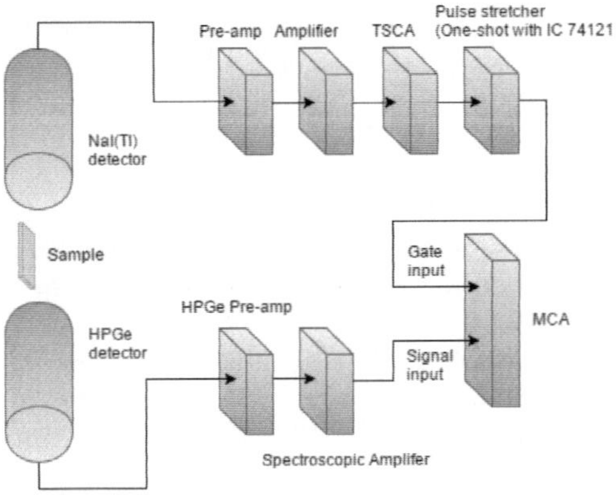

Figure 2. The NaI signal is amplified and the window corresponding to 511 keV gamma rays is selected using a timing single channel analyzer (TSCA). This signal is used as a gate for the HPGe detector.

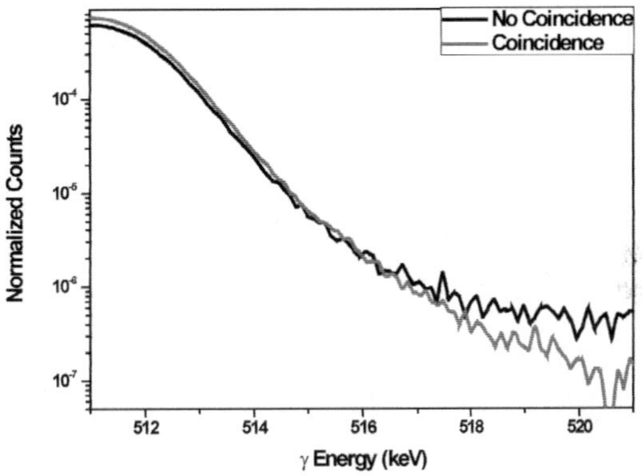

Figure 3. Collecting gamma ray spectra using coincidence mode reduces the background counts by an order of magnitude. This allows us to determine the high momentum region of the 511 keV line shape more accurately. Both spectra were normalized to their integrated intensities from 501-521 keV. Counting rates during this experiment were ~ 10 counts per second.

The system is capable of changing the kinetic energy of the positrons at the sample from 2 eV to 20 keV by appropriately biasing the sample with respect to the moderator. Figure 4 shows the positron kinetic energy measured at the entrance of the TOF tube by using it as a retarding field analyzer. Here the moderator was biased to 1 V resulting in a positron energy distribution with a FWHM of ~1.5 eV and peaking at 2.75 eV. Tests confirmed that the annihilation gammas from positrons reflected from a positively biased sample did not effect the results as shown in Figure 5. The reflected positrons retrace their path through the system and annihilate near the RGM as evidenced by the icrease in the gamma-ray counts near the RGM exit aperture with increase in positive bias on the sample.

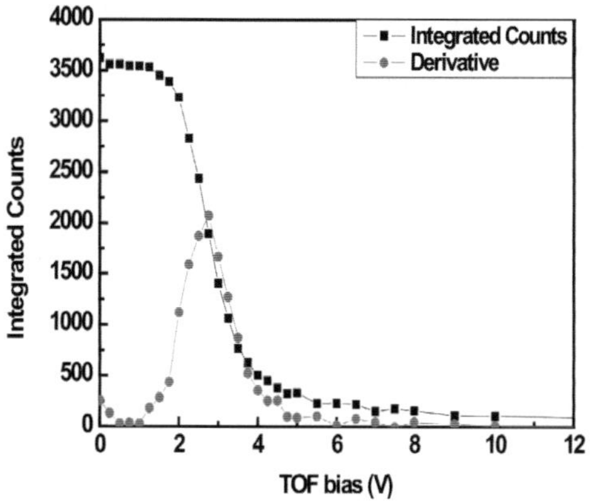

Figure 4. Positron kinetic energy distribution with a moderator bias of 1 V. Here we have plotted the integrated counts under the 511 keV peak as detected by the HPGe detector at the sample as a function of positive bias on the TOF tube. The derivative of the integrated counts give us the distribution of the positron energies, also shown in the figure. The energy is peaked at ~2.75 eV with a FWHM of ~1.5 eV.

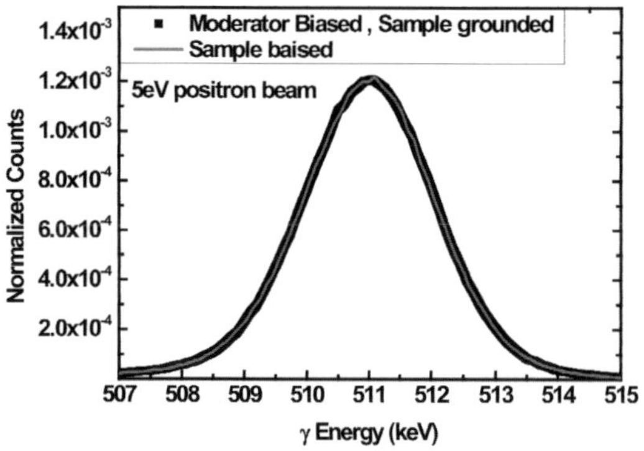

Figure 5. 511 keV gamma peak produced by biasing the system in two different ways to produce a 5 eV positron energy at the sample. The first method biases the moderator at +4 V with the sample grounded. The beam energy was determined to be 5eV at the TOF tube. In the second method, the moderator is biased to +32 V to get a 33 eV positrons at the TOF tube. The sample is then positively biased to 28 V to decrease the incident positron energy to 5 eV. Both methods are shown to produce equivalent gamma spectra.

4. Conclusions and Future Work

A positron beam system capable of simultaneously carrying out CDB and PAES measurements is under construction at UT, Arlington. The CDB spectroscopy system has been tested and utilized to investigate multilayer (6-8 layers) graphene grown on polycrystalline Cu [18]. Modifications to the system are currently underway to install a MCP electron detector and timing electronics. Electrons emitted from the sample travel through a 3 m flight path before being detected. The 3 m electric field-free region will give the new system superior energy resolution as compared to previous systems with shorter flight paths and allow for the investigation of internal and external surfaces. Additional improvements over the previous system include the addition of a HPGe detector for CDB experiments, as well as plans to digitize the data collection processes. Digitization will allow for time-correlated measurements of annihilation gammas and electrons collected at the channel plate.

Acknowledgments

This work was supported by NSF grant DMR 1508719 and DMR 1338130.

References

[1] Geim A K and Novoselov K S 2007 *Nat. Mater.* **6** 183
[2] Xu C, Wang L, Liu Z, Chen L, Guo J, Kang N, Ma X-L, Cheng H-M and Ren W 2015 *Nat. Mater.* **14** 1135

[3] Mak K F, Lee C, Hone J, Shan J and Heinz T F 2010 *Phys. Rev. Lett.* **105** 136805
[4] Wang Q H, Kalantar-Zadeh K, Kis A, Coleman J N and Strano M S 2012 *Nat. Nanotechnol.* **7** 699
[5] Qi X-L and Zhang S-C 2011 *Rev. Mod. Phys.* **83** 1057
[6] Coleman P, ed. 2000 *Positron beams and their applications* (Singapore: World Scientific)
[7] Mukherjee S, Nadesalingam M P, Guagliardo P, Sergeant A D, Barbiellini B, Williams J F, Fazleev N G and Weiss A H 2010 *Phys. Rev. Lett.* **104** 247403
[8] Seker E, Reed M L and Begley M R 2009 *Materials* **2** 2188
[9] Xu C, Xu X, Su J and Ding Y 2007 *J. Catal.* **252** 243
[10] Viswanath R N, Chirayath V A, Rajaraman R, Amarendra G and Sundar C S 2013 *Appl. Phys. Lett.* **102** 253101
[11] Frenkel A I, Vasic R, Dukesz B, Li D, Chen M, Zhang L and Fujita T 2012 *Phys. Rev. B* **85** 195419
[12] Gidley D W, Peng H G and Vallery R S 2006 *Annu. Rev. Mater. Res.* **36** 49-79
[13] Kim S, Eshed A, Goktepeli S, Sterne P A, Koymen A R, Chen W C and Weiss A H 2006 *Phys. Rev. B* **73** 014114
[14] Eshed A, Goktepeli S, Koymen A R, Kim S, Chen W C, O'Kelly D J, Sterne P A and Weiss A H 2002 *Phys. Rev. Lett.* **89** 075503
[15] Mills A P Jr. and Gullikson E M 1986 *Appl. Phys. Lett.* **49** 1121
[16] Mukherjee S, Shastry K, Anto C V, Joglekar P V, Nadesalingam M P, Xie S, Jiang N and Weiss A H 2016 *Rev. Sci. Instrum.* **87** 035114
[17] Brusa R S, Deng W, Karwasz G P and Zecca A 2002 *Nucl. Instrum. Methods Phys. Res. B* **194** 519
[18] Chirayath V A, M D Chrysler, A D McDonald, R W Gladen, A J Fairchild, A R Koymen and A H Weiss *J. Phys.: Conf. Series* in this volume

Construction of a surface positronium lifetime spectroscopy apparatus with a spin-polarized low energy positron beam

M Maekawa, K Wada, A Miyashita and A Kawasuso

Quantum Beam Science Research Directorate,
National Institutes for Quantum and Radiological Science and Technology,
1233 Watanuki, Takasaki, Gunma, 370-1292, Japan

E-mail: maekawa.masaki@qst.go.jp

Abstract. We have developed a surface positronium (Ps) lifetime spectrometer. A spin-polarized slow positron beam is generated by a sodium-22 source and a solid krypton moderator and transported to the sample chamber by electrostatic lenses. The incident energy is adjusted by a retarding potential using a deceleration tube. For the lifetime measurement, the start signal is obtained from a secondary-electron detector installed in front of the sample. The stop signal is detected by a scintillation detector for annihilation gamma rays. As a performance test, Ps lifetime measurements were carried out and clear decay curves derived from ortho-Ps self-annihilation were successfully observed. This system might be used for the estimation of the spin polarization at the surface of ferromagnetic materials.

1. Introduction

As a method to determine surface spin polarization, we have developed a spin-polarized slow positron beam apparatus [1-3]. Positrons, which are implanted near the surface and then diffuse to the surface, are emitted as positronium (Ps) when the work function of Ps is negative. By measuring 3-gamma annihilation ratio of ortho-Ps (o-Ps) decay, the spin polarization at the first surface layer can be determined. Since Ps is formed only at a vacuum region with a low-electron density, the spin polarization obtained by this method is not affected by the bulk condition. Indeed, we successfully observed novel spin phenomena, such as current-induced spin polarization [4].

The intensity of o-Ps decay can be obtained from both an energy spectrum of the annihilation gamma rays and a lifetime spectrum [5]. Since the energy distribution of 3-gamma annihilation is continuous in the range of 0-511 keV, it is not easy to estimate this precisely from the energy spectrum, which is a mixture of some effects, such as Compton scattering. From the lifetime measurement, however, it is easy to extract o-Ps intensity because of its characteristic long lifetime of 142 ns.

To obtain lifetime spectrum using a slow positron beam, a pulsed beam is often used. However, it is difficult to form a pulsed beam in our apparatus using only an electrostatic field to generate the spin-polarized positron beam. Gidley *et al.* show that the Ps lifetime measurement is possible by detecting secondary electrons when positrons with the energy range of 300 - 1600 eV are implanted onto a sample [6, 7]. In this study, to carry out a surface Ps lifetime measurement, we optimised the secondary-electron method for a spin-polarized positron beam with low energy.

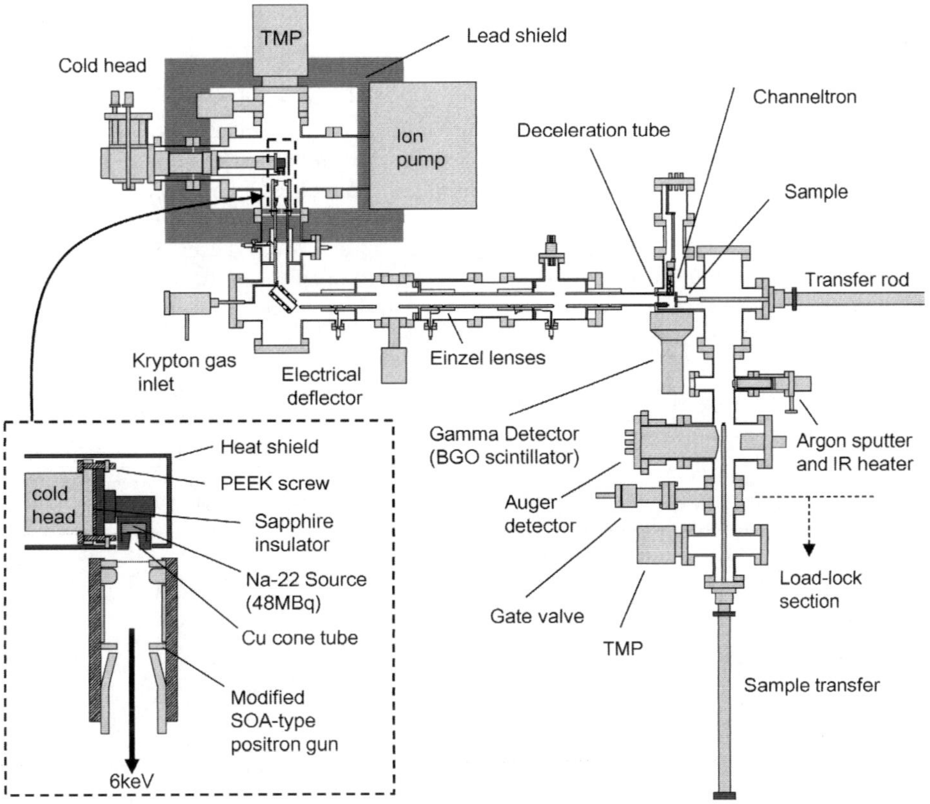

Figure 1. Schematic diagram of the apparatus developed in this work.

2. Construction of apparatus

Figure 1 shows an overview of the newly developed beam apparatus. A 48 MBq ^{22}Na sealed source and a copper cone tube with a bottom diameter of 5 mm and an angle of 5.7 degree are mounted on a 4 K refrigerator. A solid krypton moderator is formed on the inner surface of the copper cone and on the source window [8]. A slow positron beam with an energy of 6 keV is generated by a modified Soa-gun [9]. Since positrons emitted from radioisotopes are longitudinally spin-polarized [10], one can obtain spin-polarized positron beam by the electrostatic beam transportation not to rotate spin direction. The positron beam is transported by five einzel lenses and focused onto the sample. By using an electrostatic deflector, a transversely spin-polarized positron beam is formed. Spin polarization was estimated before to be approximately 27 % using the magnetic quenching method [11].

Figure 2 shows a schematic diagram of the lifetime measurement system. The energy of the incident positron beam is reduced to typically 50 eV by a deceleration tube located in front of the sample. For the detection of secondary electrons that are used as start signals, a channeltron module (Photonis, model 4831) is installed between the deceleration tube and the sample. Since the channeltron module is biased to the same electric potential of the sample, the signals are taken out through a high-voltage fast capacitor (TDK, UHV-241A). Ps which is formed on the sample surface annihilates with the lifetimes of 125 ps (para-Ps) and 142 ns (ortho-Ps). Annihilation gamma rays, which are used as stop signals, are detected by a Bismuth Germanate (BGO) scintillation detector beside the sample. These start and stop signals are collected by a digital oscilloscope (Lecroy, WavePro7100) providing the Ps lifetime spectrum.

With only the channeltron, the secondary electrons are rarely collected because the secondary electrons are pushed back to the sample by a fringing field from the deceleration tube. To collect the secondary electrons efficiently, the fringing field from the deceleration tube, that tends to push back electrons to the sample, is terminated by a tungsten mesh and the space between the mesh and the sample is covered by an electrostatic shielding consisting of an aluminium shroud. Moreover, the entrance of the channeltron is biased to +50 V relative to the sample. Figure 3(a) shows the results of numerical calculations of the electric potential around the channeltron. The trajectories of incident positrons and secondary electrons emitted from the sample surface (the energies are assumed to be 50± 3 eV and 10 ± 5 eV, respectively) are also shown in Figs. 3(b) and 3(c). These confirm that positrons hit the sample and the secondary electrons are collected efficiently at the channeltron entrance. Trajectory of incident positrons are also slightly shifted by this drawing potential, however, it is negligible because the positron beam energy is enough high.

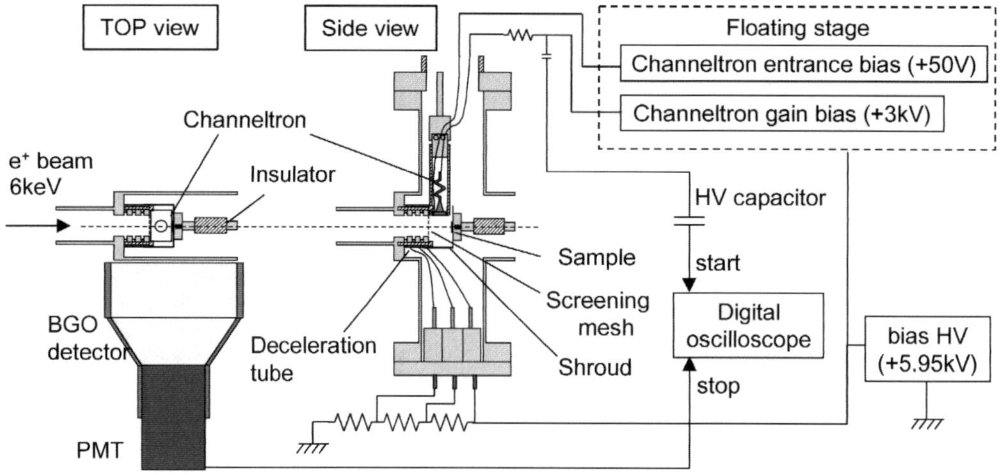

Figure 2. Schematic of the lifetime measurement system.

3. Lifetime measurements

Positonium lifetime measurements were carried out using the above apparatus. The beam intensity provided by the solid Kr moderator was estimated to be 1×10^4 e^+/s at the sample position and final counting rate of lifetime measurement was ~30 cps. By decreasing the incident energy, the counting rate was slightly increased. This is because the emission efficiency of secondary electrons tends to be higher at lower incident particle energy. Two samples were used for the test measurements. The first one was a silica aerogel block (Panasonic, SP-15) with a size of 10×15 mm and thickness of 2 mm. Figure 4(a) shows lifetime spectra when the positron beam energy was decelerated to 500 eV. Nearly the same spectrum was obtained even for an incident positron energy of 50 eV. The total spectrum area contained more than 1×10^6 events. The signal-to-noise ratio is ~50000. Short and long lifetime components are seen. From a two components analysis using the PATFIT code, the lifetimes and its intensities were estimated to be τ_1 = 20 ns (51%) and τ_2 = 120 ns (49%), respectively, with the time resolution of 11 ns. However, the lifetime of this τ_2 component tends to be underestimated due to the thermalization process [12]. From an exponential-decay fitting only in the time region of 500 ~ 1600 ns, where the thermalization effect is small enough, a lifetime of 140 ns was obtained [13, 14]. The intensity of o-Ps lifetime components for some aerogels are reported to be 40~50 % [15]. From these results, the long lifetime component is attributed to o-Ps decay. The short component τ_1 is not only from p-Ps decay because τ_1 is longer than the p-Ps lifetime convoluted with the time resolution. This component might be mixture of the o-Ps thermalization process and o-Ps pick-off annihilation on grain surfaces [16]. These results show that the apparatus is sufficient to measure the o-Ps lifetime spectrum.

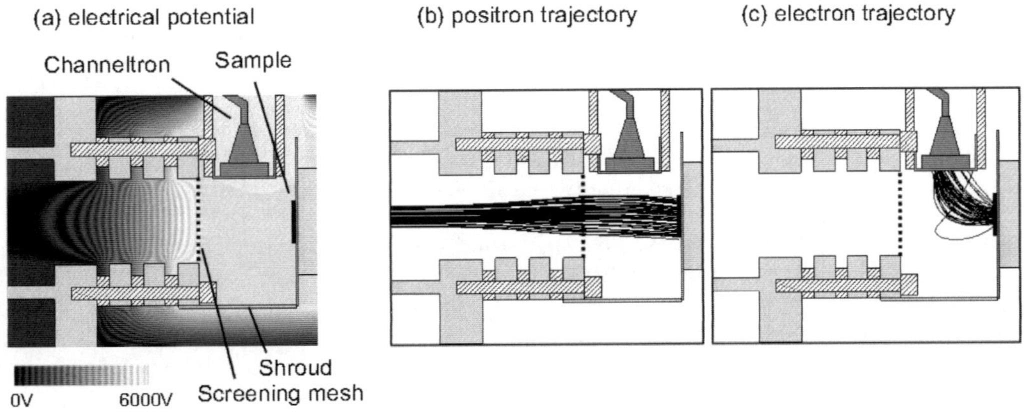

Figure 3. (a)Numerical calculations of the electric potential around the channeltron, (b)(c) simulated trajectories of the incident positron beam and secondary electrons.

The second sample was well-annealed copper plate. Figure 4(b) shows the spectrum when the positron beam energy was set to 6000 eV. The intensity of o-Ps component was decreased to ~ 6 %. For incident positron energy of E = 50 eV, the o-Ps component appears again. Its intensity was estimated to be 19 %. These results show the same tendency with the previous works [17].

4. Summary

A spin-polarized positron beam generated using a ^{22}Na source, a solid Kr moderator and electrostatic lenses was developed. A lifetime measurement system using secondary electrons as start triggers was also developed. Using this system, the lifetime spectra of o-Ps were obtained. This system will be used to determine the surface spin polarization of some ferromagnets.

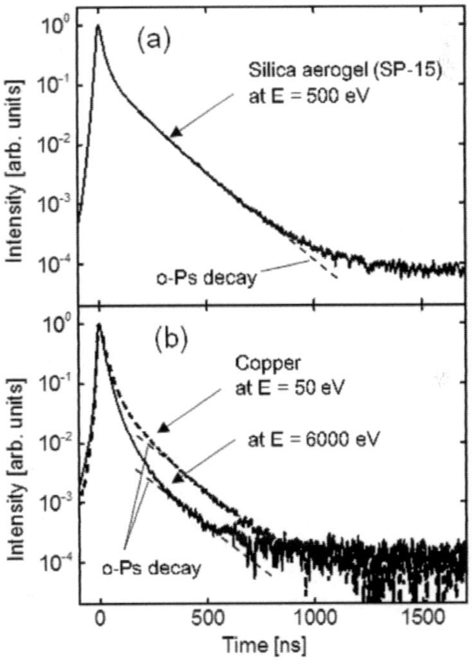

Figure 4. Positronium Lifetime spectra of (a) silica aerogel and (b) well-annealed copper.

References

[1] Maekawa M, Fukaya Y, Yabuuchi A, Mochizuki I and Kawasuso A 2013 *Nucl. Instr. Methods Phys. Res. B* **308** 9

[2] Maekawa M. Fukaya Y, Zhang H J, Li H and Kawasuso A 2014 *J. Phys.: Conf. Ser.* **505** 012033

[3] Kawasuso A, Fukaya Y, Maekawa M, Zhang H J, Seki T, Yoshino T, Saitoh E and Takanashi K 2013 *J. Mag. Mag. Mater* **342** 139

[4] Zhang H. J, Yamamoto S, Fukaya Y, Maekawa M, Li H, Kawasuso A, Seki T, Saitoh E and

Takanashi K 2014 *Sci. Rep.* **4** 4844

[5] Van Veen A, Schut H and Mijnarends P E 2000 Depth-profiling of Subsurface Regions, Interfaces and Thin Film (*Positron Beams and Their Applications)* ed P. G. Coleman (Singapore: World Scientific) p 201

[6] Gidley D W, Köymen A R and Capehart T W 1982 *Phys. Rev. Lett.* **49** 1779

[7] Rich A, Van House J, Gidley D W and Conti R S 1987 *Appl. Phys. A* **43** 275

[8] Mills Jr. A P, Voris Jr. S S and Andrew T S 1994 *J. Appl. Phys.* **76** 2556

[9] Mulligan M J and Lubell M S 1993 *Meas. Sci. Technol.* **4** 197

[10] Hanna S S and Preston R S 1957 *Phys. Rev.* **106** 1363

[11] Nagai Y. Nagashima Y, Kim J, Itoh Y and Hyodo T.2000 *Nucl. Instr. Methods Phys. Res. B* **171** 199

[12] Nagashima Y, Kakimoto M, Hyodo T, Fujiwara K, Ichimura A, Chang T, Deng J, Akahane T, Chiba T, Suzuki K, McKee B T A and Stewart A T 1995 *Phys. Rev.* A **52** pp 258

[13] Shinohara N, Suzuki N, Chang T and Hyodo T 2001 *Phys. Rev.* A **64** 042702

[14] Wada K, Saito F and Hyodo T 2010 *Phys. Rev.* A **81** 2010 062710

[15] Aghion S, Ferragut R, Moja F, Petkov M P and Jones S M 2013 *J. Phys.*: *Conf. Ser.* **443** 012064

[16] Zhou Y, Mao W, Li Q, Wang J and He C 2015 *Chem. Phys.* **459** 81

[17] Laricchia G. 1995 Positronium Beams and Surfaces (*Positron Spectroscopy of Solids, Proceedings of the International School of Physics "Enrico Fermi" vol. 125*) ed. A Dupasquier and A P Mills Jr. (Netherlands: IOS Press) p 405

IOP Conf. Series: Journal of Physics: Conf. Series **791** (2017) 012010 doi:10.1088/1742-6596/791/1/012010

Design and simulation of a pulsed positron beam at ELI-NP

N Djourelov and D Dinescu

Extreme Light Infrastructure - Nuclear Physics, Horia Hulubei National Institute for Physics and Nuclear Engineering, 30 Reactorului Street, P.O. Box MG-6, 077125 Magurele, Ilfov county, Romania

E-mail: doru.dinescu@eli-np.ro

Abstract. The design of the pulsing system for the slow e^+ beam at ELI-NP is presented. It will deliver narrow time width e^+ pulses, achieved by a combination of prebuncher, chopper and buncher. The effect of the different components on the pulse compression is simulated by a Matlab program. The results show that the best compression of the e^+ pulses, which can be achieved by the proposed pulsing system, is limited to about 100 ps (FWHM). The most effective solution, applied up to now, for minimizing the influence of the backscattered e^+ on the PALS spectra by using a bent tube filter is simulated by Comsol Multiphysics.

1. Introduction

For the development of new functional materials, the investigation of lattice defects and various atomic imperfections in solids constitutes an important step. Conventional positron annihilation methods use energetic e^+ directly emitted from radioisotopes such as ^{22}Na and are suited to measurements of bulk materials. For analyzing subsurface layers and thin films, slow e^+ beams are necessary [1]. If a radioisotope is used as the source of fast e^+ the intensity of the moderated slow e^+ is limited to ~ 10^6 e^+ s^{-1}. For creating a slow e^+ beam with a higher intensity another kind of source has to be utilized. At the European Light Infrastructure (ELI-NP), a brilliant γ-beam will produce fast e^+ in a suitable converter using the pair production mechanism. The creation of a slow e^+ beam will be achieved using a converter of tungsten foils, which also acts as the moderator [2].

One of the most powerful positron based techniques for the quantification of defects is Positron Annihilation Lifetime Spectroscopy (PALS). To perform PALS with a e^+ beam a start signal is needed. For generation of a start signal several methods have been applied: detection of secondary e^- produced by incident e^+ [3], e^+ trapping in a field and using a trigger to release the trapped e^+ [4], and the method with the best possible time resolution, injection of ultra short pulses using the timing signal of the pulsing device [5].

In the present paper we report on the designed pulsing system which will deliver short e^+ pulses. The pulsing technique by chopping and bunching consists in modulating the longitudinal velocity of e^+ by applying a time dependent electric field to accelerate/decelerate the e^+ that would arrive too late/early at the time focus. The technique is well-known and has been applied previously by a number of groups in order to develop pulsed e^+ beam from a DC e^+ source.

2. Design and simulations

The designed pulsing system is presented in Figure 1. It is composed of a prebuncher, chopper, pre-accelerator, buncher, decelerator, accelerator, and bent drift tube. We have developed a Matlab program to calculate the movement of e^+ in time varying electric fields with the goal of recording the time of arrival of the e^+ at various points of the pulsing part. The possible spherical aberrations are neglected as the time resolution (e^+ pulse width) is basically dominated by the chromatic aberrations. The system was designed for incoming e^+ with energy of 29 eV with a realistic energy distribution of 1 eV (FWHM). The e^+ will be confined close to the beam line axis by a longitudinal magnetic field of 60 G provided by solenoid coils. In the following sections we will discuss the various components of the pulsing system and their effect on the quality of the pulsing.

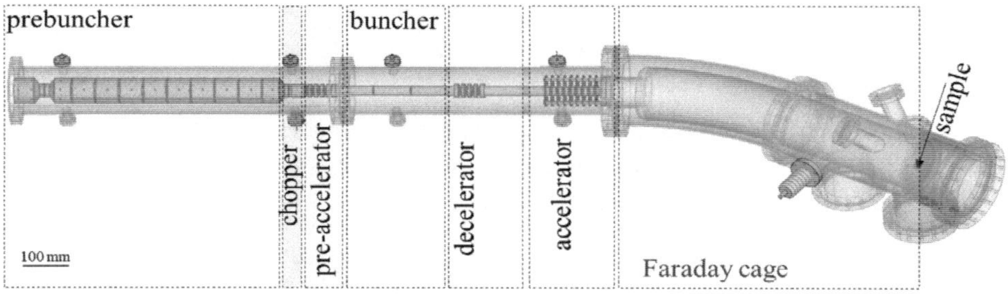

Figure 1. The 3D design of the in vacuum electrodes of the pulsing system. It consists of a prebuncher, chopper, pre-accelerator, buncher, decelerator, accelerator, and drift bent tube. The longitudinal magnetic field will be provided by a solenoid coil wound directly on the vacuum tubes and Helmholtz coils where a solenoid is not possible. The mechanical supports for the electrodes are omitted for simplicity.

2.1. Prebuncher

The purpose of the prebuncher is to bunch the e^+ beam in such a way that in the most e^+ will arrive at the main buncher gap during the linear parts of the sine wave potential which is applied to the main buncher [5].

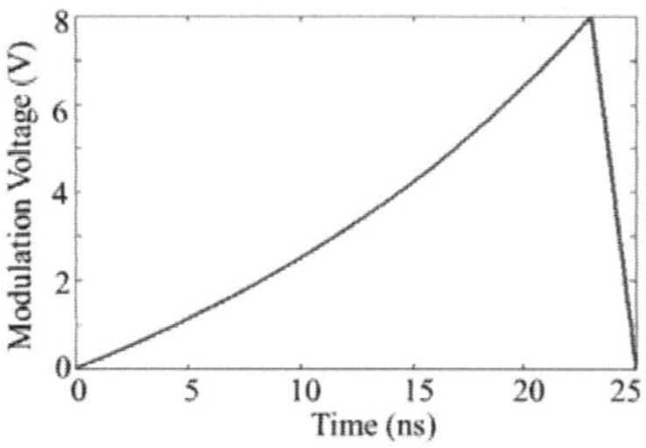

Figure 2. The periodic modulation potential to be generated by AWG and applied to the first gap of the multi-electrode prebuncher.

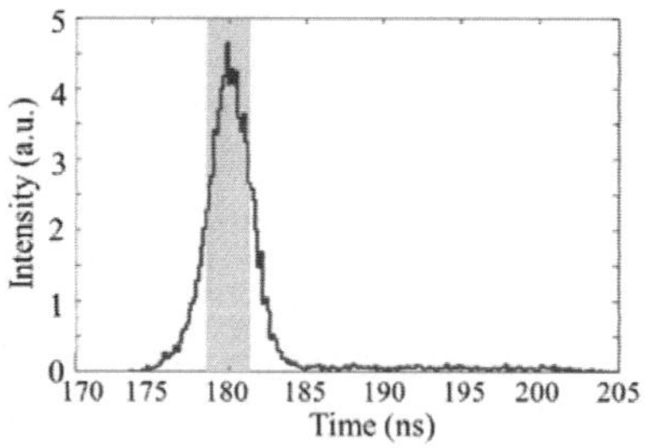

Figure 3. The simulated time compression of e^+ after the prebuncher. The shaded area represents the chopper transmission window chosen as 2.1 ns.

The ideal shape for the potential of the prebuncher electrode is a periodic saw-tooth function ($\sim t^2$). However, in a double gap prebuncher unwanted wrong energy modulation is given to e^+ when passing the second gap. One way to avoid the unwanted energy modulation is to use a multi-electrode prebuncher without drift [6]. The potential difference at the first gap determines the initial e^+ energy modulation. Then the e^+ travel in varying electric field generated by a number of electrodes. The varying potential is spread by a voltage divider over several gaps formed. If the time e^+ need to pass the prebuncher electrodes length (chosen 500 mm) is equal to one period of the saw-tooth function, the extra energy modulation averages to zero [6]. For our simulation we used a wave of 40 MHz $\sim t^2$ with an amplitude of 8 V (see Figure 2), which can be generated by a commercially available arbitrary wave generator (AWG) with slew rate of 20 V/ns. The resulting time compression after the DC e^+ beam passed the prebuncher is shown in Figure 3.

2.2. *Chopper*

The purpose of a chopper is to filter out those e^+ which will not arrive at the proper time at the buncher gaps. As a result, the chopping eliminates possible satellite peaks in the spectrum and improves the peak to background ratio. A retarding potential chopper consists of three tungsten meshes. The first and last meshes are grounded, and a chopping signal (40 MHz square-wave, 2.1 ns width and a bias voltage of 40 V) is supplied to the second mesh. In Figure 3 the transmission window of an ideal chopper is shown. The challenge in application of this type of chopper is the amplification of the square-wave. To preserve the shape of the generated square-wave, a wide bandwidth amplifier is required. Even though, in a realistic retarding potential chopper, an appearing of pulse tails due to e^+ which are almost stopped and then accelerated is a disadvantage and may deteriorate the final time resolution. These effects are not simulated in the present work. It would be worthy to try to design an E×B chopper [5, 7]. The time compression achieved by the prebuncher and the width of the chopper transmission window determines the beam intensity losses due to the pulsing system. For the signals described above the transmission efficiency of the pulsing system is 65%.

2.3. *Main buncher*

The main buncher is the second time focusing component of the pulsing system. In order to minimize the chromatic aberration in this component (due to the energy modulation introduced by the prebuncher as the first focusing component), after the chopper, the e^+ are pre-accelerated to 1.5 keV [8]. It has to be mentioned that, where technically possible, either acceleration or deceleration of a charged particle confined in longitudinal magnetic field should be done by multi electrode system as it introduces smaller extra longitudinal velocity spread as compared to one gap acceleration/deceleration. This was simulated and confirmed by Comsol multiphysics.

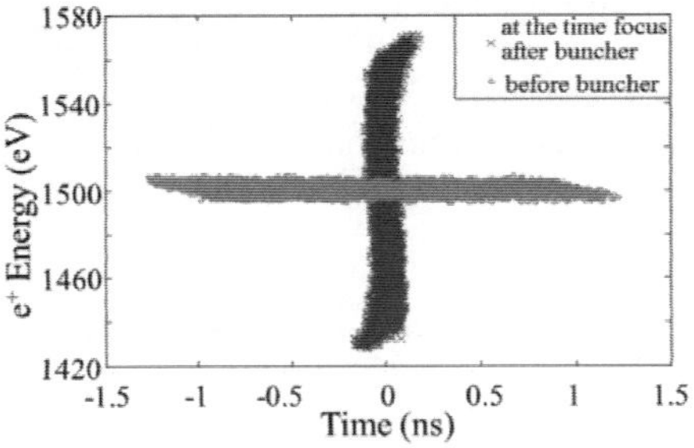

Figure 4. The phase space of e^+ before the buncher and at the target.

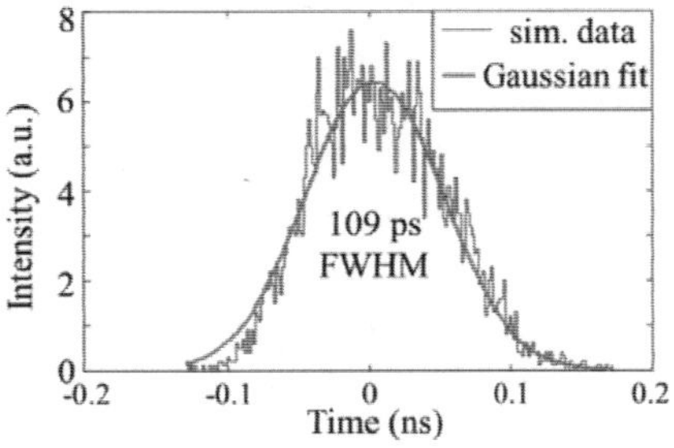

Figure 5. The time distribution of 1.5-keV e^+ at the target position. The data are fitted well by a Gaussian with FWHM=109 ps.

The designed main buncher is a simple single frequency (120 MHz) sinewave double gap buncher with a 96 mm central electrode. As a result of the Matlab optimization, we found that for obtaining the best time compression (shortest bunches at the target) a sine wave with amplitude of 47.5 V biased at 1.5 kV is necessary. The effect of the buncher on the phase space of the e^+ is shown in Figure 4. The resulted time distribution of the e^+ in a pulse is very close to a Gaussian with FWHM = 109 ps (see Figure 5). Considering these conditions, the time focus after the buncher occurs at 43.5 ns which corresponds to a drift length of ~ 1 m. By combination of the potentials applied at the decelerator and accelerator, together with small changes of the sine wave amplitude, one can adjust the time focus to occur at the target position for the different desired e^+ incident energies in the range 0.2-30 keV [9].

2.4. Simulation of the backscattered e^+

When incident e^+ hit the target some fraction of them are backscattered. If these positrons reach the accelerator they can be re-accelerated and implanted into the sample with delay from the initial e^+ bunch. This causes PALS spectrum distortions as side peaks. The best solution applied up to now uses a bent tube solenoid equipped with steering coils [7]. The longitudinal bent magnetic field guides the incident e^+ bunch through the bend and to the target, but does not allow backscattered e^+ to reach the accelerator.

In order to optimize the bend radius and the angle of the bend, we simulated this process in Comsol. As an example, Figure 6 shows the trajectories of e^+ accelerated to 4 keV in a 30° bent tube with a bend radius of R=1 m followed by a 290 mm straight tube with the target at the end. Due to Comsol limitations, the magnetic field of the bent solenoid was simulated by suitable Helmholtz coils. The bent and straight tubes form a Faraday cage for e^+ drift. The backscattered e^+ are simulated as secondary particles in Comsol, with realistic, but uncorrelated, angular and energy distributions [10]. Figure 6 shows that the backscattered e^+ are guided back into the beamline, but they do not reach the accelerator, and, instead, they annihilate (the annihilation places are represented by balls) on the tube wall far away from the target giving negligible chance for the annihilation γ-rays to be detected. The tube sizes optimization work is in progress.

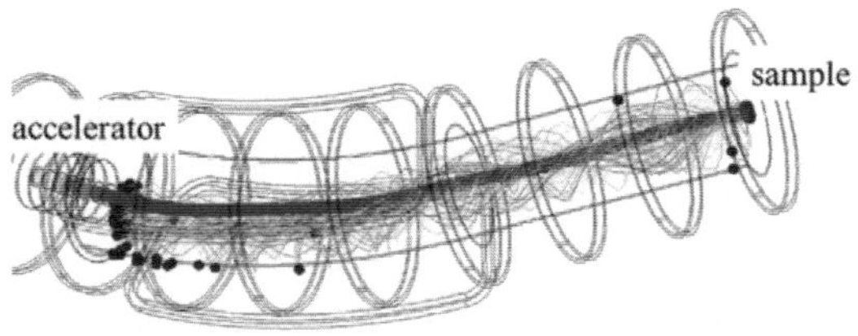

Figure 6. Comsol simulation of incident and backscattered e^+ trajectories inside the Faraday cage consisting of bent (30°, R=1 m) and straight tubes. Backscattered particles annihilate (represented by balls) on the cage wall away from the sample position.

3. Summary

The design of a pulsing system for a DC e^+ beam was presented. The simulation of the effect of the different pulsing components on the e^+ phase space was performed by a Matlab program. Optimization of the signal amplitudes showed that compression to about 100 ps (FWHM) e^+ pulses at the sample position is a limit of this setup. The most effective solution, applied up to now, of minimizing the effect of backscattered e^+ by using a bent tube filter was simulated by Comsol.

References

[1] Schultz P J and Lynn K G 1988 *Rev. Mod. Phys.* **60** 701
[2] Djourelov N, Hugenschmidt C, Balascuta S, Leca V, Oprisa A, Piochacz C, Teodorescu C and Ur C A 2016 *Rom. Rep. Phys.* **68** S735
[3] Szpala S, Petkov M and Lynn K G 2002 *Rev. Sci. Instrum.* **73** 147
[4] Gilbert S J, Kurz C, Greaves R G and Surko C M 1997 *Appl. Phys. Lett.* **70** 1944
[5] Schödlbauer D, Sperr P, Kögel G and Triftshäuser W 1988 *Nucl. Instrum. Meth. B* **34** 258
[6] Piochacz C and Hugenschmidt C 2013 *J. Phys: Conf. Ser.* **443** 012093
[7] Jungmann M, Haeberle J, Krause-Rehberg R, Anwand W, Butterling M, Wagner A, Johnson J M and Cowan T E 2013 *J. Phys: Conf. Ser.* **443** 012088
[8] A Laakso 2005 Construction of a pulsing system for low-energy positrons Phd Thesis (Helsinki, Finland)
[9] Butterling M, Jungmann M, Bondarenko V, Sachert S, Brauer G, Anwand W and Krause-Rehberg R 2011 *J. Phys: Conf. Ser.* **265** 012027
[10] Marinov H, Djourelov N, Nédélec P and Petrov L 2013 *Nucl. Instrum. Meth. A* **729** 569

Status of the project for a positron laboratory at ELI-NP

N Djourelov, A Oprisa, D Dinescu and V Leca

Extreme Light Infrastructure - Nuclear Physics, Horia Hulubei National Institute for Physics and Nuclear Engineering, 30 Reactorului Street, P.O. Box MG-6, 077125 Magurele, Ilfov county, Romania

E-mail: nikolay.djourelov@eli-np.ro

Abstract. We report on the plans of a positron laboratory to be built at the Extreme Light Infrastructure - Nuclear Physics (ELI-NP) facility. The slow e^+ source is based on e^-–e^+ pair production in a converter made of tungsten foils by a brilliant gamma beam (E_γ<3.5 MeV, $I_\gamma = 2.4\times10^{10}$ γ s^{-1}) which will be generated by Compton back–scattering of photons from a high power laser on electrons from a LINAC. Numerical simulations of γ–ray interactions with the designed converter showed that, if the tungsten foils are used for moderation of the created fast e^+, a slow e^+ beam of an intensity of ~1×10^6 s^{-1} can be obtained. Circular polarization of the γ–beam is proposed to be one of the ELI-NP's upgrades. The slow e^+ will be extracted perpendicularly to the γ–beam and will have ~ 30% transversal polarization degree. By using a transmission re-moderation stage and electrostatically turning the re–moderated beam by 90°, the transversal polarization will be changed to longitudinal. Longitudinal polarization can then be preserved in the longitudinal magnetic guiding field. The heat load to the converter was estimated at < 2 mW which implies that e^+ moderation by frozen neon is applicable.

1. Introduction

One of the methods to produce slow e^+ beams of high intensity involves the (γ, e^+e^-) reaction in high–Z materials like tungsten or platinum to obtain fast e^+ which are subsequently moderated. The γ–rays needed for the e^+–e^- pair production are obtained either from bremsstrahlung of electron beam in a target or from nuclear reactors [1-4]. Another method to produce a brilliant source of γ–rays is the Compton scattering of photons off a high-energy electron beam [5, 6]. This method will be realized for the Gamma Beam System (GBS) at the Extreme Light Infrastructure - Nuclear Physics (ELI-NP) (under construction in Magurele, Romania) to produce two brilliant γ–beams, with maximum energies of 3.5 MeV, and 19 MeV, both tunable [7]. At the very early stage of the ELI-NP project, it was considered that a γ–beam intensity of 10^{13} γ s^{-1} is achievable [8]. Based on this data and, in addition, on the presumption of a beam spot size of few tens of μm, Hugenschmidt *et al.* estimated that a slow positron beam at ELI-NP would have an intensity exceeding the present highest–intensity positron source (10^9 slow e^+ s^{-1}, NEPOMUC)[9]. However, according to the reported simulations of the GBS, the low–energy γ–beam will have an intensity of 2.4×10^{10} γ s^{-1} [7]. Although the γ–beam at the photon –e^- interaction point (IP_γ) will have an extremely low divergence, of a few μrad, the design restrictions determine that the converter for e^+ production can be placed only at a distance of ~ 4 m away from the IP_γ, where the spot size of the non-collimated γ–beam will have a diameter of ~ 6 mm (FWHM). By using these data, we have simulated the interaction of the γ-rays with the specially designed converter/

moderator assembly (CMA), made of tungsten foils, the moderation of fast e^+, the extraction of moderated e^+ from CMA, and their focusing to form a beam [10, 11]. The optimization of the CMA sizes showed that a primary slow e^+ beam can be obtained with an intensity of $1\text{-}2\times10^6$ e^+ s^{-1}.

The present paper reports on the status of the project to build a slow positron beam line at ELI-NP.

2. Positron beam line setup

Details on the planned positron spectroscopy techniques: Positron Annihilation Lifetime Spectroscopy (PALS), Coincidence Doppler Broadening Spectroscopy (CDBS), Age MOmentum Correlation spectroscopy (AMOC), and Positron annihilation initiated Auger Electron Spectroscopy (PAES) to be used at the positron laboratory at ELI-NP can be found in [11]. One of the disadvantages to generate e^+ by using a warm (non-superconducting) LINAC is the low duty cycle. In order to avoid detectors pile-up and to assure high count rates for PALS, CDBS and AMOC, application of a e^+ pulse stretcher is required [2]. The GBS machine includes a warm LINAC that will provide trains of 32 e^- bunches separated by gaps of 16 ns in a micro pulse of 0.5 μs with a repetition rate of 100 Hz [7]. By reflections in a recirculator, one and the same laser pulse will interact with all the e^- bunches in a micro pulse. Back scattered photons forms a high energy γ-beam through inverse Compton scattering with the e^- beam. When the γ-beam interacts with the CMA the created fast e^+ will have almost the same timing structure as the e^- from the LINAC. Due to different trajectory lengths of the moderated e^+ at the extraction and formation points of the primary slow e^+ beam, the fine timing structure will be smeared and the slow e^+ will feed the stretcher with ~ 0.6 μs pulses of 100 Hz repetition.

2.1. Stretcher

The e^+ pulse stretching technique is well-known and successfully applied. The simplest one is a low-field Penning-Malmberg trap with the entrance gate synchronized with the LINAC, central electrode at ground, and exit gate with potential barrier gradually decreasing between the pulses [12]. Another version of the stretcher can be realized with fixed potential for the entrance and exit electrodes, and a central electrode which potential suddenly drops when the e^+ pulse enters and is then gradually increased in each repetition cycle [13]. The advantage of the last version is that the e^+ energy can be controlled by the output electrode potential. A schematic drawing of the designed stretcher for the ELI-NP positron laboratory is shown in Figure 1.

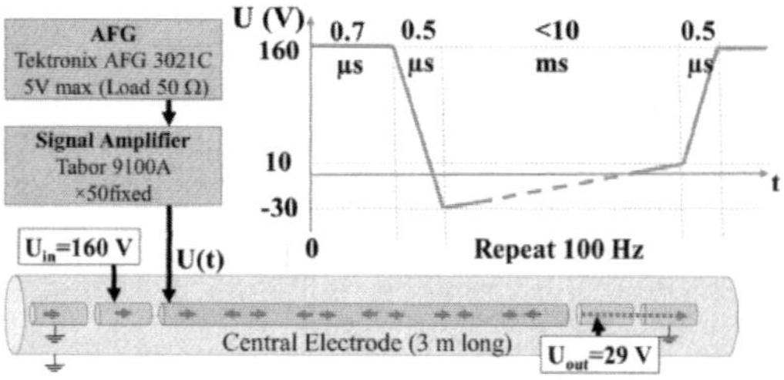

Figure 1. The designed stretcher section with a central electrode of time-varying potential and gate electrodes at fixed potentials.

The incoming e^+ come in short pulses (~ 0.6 μs) with an energy of 200 eV. The potential of 160 V of the entrance electrode will slow them down to 40 eV within the central electrode. The 3-m-long central electrode will be able to accommodate a whole pulse before the potential of the central electrode (U) drops to make a trap for e^+ between the entrance and exit potential barriers. By slowly increasing the potential U, at a maximum rate of 4 V ms^{-1}, the trapped e^+ which have enough energy to pass the decreasing potential barrier of the exit gate and will form a quasi-continuous slow e^+ beam made up of stretched pulses with length up to 10 ms. It was proven that a stretched e^+ pulse duration of 1-2 ms is enough to avoid detector pile-up problems [13].

2.2. Pulsing

Application of the PALS and AMOC techniques in combination with a continuous or quasi-continuous beam of slow e^+ requires a signal to start the electronics to measure the e^+ lifetime. One solution is the pulsing technique by which continuous beam is chopped and bunched at MHz frequency to obtain short pulses of about 100 ps at the sample position. We have designed a pulsing part consisting of multi-electrode prebuncher, retarding potential chopper, pre-accelerator, buncher, decelerator and accelerator with bent drift tube (see Figure 2).

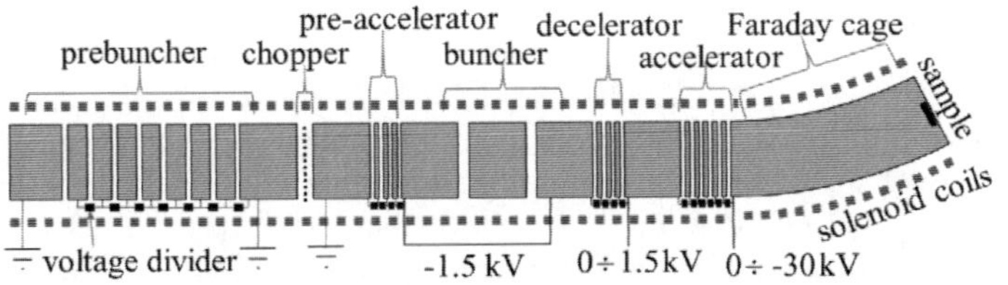

Figure 2. Scheme of the pulsing part consisting of prebuncher, chopper, pre-accelerator, buncher, decelerator, and accelerator with bent drift.

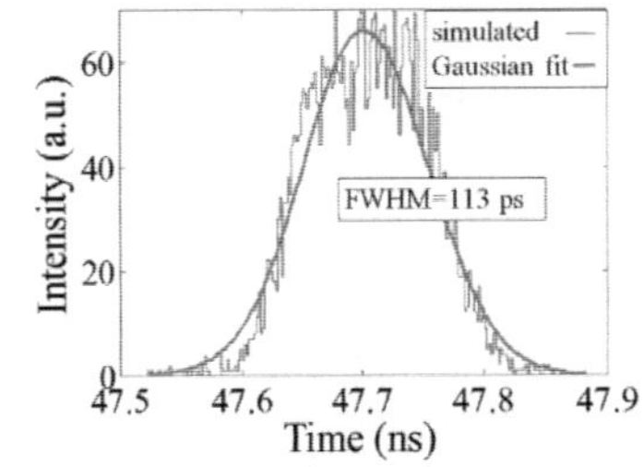

Figure 3. Simulated distribution of e^+ at the sample position.

The e^+ will enter the prebuncher with an energy of 29 eV. The e^+ will be subjected to energy modulation due to applied potential (~ t^2, 40 MHz) at the first electrode of the prebuncher. The retarding potential will have 2 ns transmission window. The chopped e^+ will be pre-accelerated to 1.5 keV before entering the double gap buncher which operates with a 120 MHz sine wave applied to the central electrode. By adjusting the potentials of the decelerator and accelerator the bunched e^+ will drift to the sample position at the desired focus time independently on their adjustable final energy (0-30 keV). The bent drift tube will be equipped with dipole coils to generate a transversal magnetic field in order to keep the incident e^+ on axis. The same magnetic field will force the backscattered e^+ from the sample to deviate from the axis and they will annihilate at the tube walls far from the sample. This method was already successfully applied to minimize the spectrum distortion due to backscattered positrons [14]. Our simplified simulation considers only chromatic aberrations and showed that the described pulsing part can compress a continuous e^+ beam to pulses of ~ 100 ps FWHM (Figure 3) with an efficiency of ~ 65%.

3. Positron polarization

The international scientific and advisory board of the ELI-NP project recommended circular polarization of the γ-beam to be the first update of the GBS. Our previous simulations showed that if γ-rays are 100% circularly polarized the slow e^+ beam may have 33% degree of spin polarization for CMA sizes optimized for best e^+ beam intensity [10]. However, close to 100% circular polarization γ-rays can be achieved only for a narrow bandwidth γ-beam. For e^+ production, we will use a wide bandwidth γ-beam (1-3.5 MeV). Simulations of the interaction of circularly polarized photons with a relativistic e^- beam have shown a strong correlation of the Stokes #3 parameter with the energy of the resulting γ-rays (see the supplementary online material of [15]). This correlation together with the cross section for e^--e^+ pair production is shown in Figure 4.

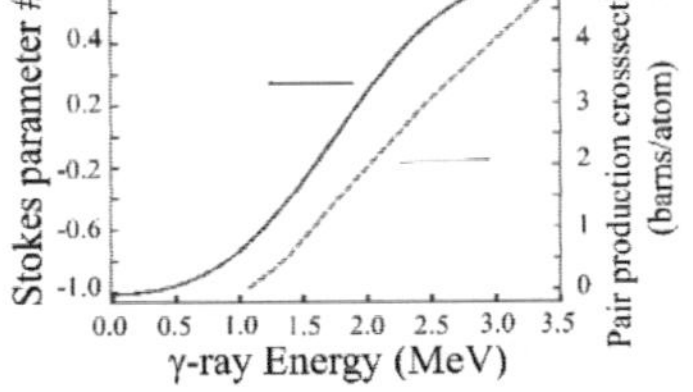

Figure 4. Circular polarization, represented by the Stokes parameter #3, of the γ-rays obtained by inverse scattering of 100% circularly polarized photons of relativistic e^- beam and the e^+- e^- pair production cross section in tungsten as a function of the γ-ray energy.

Using the correlation of the Stokes parameter #3 on the γ-ray energy we found that the e^+ spin polarization degree was previously overestimated and the new value is 30%. Actually, this overestimation was not huge due to the fact that more energetic γ-rays carry higher circular polarization and produce more e^+ (see Figure 4). The slow e^+ will be extracted perpendicularly to the γ−beam and the primary e^+ beam will have ~ 30% transversal polarization degree. By using a transmission re-moderation stage and electrostatically turning the re−moderated beam by 90°, the transversal polarization will be changed to longitudinal. Longitudinal polarization can be preserved in the longitudinal magnetic guiding field.

4. Neon moderation

Moderation of fast e^+ by a frozen neon layer due to its high efficiency is widely used in the last decade. Simulation of the interaction of the γ-ray with the CMA (with optimized sizes) showed that the heat load to the converter will be less than 2 mW [11]. This implies that e^+ moderation by frozen neon is applicable. Our plans are to freeze a layer of neon directly on the surface of the foils of the CMA. We have estimated that a commercially available cryocooler with a cooling power of 3 W at 6 K can be used. The design challenge is to modify the cold finger of the cryocooler, with a sapphire interface, in order to have an electrically insulated cold finger to be able to sustain a few kV potential difference. The high efficiency of neon moderator will compensate the intensity lost due to re-moderation.

5. Summary

We have reported on the proposed updates of the positron laboratory to be built at ELI-NP. We have re-estimated the e^+ spin polarization degree using a more realistic estimate for the circular polarization of the γ−rays from the GBS. The designs of a pulse stretcher and a pulsing part have been completed. We have proposed that the fast e^+ created in the CMA to be moderated by a neon layer, directly frozen on the CMA foils. The expected slow e^+ beam intensity, with this modification, is greater than 10^7 s^{-1} for non-polarized e^+ and greater than 10^6 s^{-1} in the case of polarized e^+.

Acknowledgments

This is work is supported by the Project Extreme Light Infrastructure –Nuclear Physics (ELI-NP) – Phase I, a project co-financed by the Romanian Government and the European Union through the European Regional Development Fund.

References

[1] Hyodo T *et al.* 2011 *J. Phys: Conf. Ser.* **262** 012026
[2] Suzuki R, Ohdaira T, Mikado T, Uedono A, Ohgaki H, Yamazaki T and Tanigawa S 1997 *Mater. Sci. Forum* **255-257** 114
[3] Krause-Rehberg R, Sachert S, Brauer G, Rogov A and Noack K 2006 *Appl. Surf. Sci.* **252** 3106
[4] Hugenschmidt C, Ceeh H, Gigl T, Lippert F, Piochacz C, Pikart P, Reiner M, Weber J and Zimnik S 2013 *J. Phys: Conf. Ser.* **443** 012079
[5] Potylitsin A P 1997 *Nucl. Instrum. Meth. A* **398** 395
[6] Omori T *et al.* 2006 *Phys. Rev. Lett.* **96** 114801
[7] Adriani O *et al.* 2014 *arXiv*:1407.3669v1 [physics.acc-ph]
[8] http://www.eli-np.ro/documents/ELI-NP-WhiteBook.pdf
[9] Hugenschmidt C, Schreckenbach K, Habs D and Thirolf P G 2012 *Appl. Phys. B* **106** 241
[10] Djourelov N, Oprisa A and Leca V 2016 *Nucl. Instrum. Meth. A* **806** 146
[11] Djourelov N, Hugenschmidt C, Balascuta S, Leca V, Oprisa A, Piochacz C, Teodorescu C and Ur C A 2016 *Rom. Rep. Phys.* **68** S735
[12] Akahane T, Chiba T, Shiotani N, Tanigawa S, Mikado T, Suzuki R, Chiwaki M, Yamazaki T and Tomimasu T 1990 *Appl. Phys. A* **51** 146
[13] Suzuki R, Mikado T, Chiwaki M, Ohgaki H and Yamazaki T 1995 *Appl. Surf. Sci.* **85** 87
[14] Jungmann M, Haeberle J, Krause-Rehberg R, Anwand W, Butterling M, Wagner A, Johnson J M and Cowan T E 2013 *J. Phys: Conf. Ser.* **443** 012088
[15] Camera F *et al.* 2016 *Rom. Rep. Phys.* **68** S539

IOP Conf. Series: Journal of Physics: Conf. Series **791** (2017) 012012 doi:10.1088/1742-6596/791/1/012012

Evaluation of the optical performance of a brightness enhancement system developed for the KUR slow positron beamline

Yoshihiro Kuzuya[1], Nagayasu Oshima[2], Atsushi Kinomura[1], Atsushi Yabuuchi[1], Koichi Sato[3] and Qiu Xu[1]

[1] Research Reactor Institute, Kyoto University, Kumatori-cho, Osaka 590-0494, Japan

[2] National Metrology Institute of Japan, National Institute of Advanced Industrial Science and Technology (AIST), Tsukuba, Ibaraki 305-8568, Japan

[3] Department of Mechanical Engineering, Kagoshima University, Kagoshima 890-0065, Japan

E-mail: kuzuya.yoshihiro.52e@st.kyoto-u.ac.jp

Abstract. A slow positron beamline using a nuclear reactor is under development at the Kyoto University Research Reactor (KUR). A brightness enhancement system was designed to reduce the initial positron beam size (~30 mm) below typical specimen sizes (< 10 mm) and keep the intensity of the beam as high as possible. After installation of the brightness enhancement system at the beamline, we tested its optical performance using an electron beam. The experimental result indicated that the system would have enough capability to obtain a small-sized beam suitable for positron measurements.

1. Introduction

Positron annihilation spectroscopy (PAS) with energy-variable positron beams is a powerful tool to investigate open-volume-type defects of atomic dimensions such as vacancies, vacancy agglomerates and micro voids near surfaces or in thin films [1]. An intense positron source is one of the key components in the slow positron beam system to achieve efficient and precise PAS measurements. We have recently developed a reactor-based, high-intensity slow positron beamline at KUR [2]. The beamline consists of a positron source, a brightness enhancement system, a pulsing system and an experiment chamber. A slow positron beam with an intensity of 1.4×10^{6} s^{-1} was successfully obtained under 1 MW reactor operation (the reactor has a maximum power of 5 MW) [2].

At KUR, the initial diameter of the positron beam is around 30 mm, and this is much larger than the size of typical samples (~10 mm) used in other slow positron beam facilities. Therefore it is desirable to reduce the initial beam size to below 10 mm and to keep the beam intensity as high as possible. We are developing a brightness enhancement system to achieve this purpose.

2. Design of a brightness enhancement system

The KUR brightness enhancement system was designed based on the system developed by Oshima et al. at AIST [3]. An initial slow positron beam generated in a solenoidal magnetic field is extracted to the outside of the magnetic field and focused on a transmission-type remoderator. Positrons re-emitted from the remoderator have a greatly reduced angular dispersion, leading to an increase in the beam brightness. At AIST, the brightness enhanced beam (remoderated beam) is then transported by electrostatic fields while in the KUR design it is transported by continuous solenoidal magnetic fields.

At KUR, we have to guide the brightness enhanced beam over a distance of ~1.5 m to the pulsing system by keeping a relatively low beam energy in the range of several eV to a few tens of eV. Electrostatic guiding through the pulsing system may be possible but would be quite complex; so solenoidal magnetic fields are used to efficiently guide the brightness enhanced beam.

In addition, when we measure large samples bigger than 30 mm, it is better to use the initial positron beam without brightness enhancement for higher count rates. The brightness enhancement system was designed taking into consideration the above points.

2.1 The characteristics of our brightness enhancement system

A schematic view of our brightness enhancement system is shown in figure 1. The system is composed of guiding coils (LA–LG), an acceleration gap, extraction coils (LC, LD, MA and MB), a magnetic lens, a transmission-type remoderator (movable) and a solenoid coil (SC). To observe the beam spot, a movable micro-channel plate (MCP) with a phosphor screen can be installed at the remoderator position. Experiments can be performed with two guiding modes in the following ways:

(1) Brightness enhancement mode: The initial positron beam magnetically transported from the positron source is accelerated by an acceleration gap (~5 keV) and extracted to a low magnetic field region by the extraction coil (MA). The large coils LE and LF are not used. The beam is focused on the transmission-type remoderator (150 nm-thick crystalline Ni) using the magnetic lens, and positrons are re-emitted from the remoderator. The brightness enhanced beam is magnetically guided to the pulsing system.

(2) Non-brightness-enhancement mode: In this case, the initial beam coming from the source is magnetically transported with the large coils (LA–LG) and without the medium and solenoid coils (MA, MB and SC). To keep a quasi-uniform magnetic field around the magnetic lens, the lens is weakly excited. The remoderator can be retracted from the center of the beamline to transport the beam without brightness enhancement.

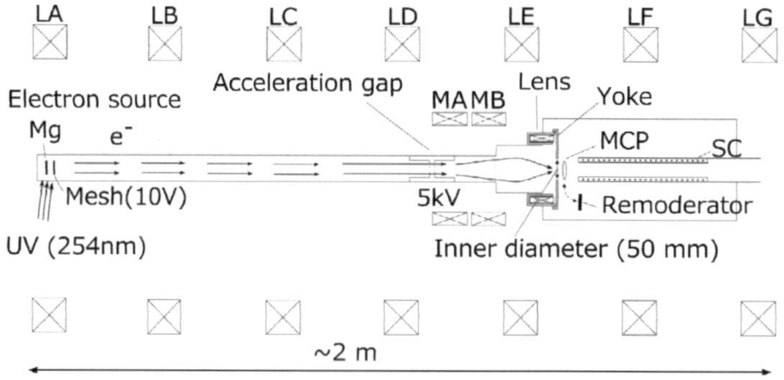

Figure 1. A schematic view of the brightness enhancement system. Different combinations of coils [LA–LG (large coils), MA and MB (medium coils), and SC (a solenoid coil)] are used depending on guiding modes. A schematic of the electron beam used for the performance test is also shown.

2.2 Guiding of the brightness enhanced beam with solenoidal magnetic fields

If the positron beam is adiabatically transported by a solenoidal magnetic field, the relationship between the magnetic field and the transverse energy is given by the following equation [4]:

$$\frac{E_{+\mathrm{r}}}{B_{\mathrm{r}}} \approx \frac{E_{+\mathrm{p}}}{B_{\mathrm{p}}}, \qquad (1)$$

where B_{r} and B_{p} are the magnetic fields at the remoderator and the pulsing system, respectively. $E_{+\mathrm{r}}$ and $E_{+\mathrm{p}}$ are the transverse energies at the remoderator and the pulsing system, respectively. Considering the law of conservation of energy ($E_{\|\mathrm{r}}+E_{+\mathrm{r}} = E_{\|\mathrm{p}}+E_{+\mathrm{p}}$) where $E_{\|\mathrm{r}}$ and $E_{\|\mathrm{p}}$ are the longitudinal energies at the remoderator and the pulsing system, respectively, the following equation is derived using equation (1):

$$E_{\parallel p} \approx E_{\parallel r} + \left(1 - \frac{B_p}{B_r}\right) E_{+r} . \tag{2}$$

Therefore the longitudinal energy spread at the pulsing system $\Delta E_{\parallel p}$ can be approximately given by the error propagation rule with the longitudinal energy spread $\Delta E_{\parallel r}$ and the transverse energy spread ΔE_{+r} at the remoderator position as follows:

$$\Delta E_{\parallel p} \approx \sqrt{\Delta E_{\parallel r}^2 + \left(1 - \frac{B_p}{B_r}\right)^2 \cdot \Delta E_{+r}^2} . \tag{3}$$

The pulse width obtained by the pulsing system is approximately proportional to $\Delta E_{\parallel p}$. Therefore, it is desirable to achieve $B_p/B_r \approx 1$ to obtain a pulsed beam with optimum time resolution. Since the designed B_p at the pulsing system is about 5 mT, the brightness enhancement system needs to have B_r of ~5 mT at the remoderator position.

2.3 Design of the magnetic lens

The following two points were considered when designing the magnetic lens.

(1) In the "brightness enhancement mode", the magnetic lens needs to focus the beam to a spot as small as possible at the remoderator. In addition, the magnetic field B_r should be ~5 mT. In order to maximize the remoderation efficiency the positron beam energy is set to ~5 keV.

(2) In the "non-brightness-enhancement mode", the positron beam with a diameter of ~30 mm must pass through the inner yoke of the lens.

The lens was designed to meet these requirements based on the magnetic lens used in AIST [3] (see figure 1). By varying the inner diameter of the lens from 10 mm to 90 mm, we calculated the magnetic field distributions for the focusing of a 5 keV positron beam at the remoderator position. In the calculation, the distance from the lens-edge to the remoderator was fixed to 10 mm and the thickness of the yoke was fixed to 5 mm. Calculation results are shown in figure 2. It is clear that the lens with an inner diameter of 50 mm meets all the requirements. Based on this calculation, the focusing lens was fabricated with an inner diameter of 50 mm and installed in the KUR system. The measured magnetic field distribution was in agreement with the calculated one.

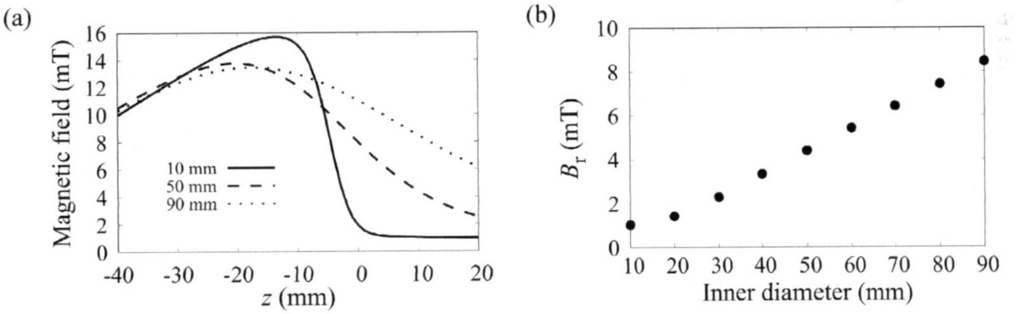

Figure 2. (a) Magnetic field distributions formed by the magnetic lens when a 5 keV positron beam is focused on the remoderator (Inner diameter of the lens: 10, 50, 90 mm). (b) Magnetic field at the remoderator position B_r as a function of the inner diameter of the lens.

3. Evaluation of the optical performance using an electron beam

We performed experiments to evaluate the optical performance of the brightness enhancement system installed at the beamline and estimated the beam diameter at the focus point. Since KUR has not been in operation for about 2 years to comply with new safety regulations, an electron beam was used (see figure 1). Electrons generated by the photoelectric effect are extracted to form a beam with a bias voltage of 10 V. The beam spot was observed using an MCP with a phosphor screen and imaged by a camera. In this experiment, the initial beam diameter of the electron source was 15 mm.

A minimum beam spot size was obtained when the current of the focusing lens was around 2.3 A [see figure 3 (a)]. We estimated the focused beam diameters (full-width at half maximum, FWHM) and confirmed that an electron beam with an initial diameter of 15 mm was focused to 2.9 mm.

We calculated the beam trajectory using the GPT [5] and Poisson Superfish [6] codes with the same parameters as those of the experiment. The cross sectional view at the focus point is shown in figure 3 (b). From this calculation, it is confirmed that a 15 mm electron beam can be focused to 3.1 mm (FWHM) with a magnetic lens current at 2.3 A. This calculation value was in good agreement with the experimental value (2.9 mm). A similar calculation indicated that a 30 mm electron beam could be focused to 5.5 mm. This implies that the demagnification factor of our focusing system is ~1/5. Considering the reemission efficiency (around 10-20%) of the 150 nm Ni remoderator [7], the initial positron beam can be at least 2.5–5.0 times brightness enhanced.

The magnetic field at the remoderator (B_r) was 4.1 mT when the current of the lens was 2.3 A. This means that the ratio $B_p/B_r \approx 1$ was almost achieved.

We also tested the "non-brightness-enhancement mode" of the system using an electron beam. The electron beam passed successfully through the system keeping the same beam diameter ~15 mm.

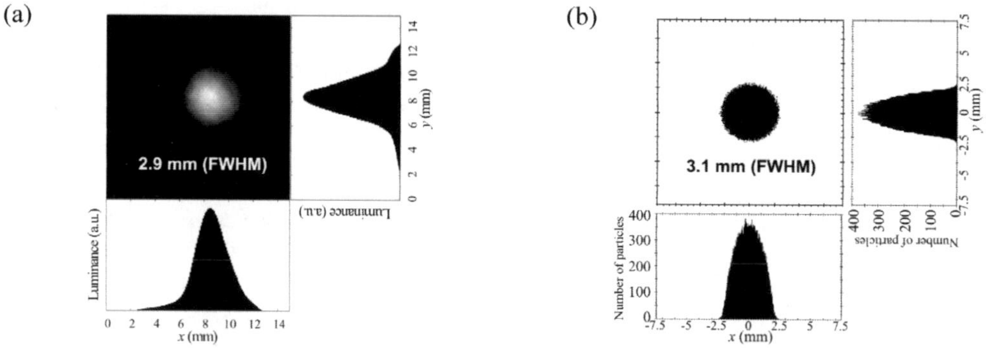

Figure 3. The intensity profiles of the beam spot obtained when the lens current was 2.3 A by (a) the experiment and (b) the calculation.

4. Summary

We designed and installed a brightness enhancement system for the KUR slow positron beamline. Then, we performed experiments to evaluate the optical properties of the system using an electron beam. A demagnification factor of 1/5 was achieved, implying that typical samples (size: ~10 mm) can be measured with our system. To obtain smaller beam sizes, further optimization of the magnetic field distribution is in progress. In the near future, we will apply this brightness enhancement system to slow positron beams.

Acknowledgements

This study was financially supported by Innovative Nuclear Research and Development Program of Ministry of Education, Culture, Sports, Science and Technology. We would like to thank colleagues of AIST, Tohoku Univ., Kyoto Univ., KEK, Tokyo Univ. of Science for their invaluable help.

References

[1] Coleman P G 2000 *Positron Beams and their Applications* (Singapore: World Scientific)
[2] Sato K et al. 2015 *Nucl. Instr. Meth. Phys. Res.* B **342** 104
[3] Oshima N et al. 2008 *J. Appl. Phys.* **103** 094916
[4] Spitzer L Jr. 1962 *Physics of Fully Ionized Gases* (New York: Interscience)
[5] van der Geer S B and de Loos M J 2001 *The general particle tracer code: Design, implementation and application* Ph.D. thesis (Eindhoven: Technische Universiteit Eindhoven)
[6] Menzel M T and Stokes H K 1987 *User's Guide for the Poisson/Superfish Group of Codes* Report LA-UR-87-115 (Los Alamos National Lab., New Mexico)
[7] Schultz P J, Gullikson E M and Mills A P Jr. 1986 *Phys. Rev. B* **34** 4421986

Evaluation of a positron-beam-pulsing system in KUR reactor-based positron beam facility

A Yabuuchi[1], A Kinomura[1], Y Kuzuya[1], K Sato[2], Q Xu[1], N Oshima[3] and B E O'Rourke[3]

[1] Research Reactor Institute, Kyoto University, Kumatori, Osaka 590-0494, Japan
[2] Department of Mechanical Engineering, Kagoshima University, Kagoshima 890-0065, Japan
[3] National Metrology Institute of Japan, National Institute of Advanced Industrial Science and Technology (AIST), Tsukuba, Ibaraki 305-8568, Japan

E-mail: yabuuchi@rri.kyoto-u.ac.jp

Abstract. The pulsing performance of the positron-beam-pulsing system of the slow positron beamline at Kyoto University research Reactor was tested using electron beams trimmed by 5-mm- or 15-mm-diameter apertures. Electron pulses with full width at half maximum of 143 ps were obtained with a 5-mm-diameter beam which corresponds to the diameter of the brightness-enhanced positron beam. This indicates the pulsing system has a sufficient pulsing performance for positron annihilation lifetime measurements.

1. Introduction

Recently, a reactor-based intense slow positron beamline has been newly constructed at Kyoto University research Reactor (KUR) which has a rated thermal power of 5 MW [1, 2]. The first positron beam extraction was achieved in 2014 with a slow positron beam intensity of 1.4×10^6 e^+/s at a reactor thermal power of 1 MW [2].

Since the KUR slow positron beamline was constructed as a part of joint usage research facilities, typical size (~10 mm) samples should be measurable with positron annihilation lifetime measurements. Thus, the KUR slow positron beamline has been designed to pulse the beam after focusing the initial 30-mm-diameter positron beam [1,2], and a positron-beam-pulsing system has been installed downstream of the brightness enhancement system [3]. From our test results using an electron beam, the initial positron beam is expected to be focused to a diameter of around 5 mm by the brightness enhancement system [4]. When a higher beam intensity is required compared with the brightness-enhanced beam, we are planning to introduce a brightness-unenhanced positron beam directly to the pulsing system. The positron beam diameter in this case was 15 mm, which is slightly smaller than the inner diameter, 20 mm, of the pulsing electrodes.

In this study, the pulsing performance of the positron-beam-pulsing system was tested using 5-mm- and 15-mm-diameter electron beams. The reactor operation was suspended for the last 2 years to comply with new safety criteria. Therefore, commissioning of the KUR slow positron beamline is performed using electrons instead of positrons at the moment.

2. Experimental method

A schematic diagram of the experimental setup of this study is shown in figure 1. A self-made electron source that consists of several magnesium (Mg) ribbons was installed upstream of the brightness enhancement system. The Mg ribbons emit photoelectrons by 254 nm (nominal) ultraviolet light irradiation. The emitted photoelectrons are extracted by a grid mesh having a 15-mm-diameter aperture, and transported to the pulsing system with an energy of 10 eV in a magnetic field of approximately 6 mT. The energy spread of the electron beam has been measured to be 1.2 eV full width at half maximum (FWHM). The electron beam was pulsed by the pulsing system which consists of reflection-type chopper, cylindrical-electrode-type sub-harmonic pre-buncher, and quarter-wave resonant-cavity-type buncher. The configuration of the pulsing system is similar to that used at the AIST LINAC-based slow positron beamline [5–8]. The resonance frequency f of the buncher RF cavity has been confirmed to be f = 122.694 MHz. The chopper, pre-buncher, and buncher were driven at $f/4$, $f/4$, and f, respectively. The pulsed electrons were detected by a single-anode microchannel plate (MCP, Hamamatsu F4655-12) with an effective diameter of 14.5 mm at the sample position.

Since electrodes of the pulsing system have an inner diameter of 20 mm, we intend to introduce the 15-mm-diameter positron beam without brightness enhancement when a higher beam intensity is required. Apertures having diameters of 5 mm or 15 mm were installed about 350 mm upstream from the chopper to demonstrate electron beams corresponding to brightness-enhanced or brightness-unenhanced positron beams. The chopper driving signals and the electron detection signals were input to a constant fraction discriminator (CFD, Ortec 935), and the time differences between the two signals were accumulated with a time-to-amplitude converter (TAC, Ortec 566) and multichannel analyzer (MCA, Laboratory Equipment MCA-Lite/M) in order to obtain time structures of the electron pulses. The time structures were obtained using the 5-mm- or 15-mm-diameter apertures, with the driving-parameters of the pulsing system optimized in each case.

3. Results and discussion

Before measuring the time structures of the beam pulses, the time resolution of the measurement circuit was confirmed by supplying the chopper driving signal through a power divider into the TAC start/stop connectors. An accumulation result of the detection-timing fluctuation of the same chopper driving signals indicated the time resolution of the measurement circuit was 44 ps FWHM which is comparable to a reported value in a similar experiment [9].

Measured time structures of electron pulses obtained using the 5-mm-diameter aperture are shown

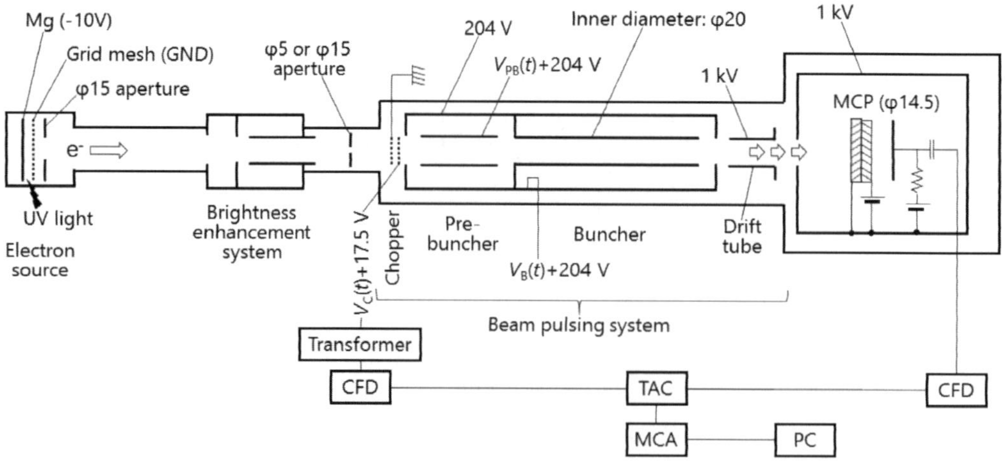

Figure 1. A schematic diagram of the electron beam pulsing test. A magnetic field of 6 mT is formed on the beamline by beam-guiding coils (not shown).

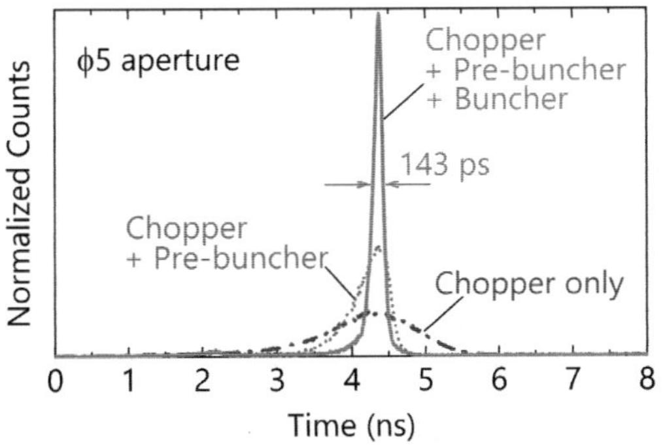

Figure 2. Time structures of electron pulses obtained with the 5-mm-diameter aperture. The dashed-dotted line (blue), dotted-line (green), and solid line (red) indicate spectra obtained with driving only the chopper, driving the chopper and pre-buncher, driving all units (chopper, pre-buncher, and buncher), respectively.

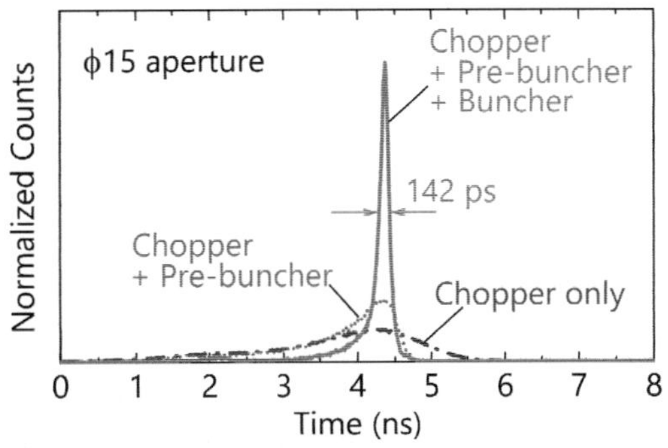

Figure 3. Time structures of electron pulses obtained with the 15-mm-diameter aperture. The dashed-dotted line (blue), dotted-line (green), and solid line (red) indicate spectra obtained with driving only the chopper, driving the chopper and pre-buncher, driving all units (chopper, pre-buncher, and buncher), respectively.

in figure 2. The result shows that the chopped electron pulses are bunched by passing through the pre-buncher and buncher. In this case, electron pulses with FWHM of 143 ps, which include the time resolution of the measurement circuit and the electron detector (MCP), are obtained. Similar electron/positron beam pulsing tests for positron annihilation lifetime measurements have been conducted using an MCP as an electron/positron detector by some other groups, and they obtained pulse widths in the range of 153–187 ps [10–12]. The pulse width in this study is sufficient for measuring positron annihilation lifetime spectra with a time resolution of a few hundred picoseconds. In our beamline, a 150-nm-thick Ni(100) single-crystal film is planned to be used as a positron remoderator at the brightness enhancement system [4]. The energy spread of positron beams emitted from Ni(100) film is reported to be 0.3 eV at FWHM [13], which is smaller than the energy spread of the electron beam used in this study (1.2 eV). When a positron beam emitted from the Ni remoderator is used, pulse widths are expected to be at least comparable or narrower than the results of this study. Therefore, our system has a sufficient pulsing performance for positron annihilation lifetime measurements at least using the brightness-enhanced positron beam.

Figure 3 shows measured time structures of electron pulses obtained using the 15-mm-diameter aperture. Even in this case, electron pulses with FWHM of 142 ps are obtained. It should be noted, however, that the actual beam diameter seems to be less than 15 mm (not uniform) in this case owing to insufficient optimization of beam trajectories. Thus, the actual beam diameter should be confirmed by an MCP phosphor screen in the future study, and the further pulsing performance tests for the trimmed 15-mm-diameter positron beam are necessary. At least, tail components of time structure curves in figure 3 are slightly increased, suggesting that beam diameters larger than 5 mm influence the time structure of the electron pulses.

Beam chopping efficiencies calculated from the measured time structures were 5–7% in the present pulsing parameter settings. Such beam duty factors are largely dependent on driving parameters of the chopper, and the duty factor could be increased to 10% or more if a degradation of time resolution can be tolerated.

From the present results, we note that the performance of the pulsing system is determined as a consequence of the balance between the beam size and the intensity. For the brightness enhanced beam (~5 mm) with the 150-nm-thick Ni remoderator, the beam intensity will be decreased to 10–20% of the initial intensity [14]. On the other hand, for the beam without the brightness enhancement, the intensity remains at 25% when the initial 30-mm-diameter beam is trimmed by the 15-mm-diameter aperture. If the large (15 mm) spot size can be tolerated, counting rates can be increased approximately by a factor of 2 with moderate pulsing performance.

4. Summary

In this study, we developed a beam-pulsing system for the KUR slow positron beamline with a brightness enhancement system, and beam-pulsing tests were conducted using an electron beam. Electron pulses with a FWHM of 143 ps were obtained from a 5-mm-diameter electron beam, which corresponds to the diameter of the brightness-enhanced positron beam. Our system was confirmed to have a sufficient pulsing performance to perform positron annihilation lifetime measurements.

Acknowledgements

This study was financially supported by Innovative Nuclear Research and Development Program of Ministry of Education, Culture, Sports, Science and Technology. We thank colleagues of Tohoku Univ., Kyoto Univ., KEK, and Tokyo Univ. of Science for their invaluable help.

References

[1] Xu Q, Sato K, Yoshiie T, Sano T, Kawabe H, Nagai Y, Nagumo K, Inoue K, Toyama T, Oshima N, Kinomura A and Shirai Y 2014 *J. Phys.: Conf. Ser.* **505** 012030
[2] Sato K, Xu Q, Yoshiie T, Sano T, Kawabe H, Nagai Y, Nagumo K, Inoue K, Toyama T, Oshima N, Kinomura A and Shirai Y 2015 *Nucl. Instrum. Methods Phys. Res. Sect. B* **342** 104
[3] Oshima N, Suzuki R, Ohdaira T, Kinomura A, Narumi T, Uedono A and Fujinami M 2008 *J. Appl. Phys.* **103** 094916
[4] Kuzuya Y, Oshima N, Kinomura A, Yabuuchi A, Sato K and Xu Q (*unpublished*)
[5] Suzuki R, Kobayashi Y, Mikado T, Ohgaki H, Chiwaki M, Yamazaki T and Tomimasu T 1991 *Jpn. J. Appl. Phys.* **30** L532
[6] Suzuki R, Kobayashi Y, Mikado T, Ohgaki H, Chiwaki M, Yamazaki T and Tomimasu T 1992 *Solid State Phenomena* **28–29** 365
[7] Suzuki R, Kobayashi Y, Mikado T, Ohgaki H, Chiwaki M and Yamazaki T 1994 *Hyperfine Interact.* **84** 345
[8] O'Rourke B E, Oshima N, Kinomura A, Ohdaira T and Suzuki R 2012 *Defect and Diffusion Forum* **331** 75
[9] Rytsölä K, Nissilä J, Kokkonen J, Laakso A, Aavikko R and Saarinen K 2002 *Appl. Surf. Sci.* **194** 260
[10] Suzuki R 1991 *Researches of the Electrotechnical Laboratory* **928** 117 (*in Japanese*)
[11] Laakso A, Pelli A, Rytsölä K, Saarinen K and Hautojärvi P 2006 *Appl. Surf. Sci.* **252** 3148
[12] Maekawa M and Kawasuso A 2012 *Nucl. Instrum. Methods Phys. Res. Sect. B* **270** 23
[13] Zafar N, Chevallier J, Laricchia G and Charlton M 1989 *J. Phys. D: Appl. Phys.* **22** 868
[14] Schultz P J, Gullikson E M and Mills A P Jr. 1986 *Phys. Rev. B* **34** 4421986

IOP Conf. Series: Journal of Physics: Conf. Series **791** (2017) 012014 doi:10.1088/1742-6596/791/1/012014

The AEgIS experiment at CERN: measuring antihydrogen free- fall in earth's gravitational field to test WEP with antimatter

R S Brusa[1, 2], C Amsler[3], T Ariga[3], G Bonomi[4,5], P Bräunig[6], L Cabaret[7], M Caccia[8,9], R Caravita[10,11], F Castelli[8,12], G Cerchiari[13], D Comparat[7], G Consolati[8,14], A. Demetrio[6], L Di Noto[10,11], M Doser[15], A Ereditato[3], C Evans[8, 14], R Ferragut[8, 14], J Fesel[15], A Fontana[5], S Gerber[15], M Giammarchi[8], A Gligorova[16], F Guatieri[1,2], S Haider[15], H Holmestad[17], T Huse[17], A Kellerbauer[13], D Krasnický[10,11], V Lagomarsino[10,11], P Lansonneur[18], P Lebrun[18], C Malbrunot[15,19], S Mariazzi[19], V Matveev[20,21], Z Mazzotta[8,12], G Nebbia[22], P Nedelec[18], M Oberthaler[6], N Pacifico[16], D Pagano[4,5], L Penasa[1,2], V Petracek[23], C Pistillo[3], F Prelz[8], M Prevedelli[24], L Ravelli[1,2], B Rienaecker[15], O M Røhne[17], A Rotondi[5,25], M Sacerdoti[8,12], H Sandaker[17], R Santoro[8, 9], P Scampoli[3,26], L Smestad[15, 27], F Sorrentino[10,11], I M Strojek[18], G Testera[11], I C Tietje[15], S Vamosi[19], E Widmann[19], P Yzombard[7], J Zmeskal[19] and N Zurlo[5,28]

[1] Department of Physics, University of Trento, via Sommarive 14, 38123 Povo, Trento, Italy

[2] TIFPA/INFN Trento, via Sommarive 14, 38123 Povo, Trento, Italy

[3] Laboratory for High Energy Physics, Albert Einstein Center for Fundamental Physics, University of Bern, 3012 Bern, Switzerland

[4] Department of Mechanical and Industrial Engineering, University of Brescia, via Branze 38, 25123 Brescia, Italy

[5] INFN Pavia, via Bassi 6, 2790 Pavia, Italy

[6] Kirchhoff-Institute for Physics, Heidelberg University, Im Neuenheimer Feld 227, 68120 Heidelberg, Germany

[7] Laboratoire Aimé Cotton, Université Paris-Sud, ENS Cachan, CNRS, Université Paris-Saclay, 91405 Orsay Cedex, France

[8] INFN Milano, via Celoria 16, 20133 Milano, Italy

[9] Department of Science, University of Insubria, Via Valleggio 11, 2290 Como, Italy

[10] Department of Physics, University of Genova, via Dodecaneso 33, 16146 Genova, Italy

[11] INFN Genova, via Dodecaneso 33, 16146 Genova, Italy

[12] Department of Physics, University of Milano, via Celoria 16, 20133 Milano, Italy

[13] Max Planck Institute for Nuclear Physics, Saupfercheckweg 1, 68117 Heidelberg, Germany

[14] Politecnico of Milano, Piazza Leonardo da Vinci 32, 20133 Milano, Italy

[15] Physics Department, CERN, 1211 Geneva 23, Switzerland

[16] Institute of Physics and Technology, University of Bergen, Allégaten 55, 5007 Bergen, Norway
[17] Department of Physics, University of Oslo, Sem Sælandsvei 24, 0371 Oslo, Norway
[18] Institute of Nuclear Physics, CNRS/IN2p3, University of Lyon 1, 68622 Villeurbanne, France
[19] Stefan Meyer Institute for Subatomic Physics, Austrian Academy of Sciences, Boltzmanngasse 3,
980 Vienna, Austria
[20] Institute for Nuclear Research of the Russian Academy of Science, Moscow 117312, Russia
[21] Joint Institute for Nuclear Research, 141980 Dubna, Russia
[22] INFN Padova, via Marzolo 8, 35131 Padova, Italy
[23] Czech Technical University, Prague, Brehová 7, 11518 Prague 1, Czech Republic
[24] University of Bologna, Viale Berti Pichat 6/2, 40126 Bologna, Italy
[25] Department of Physics, University of Pavia, via Bassi 6, 2790 Pavia, Italy
[26] Department of Physics "Ettore Pancini", University of Napoli Federico II, Complesso Universitario di Monte S. Angelo, 80126, Napoli, Italy
[27] The Research Council of Norway, P.O. Box 564, NO-1327 Lysaker, Norway
[28] Department of Civil Engineering, University of Brescia, via Branze 43, 25123 Brescia, Italy

E-mail: brusa@science.unitn.it

Abstract. The AEgIS (Antimatter Experiment: Gravity, Interferometry, Spectroscopy) experiment is designed with the objective to test the weak equivalence principle with antimatter by studying the free fall of antihydrogen in the Earth's gravitational field. A pulsed cold beam of antihydrogen will be produced by charge exchange between cold Ps excited in Rydberg state and cold antiprotons. Finally the free fall will be measured by a classical moiré deflectometer. The apparatus being assembled at the Antiproton Decelerator at CERN will be described, then the advancements of the experiment will be reported: positrons and antiprotons trapping measurements, Ps two-step excitation and a test-measurement of antiprotons deflection with a small scale moiré deflectometer.

1. Introduction

The weak equivalence principle (WEP), also known as universality of free fall (UFF), states that in a gravitational field all bodies, irrespective of their mass and composition, fall with the same acceleration. UFF of matter in the field of Earth has been tested by measuring the dimensionless parameter $\eta = \Delta a/a$ where Δa is the relative acceleration between two proof masses as they fall with an acceleration a towards the Earth; validity of UFF-WEP requires $\eta = 0$. Precision experiments with torsion balance and dropping of cold-atoms have measured η reaching values of $\cong 10^{-13}$ and $\cong 10^{-7}$, respectively [1]. Up to now, in spite of very precise experiments done with matter, none has been carried out to test UFF-WEP with antimatter.

The AEgIS experiment, set up at the Antiproton Decelerator (AD) at CERN, is designed with the primary scientific goal of measuring for the first time the free fall of antihydrogen ($\overline{H}$) in the Earth's gravitational field with percent order precision [2]. In the experiment, the measurement of the gravitational acceleration $\bar{g}$ on $\overline{H}$ will be carried out by measuring the time of flight and the vertical displacement of each $\overline{H}$ after its passage through a moiré deflectometer realized with two gratings and a position sensitive detector. The needed pulsed cold beam of $\bar{H}$ will be produced by charge exchange among excited Ps atoms and cooled antiprotons.

In this paper the current progress towards forming a cold antihydrogen beam and measuring gravity will be summarized.

2. The AEgIS experiment: method and set-up

A detailed description of the experiment can be found in [3]. The method proposed by AEgIS to form a cold $\bar{H}$ beam is sketched in Figure 1. Bunches of more than 10^8 positrons with a duration of about 10 ns are shot on a positron-positronium (Ps) converter. Collisionally cooled Ps emitted into vacuum is excited in Rydberg states. These long living Ps fly into the antiproton trap, where $\bar{H}$ can be formed in an excited state by the charge exchange reaction: $Ps^* + \bar{p} \rightarrow \bar{H}^* + e^-$. Ps excitation in high Rydberg states is necessary not only to lengthen its lifetime, but also to augment the $\bar{H}^*$ yield: the cross section of the charge exchange reaction is proportional to n^4 where n is the *Ps* principal quantum number. Finally, excited $\bar{H}^*$ are Stark accelerated towards a moiré deflectometer, the classical analogue of a matter wave interferometer. Along its travel $\bar{H}^*$ decays to ground state.

A sketch of the moiré deflectometer is shown in the right panel of Figure 1. The present design, details reported in Ref. [3], is based on a device previously used to measure the gravity acceleration of an argon beam with high sensitivity [4].

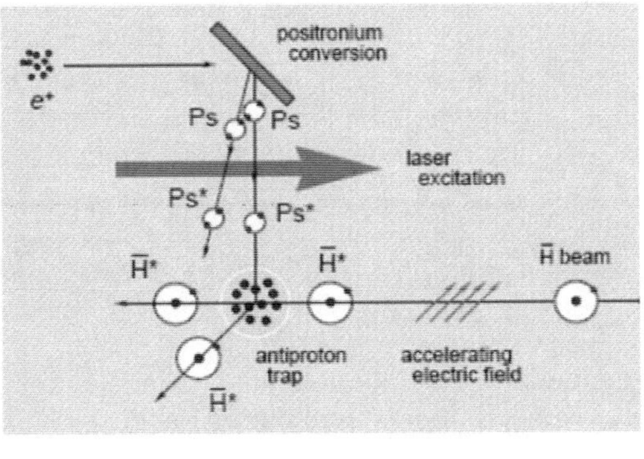

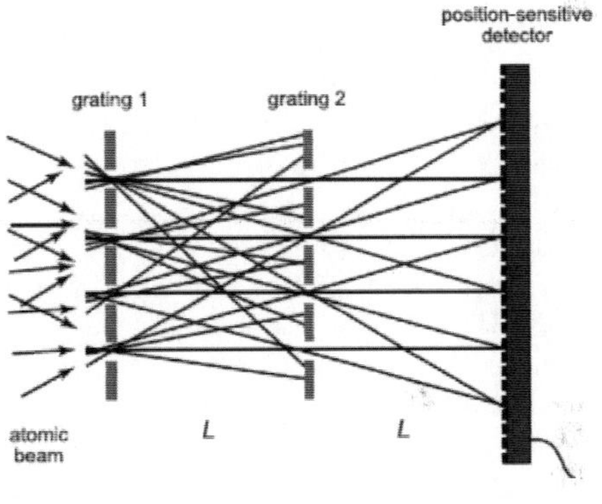

Figure 1. Left, the AEgIS method for the production of a pulsed cold $\bar{H}$ beam. Right, sketch illustrating the moiré deflectometer technique [3, 4]. A divergent $\bar{H}$ beam ($v \sim 400 - 600\ m\ s^{-1}$) propagates through two identical gratings with a period equal to the grating period (80 μm) at a distance L=40 cm. $\bar{H}$ passing the gratings follow a parabolic path and annihilate on a position sensitive detector with a downward shift that would be of the order of 10 μm considering an acting gravitational force as on a H atom.

The experimental apparatus scheme is shown in Figure 2, details in [5]. The AD delivers bunches of $\sim 3\text{x}10^7\ \bar{p}$ every $\sim$ 110 s with 5.3 MeV kinetic energy. Antiprotons passing through aluminium foils (degrader) are slowed down to a few keV and then caught in a 75 cm long set of Penning-Malmberg traps in the 5 T magnet. Trapped $\bar{p}$ are cooled to a few Kelvin by sympathetic cooling with a cloud of 10^7-10^9 electrons previously stored in a 100-150 V potential well (see section 3). Cooled $\bar{p}$ are then transferred into a second Penning-Malmberg trap in the 1T magnet and wait there for *Ps* to form $\bar{H}$.

Positrons, from a ^{22}Na radioactive source (~10 mCi) coupled to a Ne moderator [6], are cooled in a two-stage Surko buffer trap [7] and stored in a Penning-Malmberg accumulator that releases bunches of some 10^7 e^+, which are then transferred and trapped in the 5T magnet.

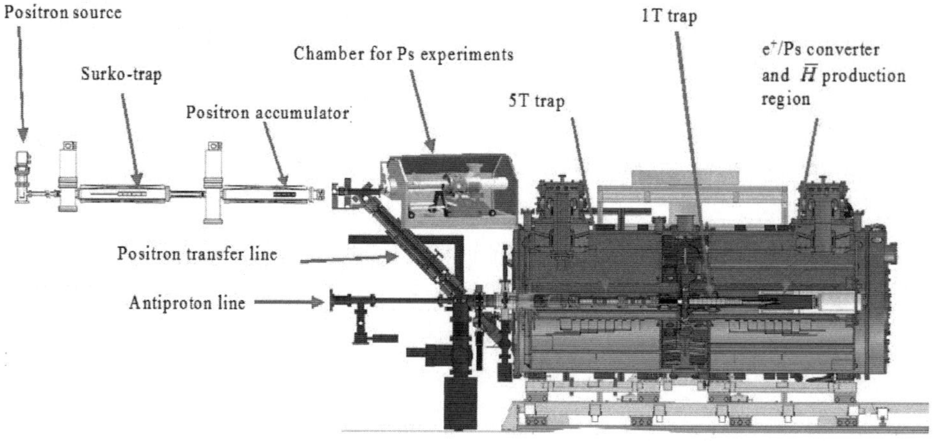

Figure 2. Scheme of the AEgIS experimental set-up.

Ps in vacuum is produced by transferring and implanting previously trapped positrons (see section 3) in a porous silica converter installed in the 1T magnet. UV (λ=205 nm) and IR (λ~1670 nm) laser light [8] is transported with glass fibers in front of the converter to perform a two-step *Ps* excitation: $1^3S \rightarrow 3^3P$, $3^3P \rightarrow$ Rydberg states (see section 4).

In addition to the main apparatus for $\bar{H}$ gravity study, a chamber to perform *Ps* formation and excitation experiments is connected to the accumulator through a 60 cm magnetic transport line. Positrons pass through a magnetic field terminator and are bunched (~ 7 ns FWHM) and focused on a *Ps* converter with a spot of less than 4 mm FWTM; bunched positron energy varies in the 3 to 8 keV range [9, 10].

3. Trapping positrons and antiprotons in the 5T magnet

A reproducible procedure for trapping e^+ and $\bar{p}$ in the 5T region was devised during the 2015 AD run [5, 11]. Bunches of e^+, with a longitudinal velocity corresponding to an energy of 300 eV, were magnetically transported in a 0.14 T field and injected into the Penning-Malmberg trap in a 4.5 T magnetic field. Positrons are cooled down by cyclotron radiation and trapped in a 50-100 eV potential well. The number of stored e^+ and e^+ lifetime in the trap were evaluated by dumping the particles on a stopper and measuring annihilation gamma-rays with plastic scintillators placed around the 1T-5T chamber. When ~2.5×10^7 e^+ are trapped, about 100% of them cool down without observed losses for storage times up to more than 30 min. Thanks to the long lifetime of e^+ in the trap, it was possible to accumulate more than 2×10^8 e^+, transferring shots of 1.8×10^7 e^+ from the accumulator, see left panel in Figure 3.

Study of trapped $\bar{p}$ was done in the same Penning-Malmberg trap used to store e^+. On the average 3.6×10^5 $\bar{p}$ were captured for each shot of 3.0×10^7 $\bar{p}$ delivered by AD, with an optimized 9 kV trapping potential. Before capturing $\bar{p}$, the trap was loaded with 10^7-10^9 electrons for $\bar{p}$ electron-cooling. The remaining hot fraction of $\bar{p}$ was firstly ejected by lowering the trapping high-voltage to a low-voltage (fraction HD-hot dump). Then the cooled fraction was dumped by lowering the low-voltage (fraction CD-cold dump). In Fig. 3, right panel, the fraction of cooled $\bar{p}$ (CD) against the total trapped $\bar{p}$ (CD+HD) is reported as a function of the time spent by $\bar{p}$ with electrons. A cooling efficiency of 90% was achieved in about 60 s with an optimum overlap of the e^-- $\bar{p}$ clouds.

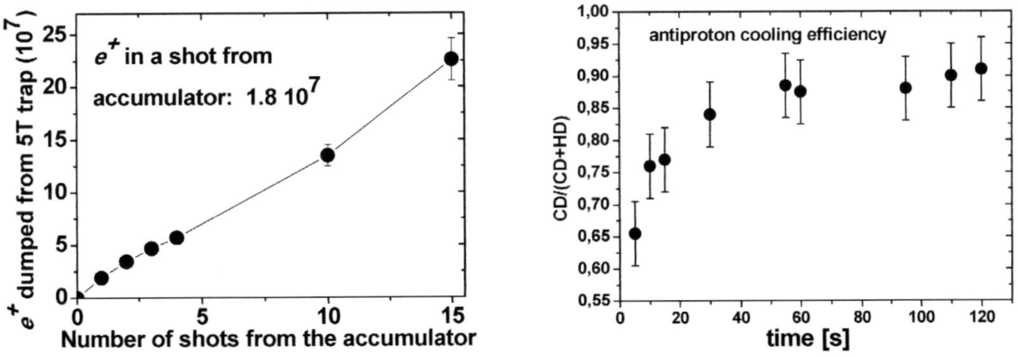

Figure 3. Left: Trapped e^+ in the 5T region as a function of number of shots from the accumulator; errors are statistical and due to the fluctuation in the number of e^+ from the accumulator. Right: fraction of cold $\bar{p}$ as a function of time spent in the trap with electrons; errors are the statistical and due to fluctuation in the number of $\bar{p}$ delivered by AD. Reproduced from Ref. [5] and Ref. [11].

4. Positronium formation and excitation in the secondary chamber

Ps formation, emission into vacuum and excitation in Rydberg states were proved and verified in the "chamber for Ps experiments" [10] in line with the Surko trap and the accumulator, see Figure 2.

High yield of Ps in vacuum was observed [10] using a Si target with oriented oxidized nanochannels [12, 13] as e^+/*Ps* converters. *Ps* excitation was measured by the SSPALS (Single Shot Positron Annihilation Lifetime Spectroscopy) method firstly introduced by Cassidy and Mills [14] to study two-step, $1^3S \rightarrow 2^3P$, $2^3P \rightarrow$ Rydberg excitation [15], Ps-Ps interaction [16] and to observe Ps_2 molecules [17].

The lifetime spectra from the single gamma ray shots were acquired with a $PbWO_4$ scintillator coupled to a Hamamatsu R11265-100 photomultiplier tube. A laser system to perform the two-step *Ps* excitation $1^3S \rightarrow 3^3P$, $3^3P \rightarrow$ Rydberg was designed and set up [8].

A UV laser pulse (205 nm, energy 54 μJ, pulse width ~ 1.5 ns) was used to excite *Ps* from ground to n=3 state and simultaneously an IR laser (λ=1064 nm, 50 mJ, 10 ns temporal pulse width) was shot to ionize the excited *Ps*.

An IR laser (tuneable wavelength in the ~ 1650- ~1720 nm range, energy ~ 1mJ, pulse width ~4 ns) was pulsed at the same time with the UV laser to excite *Ps* from n=3 to Rydberg levels.

Results are shown in Figure 4. Experiment details and analysis of the data are extensively reported in [18].

The black lines in Figure 4, after the annihilation prompt peak, correspond to the *Ps* decaying in vacuum. When the UV+IR (λ=1064 nm) lasers are shot on the *Ps* cloud, *Ps* population is decreased by the fraction of ionized *Ps* atoms, as shown by the grey curve in the left panel of Figure 4. The decrease is evaluated with the parameter $S = (f_{off}-f_{on})/f_{off}$ where f_{off} and f_{on} are the areas of the SSPAL spectra between 50 and 250 ns with the laser *off* and *on*. In the inset (left panel Fig.4) the $1^3S \rightarrow 3^3P$ excitation linewidth obtained measuring $S\%$ as a function of the UV wavelength is shown.

The lifetime of *Ps*, when excited in Rydberg states, increases up to microseconds allowing *Ps** to reach the walls of the vacuum chamber. In this case the SSPALS spectrum shows a decrease of annihilations after the prompt peak due to the *Ps**formation (as shown in the grey curve up 300 ns in the right panel of Figure 4) and an increase of annihilations when *Ps** start reaching the chamber walls, as shown in the grey curve after 300 ns in the right panel of Figure 4. A scan of the IR laser wavelength, while keeping the UV laser wavelength constant on n=3 resonance (λ=205.05 nm), was carried out to resolve the n=15 Rydberg line, and partially the n=16-17 lines due to the excessive broadening. The $-S\%$ calculated between 300-600 ns is shown in the inset (right panel Figure 4) as a function of the IR wavelength.

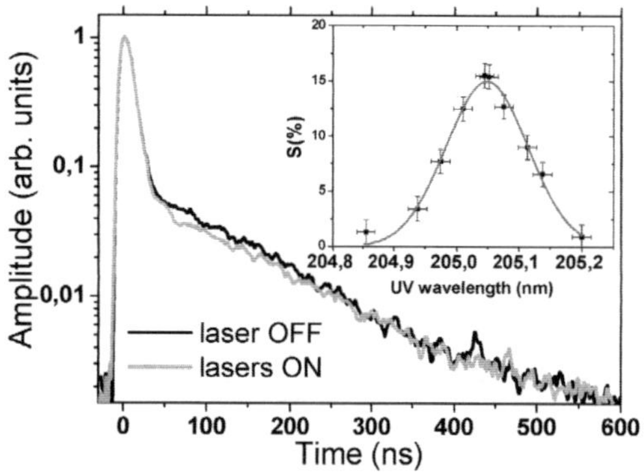

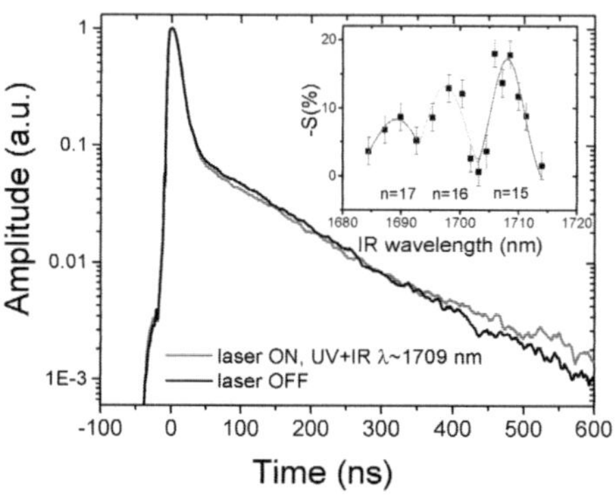

Figure 4. SSPALS spectra. Left: *Ps* in vacuum, laser off , black line; laser (UV+IR for ionization) on, grey line; the inset shows the scan of the UV wavelength, showing the n=3 excitation line. Right: *Ps* in vacuum, laser off , black line; laser (UV+IR for excitation) on, grey line; the inset shows the scan with the IR wavelength, showing the *Ps* excitation in the n=15-17 Rydberg states; the continuous line is for eye guide. Reproduced from Ref. 18.

5. Moiré deflectometer: concept proof with antiprotons

To prove the basis of the technique, a small scale moiré deflectometer was realized and tested with a beam of antiprotons with 106 keV mean energy [19]. The distance L between the gratings and the detector was set L=25 mm, see Figure1 for the scheme. The slits in the 100 μm thick silicon grating were manufactured with a 12 μm width and a 40 μm periodicity, ensuring a classical regime when these dimensions are compared to the de Broglie wavelength of the antiprotons. The annihilation positions of $\bar{p}$, which have passed the grating, were detected by an emulsion detector [20, 21] with a 2 μm resolution. The moiré deflectometer and the emulsion detector were mounted at the end of the two main magnets (1T and 5T, Figure 2) in a dedicated vacuum chamber. A moiré pattern of 241 antiproton annihilation events was recorded and the absolute fringe pattern shift was determined by comparing with a reference Talbot-Laue pattern obtained by illuminating the deflectometer with light. The shift is given by $\Delta y = F\tau^2/m$, where F is the force perpendicular to the slits, τ is the time of flight between the two gratings and m the $\bar{p}$ mass. The observed upward shift, $\Delta y = 9.8$ μm $\pm$ 0.9 μm (stat.) $\pm$ 6.4 μm (syst.) (Figure 5), was found to be consistent with a mean force of 530 aN $\pm$ 50 aN (stat.) $\pm$ 350 aN (syst.) acting on the antiprotons. This force can be caused by a magnetic field component of ~7.4 G, compatible with a magnetic field of ~ 10 G measured in the position of the deflectometer.

The gravitational force acting on a $\bar{H}$ atom will be ten order of magnitude smaller than the force measured in this test with the small scale moiré, nevertheless this experiment proves that the detection of the fall under gravity of antihydrogen atoms can be achieved if the small moiré deflectometer is scaled for increasing the time of flight τ. As a matter of fact, with a shift of the order of 10 μm, a better sensitivity of 11 orders of magnitude can be reached forming a $\bar{H}$ beam with 500 m/s velocity and increasing at 1 m the distance between the two gratings.

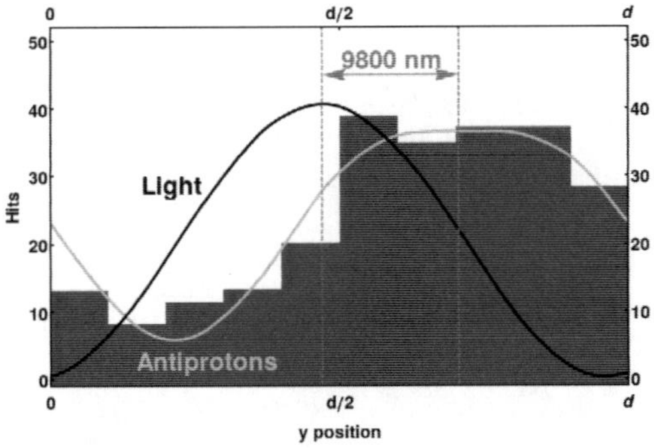

Figure 5. The light and $\bar{p}$ patterns showing the observed $\bar{p}$ shift of antiprotons with the moiré deflectometer. Reproduced from Ref. [19].

6. Conclusions

The design of the AEgIS experiment requires the development of several techniques: first to manipulate antiprotons, positrons and positronium for the production of an antihydrogen beam and then, to manipulate antihydrogen for studying its free fall in the Earth's gravitational field. In this paper the last achievements in this development have been reported. They can be summarized as follow.

At present, about 3.6x10^5 antiprotons are captured per shot of antiprotons (3.0x10^7) delivered by AD and 90% of them are electron-cooled in the 4.5 T trap. Although working with a 10 mCi ^{22}Na source, up to 2.2x10^8 positrons can be routinely trapped and cooled in the 4.5 T magnet and can be there stored for tens of minutes without significant losses. With a stronger source it is expected to increase the number of trapped e^+ and decrease the time to fill the trap.

A laser system was set up to perform a two-step $1^3S \rightarrow 3^3P$, $3^3P \rightarrow$ Rydberg Ps excitation: a fundamental advancement towards the antihydrogen production by the charge exchange reaction between antiprotons and excited Ps. Clouds of Ps atoms were obtained in vacuum by implanting positron bunches in an efficient e^+/Ps nanochannelled silicon converter allocated in a dedicated chamber for performing Ps experiments. Positrons from a two-stage Surko trap were transferred into an accumulator and then dumped and re-bunched on the Si converter. The excitation experiment was performed detecting the Ps excited fraction by employing the SSPALS technique.

Finally the feasibility of the free fall measurement of $\bar{H}$ by a moiré deflectometer was proved by measuring the deflection of an antiproton beam with a small scale moiré deflectometer and an emulsion detector with 2 μm resolution. It was shown that shift of ten microns, as expected in a free fall experiment with $\bar{H}$, can be effectively measured.

References

[1] Nobili A M 2016 *Phys. Rev.* A **93** 023617
[2] Drobychev G et al 2007 CERN-SPSC-2007-017, http://cdsweb.cern.ch/record/1037532
[3] Doser M *et al.* 2012 *Class. Quantum Grav.* **29** 184009
[4] Oberthaler M K, Bernet S, Rasel E M, Schmiedmayer J and Zeilinger A 1996 *Phys. Rev. A* **54** 3165
[5] Mariazzi S *et al.* 2016 *J. Phys: Conf. Ser.* in print
[6] Mills A P Jr and Gullikson E M 1986 *Appl. Phys. Lett.* **49** 1121
[7] Danielson J R, Dubin D H E, Greaves R G and Surko C M 2015 *Rev. Mod. Phys.* **87** 247
[8] Cialdi S, Boscolo I, Castelli F, Villa F, Ferrari G and Giammarchi M G 2011 *Nucl. Instrum. Methods Phys. Res. B* **269** 1527
[9] Penasa L, Di Noto L, Bettonte M, Mariazzi S, Nebbia G and Brusa R S 2014 *J. Phys: Conf. Ser.* **505** 012031
[10] Aghion S *et al.* 2015 *Nucl. Instrum. Methods Phys. Res. B* **362** 86

[11] Caravita R *et al.*2016 *WAG proceeding* in print
[12] Mariazzi S, Bettotti P, Larcheri S, Toniutti L and Brusa R S 2010 *Phys. Rev. B* **81** 235418
[13] Mariazzi S, Bettotti P and Brusa R S 2010 *Phys. Rev. Lett.***104** 243401
[14] Cassidy D B, Deng S H M, Tanaka H K M and Mills A P Jr 2006 *Appl. Phys. Lett.* **88** 194105
[15] Cassidy D B, Hisakado T H, Tom H W K and Mills A P Jr 2012 *Phys. Rev. Lett.* **108** 043401
[16] Cassidy D B and Mills A P Jr 2008 *Phys. Rev. Lett.* **100** 013401
[17] Cassidy D B and Mills A P Jr 2007 *Nature* **449** 195
[18] Aghion S *et al.* 2016 *Phys. Rev. A* **94** 012507
[19] Aghion S *et al.* 2014 *Nat. Commun.* **5** 4538
[20] M. Kimura *et al.* 2013 *Nucl. Instrum. Methods Phys. Res. A* **732** 325
[21] Aghion S *et al.* 2013 *JINST* **8** P08013

IOP Conf. Series: Journal of Physics: Conf. Series **791** (2017) 012015 doi:10.1088/1742-6596/791/1/012015

Effects of vibrational anharmonicity and inter-mode couplings on the binding energy of a positron to molecules

Y Kita and M Tachikawa

Quantum Chemistry Division, Yokohama City University, Seto 22-2, Kanazawa-ku, Yokohama 236-0027, Japan

E-mail: (YK) ykita@yokohama-cu.ac.jp, (MT) tachi@yokohama-cu.ac.jp

Abstract. We theoretically analyzed the effect of the anharmonicity and the inter-mode couplings of molecular vibrations on a positron affinity (PA), which is the binding energy of a positron, of HCN and CH_2O molecules. Based on the different theoretical approaches for describing vibrational wave functions, we confirmed that (i) not only the anharmonicity but also the inter-mode couplings on potential energy surface are indispensable for the accurate calculation of PA values, and (ii) the effect of inter-mode couplings on *a positron affinity surface* is sufficiently weaker than that on potential energy surface.

1. Introductions

Positrons are widely used in both scientific and technological areas such as physics, chemistry, material science, medicine, and their interdisciplinary areas [1, 2]. The understanding of the fundamental processes of positrons in materials is expected to provide a useful information for the interaction of the positron with surfaces, bulk materials, and polymers, as investigated using techniques such as Doppler broadening of positron annihilation radiation, angular correlation of annihilation radiation, positron annihilation lifetime spectroscopy, etc.

A positron affinity (PA), which is a binding energy of a positron to an atom or molecule, is one of the most important properties for positron-attached systems as well as an annihilation rate. Surko and co-workers have succeeded in measuring the experimental PA values for many molecular species such as hydrocarbons, alcohols, nitriles, aldehydes, etc. using the vibrational Feshbach resonance (VFR) technique [3-6]. In VFR processes, a positron molecular complex can be formed at vibrational excited states of the molecule by absorbing the excess energy of an incident positron. Thus, the effect of molecular vibrations could have an important role on the binding of a positron in this phenomenon.

Recently, we have theoretically demonstrated the effect of molecular vibrations on the PA values of hydrogen cyanide (HCN) and formaldehyde (CH_2O) molecules [7, 8] using *ab initio* multi-component molecular orbital and vibrational quantum Monte Carlo methods. For both molecules, vibrational excitations cause a drastic change in molecular PA values, and such PA variations mainly arises from the change in molecular permanent dipole moments by the excitations. The details of anharmonicity and inter-mode couplings of molecular vibrations, however, have not discussed from a quantitative point of view, yet. In the present study, we discussed the PA values of HCN and CH_2O molecules at vibrational excited states using different theoretical approaches for describing vibrational wave functions, in order to elucidate the effect of the anharmonicity and the inter-mode couplings of molecular vibrations on the positron binding to the molecules.

IOP Conf. Series: Journal of Physics: Conf. Series **791** (2017) 012015 doi:10.1088/1742-6596/791/1/012015

2. Method

In order to estimate molecular PA values including the effect of molecular vibrations, we employed the vibrational averaged scheme [7]. The vibrational averaged PA value (PA_v) are defined as

$$PA_v = \int PA(\mathbf{Q})|\Psi_v(\mathbf{Q})|^2 d\mathbf{Q} \Big/ \int |\Psi_v(\mathbf{Q})|^2 d\mathbf{Q}, \tag{1}$$

where $\mathbf{Q} = (Q_1, Q_2, \ldots)$ means vibrational normal mode coordinates, Ψ_v the v th vibrational wave function of a parent molecule, and $PA(\mathbf{Q})$ the following vertical PA at a given $\mathbf{Q}$:

$$PA(\mathbf{Q}) \equiv E^{M}(\mathbf{Q}) - E^{[M;e^+]}(\mathbf{Q}), \tag{2}$$

where E^{M} and $E^{[M;e^+]}$ are total energies of a parent molecule (M) and its positronic complex ([M;e^+]), respectively. The PA_v value corresponds to the weighted average of vertical PA values with vibrational probability density function ($|\Psi_v|^2$). The vibrational wave function is assumed to be the form of products of modal functions ϕ_i as $\Psi_v = \prod_i^{\text{all modes}} \phi_i$. We employed three different methods to obtain modal functions: (i) vibrational quantum Monte Carlo (VQMC) method [7], (ii) Störmer-Levy-Oshida (SLO) method [10], and (iii) harmonic approximation (HA). We note here that VQMC method can take both the anharmonicity and the inter-mode couplings on potential energy surface (PES) into account under the mean-field approximations. The SLO method is one of numerical solvers to obtain the accurate solution of one-dimensional Schrödinger equation, and thus the solutions includes the effect of one-dimensional anharmonicity along each vibrational coordinate. The HA method includes neither the anharmonicity nor the inter-mode couplings. In order to calculate PA(**Q**) values, the multi-component molecular orbital (MC_MO) method [9] was used in this study. Since a PA value is defined at a given position in a multi-dimensional space (**Q**-space), PA(**Q**) in Eq. (2) has the form of multi-dimensional surface like PES, and contains multi-dimensional contributions (or inter-mode couplings). In order to discuss the inter-mode couplings on PA(**Q**) surface (*positron affinity surface, PAS*) as well as PES, we also calculated PA values containing only one-dimensional contributions along each vibrational coordinate with VQMC wave functions.

In the vertical PA calculations, we employed the truncated singly and doubly excited configuration interaction (CISD) method for a positron-attached molecule, where three types of excitations, (i) single electronic, (ii) single positronic, and (iii) simultaneous single electronic - single positronic excitation configurations are included. The 6-31++G(*2df,2pd*) and [15*s*15*p*3*d*2*f*1*g*] Gaussian-type functions (GTFs) were employed for electronic and positronic basis sets, respectively, for [HCN;e^+] system. For [CH_2O;e^+] system, the electronic and positronic basis sets are the 6-31+G(3*d*2*f*) and [12*s*8*p*6*d*2*f*] GTFs, respectively. The exponents of the positronic GTFs were determined by the even-tempered scheme. These basis sets were constructed as reproducing the accurate vertical PA values obtained with diffusion Monte Carlo calculations at equilibrium geometries of HCN and CH_2O molecules [8, 11].

In the potential energy calculations, CCSD(T) (coupled-cluster singles and doubles with perturbative triples) with Dunning's augmented correlation consistent polarized valence triple-zeta (aug-cc-pVTZ) basis set was used for HCN molecule. The four-dimensional PES was construct by a potential energy table consisting of 14,695 grid points. For CH_2O molecule, the CCSD/aug-cc-pVTZ level of *ab initio* calculation was used. A six-dimensional PES of CH_2O molecule was truncated by 3-mode representation technique [12] which includes inter-mode couplings up to third order. The total number of the grid points is 33,476.

3. Results and Discussion

The PA_v values of HCN molecule obtained with different vibrational wave functions are summarized in Table 1. As reported in our previous paper [7], VQMC results show that the excitations of the CN and CH stretching modes enhance the PA value compared to the value at the ground state, while that of the bending mode deenhances it. The largest PA enhancement is found for the excited states of the CH stretching mode. Such PA_v variations can be described qualitatively even with the vibrational wave functions by SLO and HA methods. From a quantitative point of view, however, HA method significantly underestimates the PA_v variations in all vibrational excitations. For instance, the PA_v variation from the ground state value (ΔPA) by the lowest CH stretching excitation is calculated as +3.07 meV with VQMC method, while the variation is estimated as only +0.42 meV with HA method. The vibrational wave function by SLO method can improve the quantitative accuracy of PA_v variations, but still considerably underestimate PA_v variations by the vibrational excitations of the bending modes. The ΔPA by the lowest bending excitation is calculated as −1.72 meV with VQMC method, while the variation is estimated as −0.31 meV with SLO method. Using the linear regression analysis (LRA) used in our previous study [7], we also confirmed that these PA variations with SLO and HA methods are mainly due to the change in the permanent dipole moment by vibrational excitations. Although the quantitative accuracy of SLO and HA methods is lower than that of VQMC method, the origin of such PA variations is consistent with the previous VQMC results.

The mean error (ME) and mean absolute error (MAE) from VQMC results for SLO and HA methods are also given in Table 1. The ME and MAE are here defined by the average of $PA_v^{(SLO/HA)} - PA_v^{(VQMC)}$ and its absolute value over all states, respectively. The ME of SLO method is the positive value of +1.35 meV meaning that the method tends to overestimate PA_v values totally, while the HA gives the negative ME value of −1.38 meV meaning the underestimation of PA_v values. The MAE with SLO method is lower than that of HA method, but is greater than 1 meV. These results mean that not only the anharmonicity but also the inter-mode couplings on potential energy surface are indispensable for the quantitative estimation of PA values of HCN molecule. Table 1 also show the PA_v value including only one-dimensional contributions on PAS with VQMC wave function ($PA_v^{(1D)}$). The ME and MAE for $PA_v^{(1D)}$ values are calculated as $\approx$0.00 meV (less than 1μeV) and 0.14 meV, respectively. These small ME and MAE values mean that the effect of inter-mode couplings on PAS is sufficiently weaker than that on PES.

Table 1. Vibrational averaged positron affinity (PA_v) values of HCN molecule at the ground (GS), fundamental tone (X_1), and lowest overtone (X_2) states using the vibrational wave functions obtained with variational Monte Carlo method (VQMC), Störmer-Levy-Oshida method (SLO), and harmonic approximation (HA). The PA_v variation from each ground state (ΔPA) are also shown. Unit in meV.

State	VQMC				SLO		HA	
$X_{1,2}$ [a]	$PA_v^{(VQMC)}$ [b]	ΔPA	$PA_v^{(1D)}$ [c]	ΔPA	$PA_v^{(SLO)}$	ΔPA	$PA_v^{(HA)}$	ΔPA
GS	39.95 ±0.01	–	39.95	–	40.96	–	39.51	–
1_1	38.23 ±0.01	−1.72	38.29	−1.66	40.65	−0.31	39.14	−0.38
2_1	40.88 ±0.01	+0.93	40.92	+0.98	42.09	+1.13	39.80	+0.29
3_1	43.02 ±0.01	+3.07	42.79	+2.84	43.42	+2.47	39.94	+0.42
1_2	36.99 ±0.01	−2.96	37.40	−2.55	40.41	−0.54	38.82	−0.69
2_2	41.93 ±0.01	+1.98	41.88	+1.93	43.26	+2.30	40.09	+0.58
3_2	46.34 ±0.02	+6.39	46.12	+6.17	45.99	+5.03	40.35	+0.84
Mean error from VQMC results			$\approx$0.00		+1.35		−1.38	
Mean absolute error from VQMC results			0.14		1.45		2.16	

[a] X denotes a vibrational mode: "1" means the bending mode, "2" CN stretching mode, and "3" CH stretching mode.
[b] All the data from Ref. [7]. The error means the standard error of Monte Carlo samplings.
[c] The inter-mode couplings of PA surface is not included.

The PA_v values of CH_2O molecule obtained with different vibrational wave functions are summarized in Table 2. In the case of CH_2O molecule, the excitations of modes 2 (C=O stretching), 3

(CH_2 bending), and 6 (CH_2 rocking) tend to enhance the vibrational PA_v value, while those of modes 1 (symmetric C–H stretching), 4 (out-of-plane bending), and 5 (anti-symmetric C–H stretching) tend to deenhance it. Among all vibrational modes, the C=O stretching mode gives the largest contribution to the enhancement of PA_v value [8]. The ΔPA values with the SLO and HA methods are qualitatively consistent with those with VQMC method for the excited states of modes 1, 2, 4, and 5 which have the absolute ΔPA values greater than about 1 meV.

The HA method gives the negative ME values as −1.05 meV which means the method totally tends to underestimate PA_v values. Although the SLO method gives the small ME value of −0.10 meV, such tendency might be due to an accidental cancellation of errors. In fact, the MAE value of SLO method is close to 1 meV. As in the case of HCN molecule, the smallest MAE value is found for $PA_v^{(ID)}$ values. From these results, we again confirmed that (i) the both the anharmonicity and the inter-mode couplings on PES are indispensable for the quantitative estimation of PA values of CH_2O molecule, and (ii) the effect of inter-mode couplings on PA surface is sufficiently weaker than that on PES.

Finally, we make a comment on the origin of PA variations at different theoretical level. As shown in our previous work [8], the PA enhancement/deenhancement of CH_2O molecule mainly arise from the change in the permanent dipole moments by vibrational excitations. The $PA_v^{(HA)}$ variations, however, could not be described by the change in the dipole moment (the R^2 value in LRA is less than 0.2). Thus, the potential anharmonicity must be essentially indispensable for describing the vibrational states of CH_2O molecule. Although the LRA results for $PA_v^{(SLO)}$ and $PA_v^{(ID)}$ values are consistent with VQMC results, the SLO method, in which the inter-mode couplings on PES is neglected, qualitatively fail to describe PA variations at the excited states of CH_2 bending and rocking modes as mentioned above.

Table 2. Vibrational averaged positron affinity values of CH_2O molecule. All abbreviations are defined in Table 1. Unit in meV.

State	VQMC				SLO		HA	
$X_{1,2}$ [a]	$PA_v^{(VQMC)}$ [b]	ΔPA	$PA_v^{(ID)}$ [c]	ΔPA	$PA_v^{(SLO)}$	ΔPA	$PA_v^{(HA)}$	ΔPA
GS	25.78±0.03	–	25.66		25.67	–	25.06	–
1_1	24.51±0.02	−1.27	24.34	−1.32	24.54	−1.12	25.02	−0.04
2_1	30.69±0.06	+4.91	30.54	+4.88	30.40	+4.73	27.28	+2.22
3_1	26.03±0.03	+0.25	25.95	+0.29	25.31	−0.36	24.77	−0.29
4_1	24.88±0.03	−0.90	24.82	−0.84	24.50	−1.16	23.87	−1.19
5_1	23.60±0.02	−2.18	23.33	−2.34	24.95	−0.72	24.31	−0.75
6_1	25.91±0.03	+0.13	25.81	+0.15	25.47	−0.19	24.86	−0.19
1_2	23.57±0.02	−2.21	23.31	−2.35	23.45	−2.22	24.96	−0.09
2_2	36.07±0.08	+10.29	35.87	+10.21	35.16	+9.50	29.50	+4.44
3_2	26.06±0.04	+0.28	26.02	+0.36	24.99	−0.67	24.51	−0.55
4_2	24.08±0.03	−1.70	24.07	−1.59	23.43	−2.23	22.75	−2.31
5_2	21.43±0.01	−4.35	20.87	−4.79	24.28	−1.38	23.58	−1.48
6_2	26.14±0.03	+0.36	26.08	+0.42	25.29	−0.37	24.68	−0.38
Mean error from VQMC results			−0.16		−0.10		−1.05	
Mean absolute error from VQMC results			0.16		0.75		1.78	

[a] X denotes a vibrational mode: "1" symmetric CH stretching, "2" CO stretching, "3" CH_2 bending, "4" out-of-plane bending, "5" anti-symmetric CH stretching, and "6" CH_2 rocking.
[b] All data from Ref. [8]. The error means the standard error of Monte Carlo samplings.
[c] The inter-mode couplings of PA surface is not included.

4. Conclusion

In order to elucidate the effect of the anharmonicity and the inter-mode couplings of molecular vibrations on the positron binding to HCN and CH_2O molecules, we theoretically analyzed the vibrational averaged positron affinity (PA) of both molecules using the vibrational wave functions obtained with three different theoretical methods: (i) vibrational quantum Monte Carlo method which includes the effects of both anharmonicity and inter-mode couplings under the mean-field approximation, (ii) Störmer-Levy-Oshida method which takes the effect of only the one-dimensional anharmonicity along each vibrational coordinate into account, and (iii) the harmonic approximation which does not include both effects. For both molecules, our results show that not only the anharmonicity but also the inter-mode couplings on potential energy surface are indispensable for the quantitative estimation of PA values. We also confirmed that the effect of inter-mode couplings on *a positron affinity surface* are sufficiently weaker than that on a potential energy surface.

Acknowledgments
Financial support was provided by Grant-in-Aid for Scientific Research and for the priority area by Ministry of Education, Culture, Sports, Science and Technology, Japan for YK and MT.

References

[1] Coleman P G 2000 *Positron Beams and Their Applications* (WorldScientific: Singapore)
[2] Charlton M, Humberston J W 2001 *Positron Physics* (Cambridge University Press: Cambridge)
[3] Gribakin G F, Young J A, Surko C M 2010 *Rev. Mod. Phys.* **82** 2557
[4] Danielson J R, Gosselin J J, Surko C M 2010 *Phys. Rev. Lett.* **104** 233201
[5] Jones A C L, Danielson J R, Gosselin J J, Natisin M R, Surko C M 2012 *New J. Phys.* **14** 015006
[6] Danielson J R, Jones A C L, Natisin M R, Surko C M 2012 *Phys. Rev. Lett.* **109** 113201
[7] Kita Y, Tachikawa M 2014 *Eur. Phys. J. D* **68** 116
[8] Yamada Y, Kita Y, Tachikawa M 2014 *Phys. Rev. A* **89** 062711
[9] Tachikawa M 2001 *Chem. Phys. Lett.* **350** 269
[10] Saitoh T, Mori K, Itoh R 1981 *Chem. Phys.* **60** 161
[11] Kita Y, Maezono R, Tachikawa M, Towler M, Needs R J 2009 *J. Chem. Phys.* **131** 134310
[12] Carter S, Culik S J, Bowman J M 1997 *J. Chem. Phys.* **107** 10458

Advanced characterization of lithium battery materials with positrons

Bernardo Barbiellini[1] **and Jan Kuriplach**[2]
[1]Department of Physics, Northeastern University, Boston MA 02115, USA
[2]Department of Low Temperature Physics, Faculty of Mathematics and Physics, Charles University, V Holešovičkách 2, CZ-180 00 Prague, Czech Republic
E-mail: [1]b.amidei@neu.edu, [2]jan.kuriplach@mff.cuni.cz

Abstract. Cathode materials are crucial to improved battery performance, in part because there are not yet materials that can maintain high power and stable cycling with a capacity comparable to that of anode materials. Our parameter-free, gradient-corrected model for electron-positron correlations predicts that spectroscopies based on positron annihilation can be deployed to study the effect of lithium intercalation in the oxide matrix of the cathode. The positron characteristics in oxides can be reliably computed using methods based on first-principles. Thus, we can enable a fundamental characterization of lithium battery materials involving positron annihilation spectroscopy and first-principles calculations. The detailed information one can extract from positron experiments could be useful for understanding and optimizing both battery materials and bi-functional catalysts for oxygen reduction and evolution.

1. Introduction

Lithium-ion batteries have played a major role in the development of portable electronics involving mobile phones, tablets and laptops because of their high energy density [1]. Clearly, design and optimization of materials are important to enable further improvements and new applications of next-generation rechargeable batteries [2]. These advances depend on studying novel compounds and gaining fundamental understanding needed to guide applied research. High throughput first-principles calculations can play a major role in characterizing and predicting the structures and properties of complex materials at the atomic scale [3]. An important goal of this effort is to establish a robust feedback loops between first-principles computations and advanced characterization based on modern spectroscopy for a rational and speedy development of lithium-ion batteries. For example, soft and hard x-ray studies performed at state-of-the art synchrotron facilities have recently revealed useful details of lithium-ion battery function [4, 5, 6, 7, 8]. Here we show that positrons can provide unique insights in cathode materials for Li-ion batteries as well.

Positrons spectroscopy in materials science has been thoroughly described in Positron Spectroscopy of Solids [9], reviewed by Puska and Nieminen [10] and more recently by Tuomisto and Makkonen [11]. New studies involving positron spectroscopy in materials science include different systems such as topological insulators [12], PbSe quantum dots [13], tungsten surfaces coated with sodium [14], metal organic frameworks [15, 16] and also cathode materials for lithium-ion batteries [17, 18].

IOP Conf. Series: Journal of Physics: Conf. Series **791** (2017) 012016 doi:10.1088/1742-6596/791/1/012016

Layered lithium transition-metal oxide $LiCoO_2$ (LCO) and olivine-type $LiFePO_4$ (LFP) are important cathode materials. In $LiFePO_4$, the FeO_6 octahedron determines important features of the electronic structure related to Fe $3d$ electrons [4]. The CoO_6 octahedron plays an important role as well in the $LiCoO_2$ electronic structure. However, the O $2p$ character provides the dominating role for the Li_xCoO_2/CoO_2 in the redox orbitals [5] while the Fe $3d$ character dominates in the redox orbital of the $LiFePO_4/FePO_4$ reaction [4, 6]. When the lithium atoms are removed from the oxide matrix, Li_xCoO_2 displays complex physics involving metal/insulator transition and Li-vacancy ordering [17]. In the case of $LiFePO_4/FePO_4$, nanoscale phase separation is always observed. The most desired properties of cathode materials involve high redox potential, high lithium capacity and high electron and Li-ion conductivity.

2. Theory and computational method

2.1. Electron-positron correlations

The density functional theory (DFT) reduces the quantum-mechanical many-body problem to a set of manageable self-consistent one-body Schrödinger-like equations. DFT solves the electronic structure of a system in its ground state so that the electron density is the basic quantity. The DFT is generalized to positron-electron systems by including the positron density as well in the scheme is two-component DFT (TCDFT) [10]. When a positron impurity is immersed in an electron gas with density $\rho(\boldsymbol{R})$, the Coulomb attraction produces a screening cloud around the positron. The corresponding potential seen by the positron is constructed as

$$V_{pos}(\boldsymbol{r}) = V_{test}(\boldsymbol{r}) + V_c(\boldsymbol{r}) \ , \tag{1}$$

where V_{test} is the potential for a positron as a test charge and where V_c is the polarization or correlation potential describing the positron perturbation. The correlation potential can be obtained via the Hellmann-Feynman theorem as

$$V_c(\boldsymbol{r}) = -\int_0^1 dZ \int d^3\boldsymbol{R}\, \frac{\rho(\boldsymbol{R})[g(\boldsymbol{r},\boldsymbol{R},Z)-1]}{|\boldsymbol{r}-\boldsymbol{R}|} \ , \tag{2}$$

where $\rho(\boldsymbol{R})[g(\boldsymbol{r},\boldsymbol{R},Z)-1]$ is the screening cloud density around a positive particle with charge Z. The local density approximation (LDA) based on accurate quantum Monte Carlo computations [19] describes the screening cloud around a delocalized positron impurity in an electron gas given by a density that varies slowly over space. The effect of the density gradient on the correlation energy can be deduced from the distortion of the polarization cloud due to this gradient. In the high density limit, the lowest order gradient correction similar to Eq. (4) of Ref. [20] reads

$$\Delta V_c = \frac{1}{4}\frac{e^2}{a^*} \times \beta \times f \times \epsilon \, . \tag{3}$$

In Eq. (3), the constant $\beta = 0.066725$ is calculated with the use of diagrammatic techniques in many body perturbation theory [21], the quantity $\epsilon = |\nabla \ln \rho|^2/(q_{TF})^2$ is the square of the reduced gradient[1] (which depends on the ratio of the inhomogeneity length $1/|\nabla \ln \rho|$ and the Thomas-Fermi length $1/q_{TF}$), e is the electron charge and a^* is the effective Bohr radius for the electron-positron liquid. The distance a^* is given by the Bohr radius a_B renormalized by an effective mass $a^* = a_B/\mu$. For a liquid where the positron density equals the electron one [22] the effective mass is given by $\mu = 1/2$ and this relationship remains approximately correct also in the limit of the positron impurity [23, 24, 25]. The factor f depends on the fraction of positrons in the electron-positron liquid and is similar for the rescaling for spin polarization used in Ref. [20]. When the positron is an impurity, $f = 1/2$. Whereas equality of the positron and electron densities implies $f = 1$. Therefore, the difference between the electronic gradient

[1] In fact we employ $\epsilon = 4t^2$, t being the reduced gradient as it is usually defined (see Ref. [20]).

correction of Eq. (4) in Ref. [20] and Eq. (3) is given by the factor $f \times \mu = 1/4$. In order to interpolate to the case of rapid density variations (i.e. large ϵ), we use the formula of the parameter-free generalized gradient approximation (GGA) [26]

$$V_c(\boldsymbol{r}) = V_c^{LDA}(\boldsymbol{r}) \exp\left[-\alpha(\boldsymbol{r})\,\epsilon(\boldsymbol{r})/3\right] , \tag{4}$$

where α is a function of the local density given by

$$\alpha(\boldsymbol{r}) = -\frac{3}{16}\frac{\beta}{V_c^{LDA}(\boldsymbol{r})}\frac{e^2}{a_B} . \tag{5}$$

The corresponding electron-positron enhancement factor is

$$\gamma - 1 = (\gamma_{LDA} - 1)\exp(-\alpha\epsilon) . \tag{6}$$

2.2. Calculations of electronic structure and positron properties

The electronic structure of the studied Li_xCoO_2, $NaCoO_2$ and Li_xFePO_4 materials was determined using the WIEN2k code [27]. This code is an implementation of the augmented plane-wave plus local orbital (APW+lo) method, which is considered to be one of the most accurate techniques to calculate electronic structure of solids. In these calculations, a GGA exchange-correlation functional for electrons after Perdew, Burke, and Ernzerhof [20] was employed. In the case of Li_xFePO_4 system, we investigated the influence of the Hubbard U parameter on the electronic structure and took its values values from Ref. [28] performing thereby GGA+U calculations. The selfconsistent electron density and Coulomb potential were taken from electronic structure calculations in order to construct the positron potential (including its correlation part discussed above). The Schrödinger equation for positron is then solved on a 3D regular mesh covering the unit cell of studied materials. Full computational details are given in Ref. [29].

$LiCoO_2$ has a layered structure of $O3$-type. The structural parameters for the corresponding rhombohedral unit cell were taken from Ref. [30]. $NaCoO_2$, which we have examined too as a possible cathode material, is isostructural with $LiCoO_2$, and we used structural parameters from Ref. [31]. The structure of $LiCoO_2$ and $NaCoO_2$ is shown in figure 1 where alternating Li (or Na), O and Co layers are well seen. The Co, O and alkali ions (Li or Na) are represented by blue, red and yellow spheres, respectively. For the sake of simplicity, we employed the same structure for partially delithiated system $Li_{0.5}CoO_2$, removing just some Li atoms to represent vacancies. No structure relaxation was performed.

Following our study on oxides based on the parameter-free GGA [32], we have calculated the sum A^+ of the electron and positron chemical potentials (called usually positron affinity) and the positron lifetime, τ, in Li_xCoO_2, $NaCoO_2$ and Li_xFePO_4 within the TCDFT framework. In particular, τ was calculated via the positron annihilation rate ($1/\tau$) according to the formula

$$\frac{1}{\tau} = \pi r_0^2 c \int \mathrm{d}\boldsymbol{r}\, \rho^+(\boldsymbol{r})\, \rho(\boldsymbol{r})\, \gamma[\rho(\boldsymbol{r}), \epsilon(\boldsymbol{r})] , \tag{7}$$

where r_0 and c are the classical electron radius and speed of light, respectively. The spatial integration proceeds over the unit cell. The positron density is obtained simply as $\rho^+ = |\psi^+|^2$ from the properly normalized positron ground state wave function ψ^+.

3. Results

In our previous study on oxides [32], we have shown the crystal structures of $LiFePO_4$ and $FePO_4$ together with the corresponding isosurface occupied by the positron density distribution (PDD) $\rho^+(\boldsymbol{r}) = |\psi^+(\boldsymbol{r})|^2$. We have observed that in $FePO_4$ the positron density occupies one-dimensional channels where Li is located in $LiFePO_4$. Therefore, positrons studies should

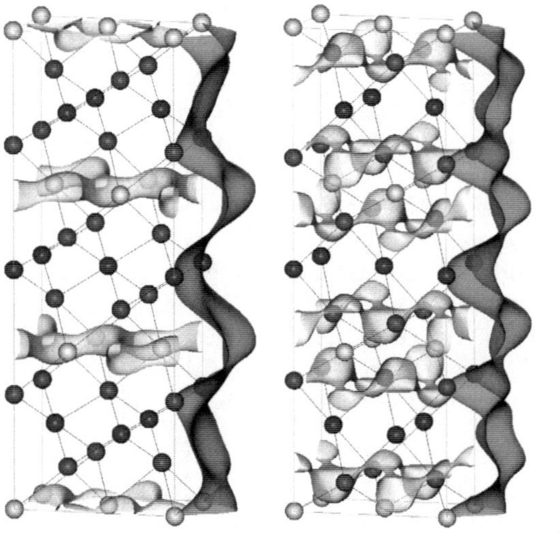

Figure 1. The structure and PDD for $LiCoO_2$ (left) and $NaCoO_2$ (right). See the text for the color code.

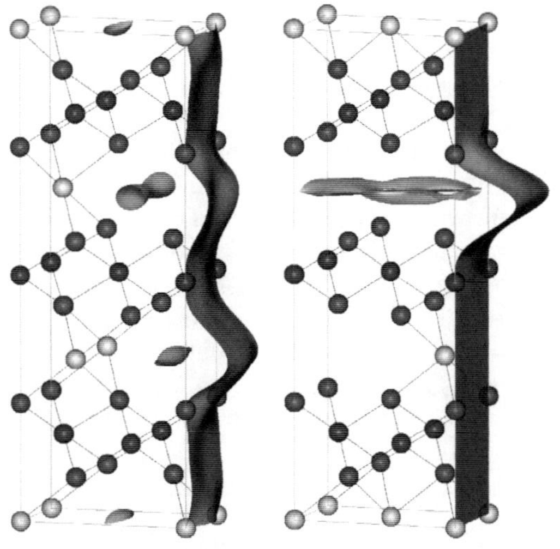

Figure 2. PDD in $Li_{0.5}CoO_2$. Li vacancies are distributed evenly in each Li plane (left) or cluster more in one of them (right).

facilitate the understanding of the Li atom transport. We provide here a similar visualization on the positron density in the CoO_2 matrix with Li and Na intercalation. In figures 2-3, the Co, O and alkali ions (Li or Na) are represented in the same way as in figure 1, whereas the isosurfaces of the positron density are colored in dark yellow (all figures). The 'vertical' red surfaces reflect the variation of ρ^+ along a vertical plane at the corresponding simulation box side.

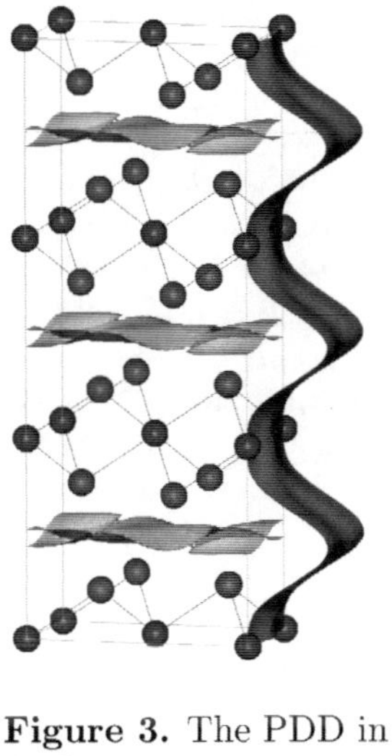

Figure 3. The PDD in CoO_2.

The PDD in $LiCoO_2$ shown in figure 1 (left part) is more concentrated in the Li planes compared to the PDD of $NaCoO_2$ displayed also in figure 1 (right part). In $NaCoO_2$ the positron samples more the adjacent O planes. In the case of $LiCoO_2$, $\tau = 131.0$ ps and $A^+ = -5.30$ eV while $\tau = 119.7$ ps and $A^+ = -6.61$ eV for $NaCoO_2$. Therefore the substitution of Li by Na gives significant differences both for the lifetime (a decrease by $\sim$ 10 ps) and A^+ (a decrease by $\sim$ 1.3 eV). It is surprising to see that positrons have less affinity for $LiCoO_2$ and at the same time live longer in this material, which is due to positron redistribution compared to $NaCoO_2$.

The positron wave function is significantly modified by delithiation. This effect corresponding to $x = 0.5$ is demonstrated in figure 2 for two cases. When the Li vacancies do not cluster and are evenly distributed in each Li plane, positrons occupy equally each such plane (see the left part of figure 2). In this case, $\tau = 179.2$ ps and $A^+ = -8.23$ eV. Oppositely, ρ^+ is distributed mainly in the Li vacancy cluster and not in planes with no or smaller amount of Li vacancies, as it is illustrated in the right part of figure 2. Consequently, the lifetime increases to 214.0 ps and A^+ decreases to -9.99 eV.

It is also instructive to inspect what happens if Li is completely removed from the lattice. The Li_xCoO_2 (and Na_xCoO_2) systems become pure CoO_2. In such a case, the lattice dimension perpendicular to Li (or Na) layers somewhat shrinks and the amount of open space in the structure decreases. Nevertheless, the positron lifetime remains quite long (151.0 ps) and A_+ value −8.98 eV is about in the middle between $x = 0.5$ cases just discussed. This is caused by quite large 'empty' planes between CoO_2 layers, which can be deduced from figure 3 where positron density distribution is plotted. Complete delithiation does not occur in commercial LCO batteries, but it is possible to achieve it in a laboratory experiment [33] (from where we also took lattice parameters for CoO_2).

Finally, table 1 gives lifetimes τ and chemical potentials sums A^+ for various magnetic configurations of $LiFePO_4$ and $FePO_4$. The positron lifetime changes dramatically with lithiation but remains rather insensitive to the magnetic structure. However, as illustrated by table 1, measurements of A^+ could be very useful to benchmark theoretical models involving various magnetic configurations and the Hubbard parameter U. Moreover, the variation $\Delta A^+ = A^+(x{=}0) - A^+(x{=}1)$ upon delithiation (3.5 eV for the AFM case) appeared to be related to the redox potential [28]. Since the positron chemical potential does change significantly with Li intercalation, ΔA^+ yields mostly the variation of the electron chemical potential.

Table 1. Positron characteristics for Li_xFePO_4 as a function of different magnetic structures: nonmagnetic (NM), ferromagnetic (FM), antiferromagnetic (AFM), ferromagnetic with a U parameter (FM+U). $U(x{=}1) = 3.71$ eV and $U(x{=}0) = 4.90$ eV [28].

x	Magnetic structure	Lifetime τ (ps)	A^+ (eV)
1	NM	169.4	-6.32
1	FM	169.1	-5.75
1	AFM	169.7	-5.70
1	FM+U	169.1	-6.87
0	NM	205.0	-9.00
0	FM	205.8	-9.13
0	AFM	207.2	-9.37
0	FM+U	206.4	-9.91

4. Conclusion

The parameter-free GGA method predicts that spectroscopies based on positron can be deployed to detect the elusive lithium in the oxide matrix of Li-batteries. The positron lifetime τ changes dramatically with lithiation and provide information about the space available for Li migration. Nevertheless, the interpretation of existing lifetime data [17, 18] for both Li_xCoO_2 and Li_xFePO4 is challenging due to the complicated microstructure of the studied samples. Measurements of A^+ could be very useful to benchmark various theoretical models and the variation of A^+ upon delithiation could be used to determine redox potentials. Similar advanced characterization can be deployed in related materials used as bifunctional catalysts for oxygen reduction and oxygen evolution reactions [34].

Acknowledgments

The work at Northeastern University was supported by the US Department of Energy (DOE), Office of Science, Basic Energy Sciences grant number DE-FG02-07ER46352 (core research), and benefited from Northeastern University's Advanced Scientific Computation Center (ASCC), the NERSC supercomputing center through DOE grant number DE-AC02-05CH11231, and support (applications to layered materials) from the DOE EFRC: Center for the Computational Design of Functional Layered Materials (CCDM) under DE-SC0012575. J. K. acknowledges the support by the Ministry of Education, Youth and Sports of the Czech Republic from the Large Infrastructures for Research, Experimental Development and Innovations project "IT4Innovations National Supercomputing Center – LM2015070".

References

[1] Goodenough J B and Kim Y 2010 *Chem. Mater.* **22** 587
[2] Islam M S and Fisher C A J 2014 *Chem. Soc. Rev.* **43** 185
[3] Ceder G and Persson K 2013 How supercomputers will yield a golden age of materials science *Scientific American* **309(6)** 34
[4] Liu X, Liu J, Qiao R, Yu Y, Li H, Suo L, Hu Y-S, Chuang Y-D, Shu G, Chou F, Weng T-C, Nordlund D, Sokaras D, Wang Y J, Lin H, Barbiellini B, Bansil A, Song X, Liu Z, Yan S, Liu G, Qiao S, Richardson T J, Prendergast D, Hussain Z, de Groot F M F and Yang W 2012 *J. Am. Chem. Soc.* **134** 13708
[5] Mizokawa T, Wakisaka Y, Sudayama T, Iwai C, Miyoshi K, Takeuchi J, Wadati H, Hawthorn D G, Regier T Z and Sawatzky G A 2013 *Phys. Rev. Lett.* **111** 056404
[6] Liu X, Wang Y J, Barbiellini B, Hafiz H, Basak S, Liu J, Richardson T, Shu G, Chou F, Weng T-C, Nordlund D, Sokaras D, Moritz B, Devereaux T P, Qiao R, Chuang Y-D, Bansil A, Hussain Z and Yang W 2015 *Phys. Chem. Chem. Phys.* **17** 26369
[7] Suzuki K, Barbiellini B, Orikasa Y, Go N, Sakurai H, Kaprzyk S, Itou M, Yamamoto K, Uchimoto Y, Wang Y J, Hafiz H, Bansil A and Sakurai Y 2015 *Phys. Rev. Lett.* **114** 087401
[8] Suzuki K, Barbiellini B, Orikasa Y, Kaprzyk S, Itou M, Yamamoto K, Wang Y J, Hafiz H, Uchimoto Y, Bansil A, Sakurai Y and Sakurai H 2016 *J. Appl. Phys.* **119** 025103
[9] 1995 *Positron Spectroscopy of Solids* ed A Dupasquier and A P Mills Jr (Amsterdam: IOS Press)
[10] Puska M J and Nieminen R M 1994 *Rev. Mod. Phys.* **66** 841
[11] Tuomisto F and Makkonen I 2013 *Rev. Mod. Phys.* **85** 1583
[12] Callewaert V, Shastry K, Saniz R, Makkonen I, Barbiellini B, Assaf B A, Heiman D, Moodera J S, Partoens B, Bansil A and Weiss A 2016 *Phys. Rev.* B **94** 115411
[13] Shi W, Eijt S W H, Sandeep Suchand C S, Siebbeles L D A, Houtepen A J, Kinge S, Brück E, Barbiellini B and Bansil A 2016 *Appl. Phys. Lett.* **108** 081602
[14] Terabe H, Iida S, Yamashita T, Tachibana T, Barbiellini B, Wada K, Mochizuki I, Yagishita A, Hyodo T and Nagashima Y 2015 *Surf. Sci.* **641** 68
[15] Crivelli P, Cooke D, Barbiellini B, Brown B L, Feldblyum J I, Guo P, Gidley D W, Gerchow L and Matzger A J 2014 *Phys. Rev.* B **89** 241103(R)
[16] Jones A C L, Goldman H J, Zhai Q, Feng P, Tom H W K and Mills A P Jr 2015 *Phys. Rev. Lett.* **114** 153201
[17] Parz P, Fuchsbichler B, Koller S, Bitschnau B, Mautner F-A, Puff W and Würschum R 2013 *Appl. Phys. Lett.* **102** 151901
[18] Zhang P, Wang Y, Lin M, Zhang D, Ren X and Yuan Q 2012 *J. Electrochem. Soc.* **159** A402
[19] Drummond N D, López Ríos P, Needs R J and Pickard C J 2011 *Phys. Rev. Lett.* **107** 207402
[20] Perdew J P, Burke K and Ernzerhof M 1996 *Phys. Rev. Lett.* **77** 3865
[21] Ma S-K and Brueckner K A 1968 *Phys. Rev.* **165** 18
[22] Brinkman W F, and Rice T M 1973 *Phys. Rev.* B **7** 1508
[23] Leung C H, Stott M J and Almbladh C O 1976 *Phys. Lett.* A **57** 26
[24] Jarlborg T and Singh A K 1987 *Phys. Rev.* B **36** 4660
[25] Drummond N D, López Ríos P, Pickard C J and Needs R J 2010 *Phys. Rev.* B **82** 035107
[26] Barbiellini B and Kuriplach J 2015 *Phys. Rev. Lett.* **114** 147401
[27] Blaha P, Schwarz K, Madsen G K H, Kvasnicka D and Luitz J 2001 *WIEN2k: An Augmented Plane Wave and Local Orbitals Program for Calculating Crystal Properties* (Vienna University of Technology, Vienna, Austria).
[28] Zhou F, Cococcioni M, Marianetti C A, Morgan D and Ceder G 2004 *Phys. Rev.* B **70** 235121
[29] Kuriplach J and Barbiellini B 2014 *Phys. Rev.* B **89** 155111
[30] Takahashi Y, Kijima N, Dokko K, Nishizawa M, Uchida I and Akimoto J 2007 *J. Sol. State Chem.* **180** 313
[31] Viciu L, Bos J W G, Zandbergen H W, Huang Q, Foo M L, Ishiwata S, Ramirez A P, Lee M, Ong N P and Cava R J 2006 *Phys. Rev.* B **73** 174104
[32] Kuriplach J and Barbiellini B 2016 Parameter-free gradient correction for positron states in oxides *Mater. Sci. Forum* (accepted, Proceedings of ICPA-17)
[33] Venkatraman S and Manthiram A 2002 *Chem. Mater.* **14** 3907
[34] Maiyalagan T, Jarvis K A, Therese S, Ferreira P J and Manthiram A 2014 *Nat. Commun.* **5** 3949

IOP Conf. Series: Journal of Physics: Conf. Series **791** (2017) 012017 doi:10.1088/1742-6596/791/1/012017

Point defects in ZnO crystals grown by various techniques

J Čížek[1], M Vlček[1], P Hruška[1], F Lukáč[1], O Melikhova[1], W Anwand[2], F Selim[3], Ch Hugenschmidt[4] and W Egger[5]

[1]Faculty of Mathematics and Physics, Charles University, V Holešovičkách 2, 18000, Praha 8, Czech Republic
[2]Institute of Radiation Physics, Helmholtz-Zentrum Dresden-Rossendorf, Bautzner Landstr. 400, 01328 Dresden, Germany
[3]Center for Photochemical Sciences, Bowling Green State University, Bowling Green, Ohio 43403, USA
[4]Technical University München, Lichtenbergstraße 1, 85747 Garching, Germany
[5]Bundeswehr München, Werner-Heisenberg-Weg 39, 85577 Neubiberg, Germany

E-mail: jakub.cizek@mff.cuni.cz

Abstract. In the present work point defects in ZnO crystals were characterized by positron lifetime spectroscopy combined with back-diffusion measurement of slow positrons. Defects in ZnO crystals grown by various techniques were compared. Hydrothermally grown ZnO crystals contain defects characterized by lifetime of ≈181 ps. These defects were attributed to Zn vacancies associated with hydrogen. ZnO crystals prepared by other techniques (Bridgman, pressurized melt growth, and seeded chemical vapour transport) exhibit shorter lifetime of ≈165 ps. Positron back-diffusion studies revealed that hydrothermally grown ZnO crystals contain higher density of defects than the crystals grown by other techniques. The lowest concentration of defects was detected in the crystal grown by seeded chemical vapor transport.

1. Introduction

Zinc oxide (ZnO) is a wide band gap (3.4 keV) semiconductor with a high exciton binding energy of 60 meV. Nowadays ZnO attracts large interest as a promising material for blue/UV light emitters and optoelectronic devices [1]. Because of the progress in the growth techniques high quality ZnO crystals became available [2]. At present there are several techniques capable of production of high quality ZnO crystals, namely seeded chemical vapor transport growth (CVT) [3], pressurized melt growth (PMG) [4], Bridgman growth (BG) [5] and hydrothermal growth (HTG) [6].

The main obstacle for widespread use of ZnO consists in doping asymmetry [1]: practically all ZnO crystals exhibit n-type conductivity but it is hard to achieve stable and reproducible p-type doping. A solution of this problem requires understanding of point defects which govern optical and electrical properties of ZnO crystals and ability to control their population. To contribute to this research in the present work point defects in ZnO crystals grown by various methods were characterized by positron lifetime spectroscopy combined with positron back-diffusion measurement.

2. Experimental details

A variety of ZnO single crystals were studied, namely (i) HTG ZnO crystals supplied by the following producers: MaTecK (Germany), Crystec (Germany), MTI (USA), Altramet (USA); (ii) PMG crystals provided by Cermet (USA), (iii) BG crystals grown at the Institut für Kristallzüchtung (IKZ) in Berlin (Germany); and (iv) CVT crystals grown at the Oak Ridge National Laboratory (ORNL) in USA.

Bulk positron lifetime investigations were performed using a 1.5 MBq ^{22}Na radioisotope deposited on a 2 μm thick mylar foil employing a digital positron lifetime spectrometer [7] with time resolution of 145 ps (FWHM of resolution function). The source contribution consisted of two components with lifetimes of 368 ps and 1.5 ns and relative intensities of 7 and 1%. Measurement of back-diffusion of slow positrons was carried out on a continuous magnetically guided slow positron beam [8] with energy of incident positrons adjustable in the range from 0.03 up to 35 keV. Doppler broadening of annihilation peak was measured by a HPGe detector with energy resolution of 1.09 keV at 511 keV and was evaluated using the line-shape *S* parameter. Selected samples were investigated on a pulsed slow positron beam PLEPS [9] at the intense positron source NEPOMUC. The energy of incident positrons in PLEPS was varied in the range 0.5-18 keV and the time resolution was 300 ps.

3. Results and discussion

3.1. As-grown crystals

All as-grown ZnO single crystals studied exhibited a single component positron lifetime spectrum (except of the source contribution). Chemical analysis revealed different concentrations of impurities not only in ZnO crystals provided by various producers but also in crystals from the same producer but supplied in different years [10]. It is therefore important to check whether crystals provided by single supplier in various years can be considered to be the same from the point of view of intrinsic defects. Hence, at first we compared batches of crystals fabricated by MaTecK and Cermet companies in various periods. Figure 1 shows that lifetimes measured for HTG ZnO crystals supplied by MaTecK in various years are practically the same. Similarly lifetimes for PMG ZnO crystals provided by Cermet in different years are close to each other. Thus, regarding defects, the crystals prepared in various periods are comparable. However there is a striking difference between HTG and PMG crystals.

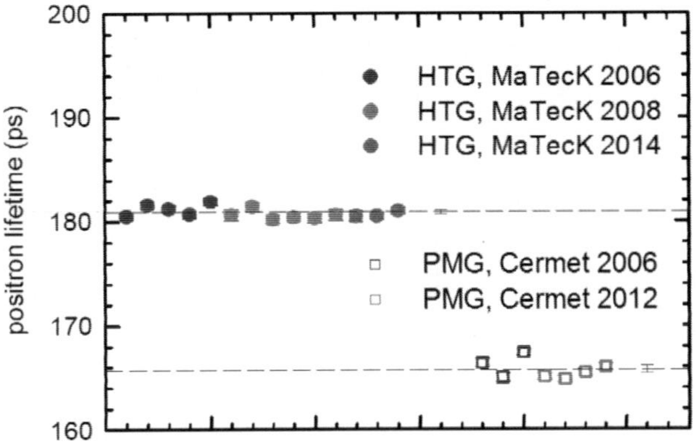

Figure 1. Results of bulk positron lifetime measurements for batches of HTG and PMG ZnO crystals supplied by MaTecK and Cermet companies in various years.

Figure 2 shows results of bulk positron lifetime measurements of ZnO crystals prepared by various methods. All HTG crystals exhibit similar lifetime around 181 ps independently on the supplier. On the other hand, crystals prepared by other techniques (PMG, BG and CVT) are characterized by a remarkably shorter lifetime around 165 ps. Positron lifetimes reported for ZnO crystals in literature [11-15] are plotted in the figure as well. Positron lifetimes measured on various setups may slightly differ. Scatter of literature data is, therefore, higher due to systematic errors but the literature data are basically in accordance with the aforementioned picture.

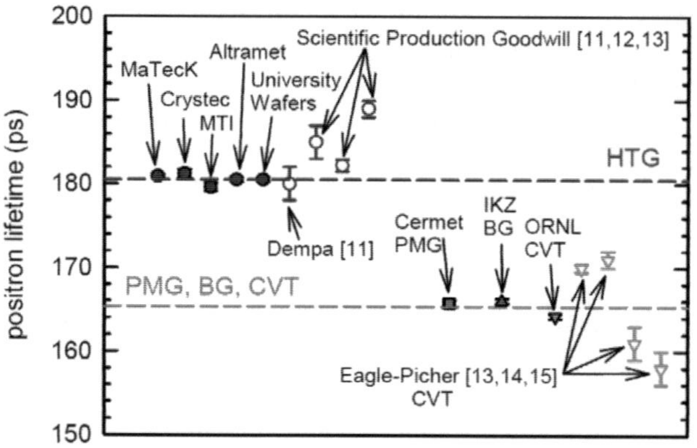

Figure 2. Positron lifetimes for various ZnO crystals. Data measured in the present work and taken from the literature [11-15] are plotted by full and open symbols, respectively

Examples of S parameter curves for as-grown ZnO crystals prepared by various techniques are plotted in Figure 3a. The $S(E)$ curves were fitted by VEPFIT code [16] using a single layer model. The mean positron diffusion length L_+ and the S parameter for ZnO bulk obtained from fitting are plotted in Figure 3b and in the inset of Figure 3a, respectively. The CVT ZnO crystal exhibits the longest L_+ among all samples studied. PMG crystals are characterized by shorter L_+ and the shortest L_+ values were measured for HTG ones. The S parameters exhibit opposite behavior. It indicates that the CVT crystal contains the lowest density of defects among the samples studied; the concentration of defects in PMG crystals is slightly higher and HTG crystals exhibit the highest density of defects. Note that in contrast to other samples CVT ZnO crystals are completely colorless indicating very low level of impurities.

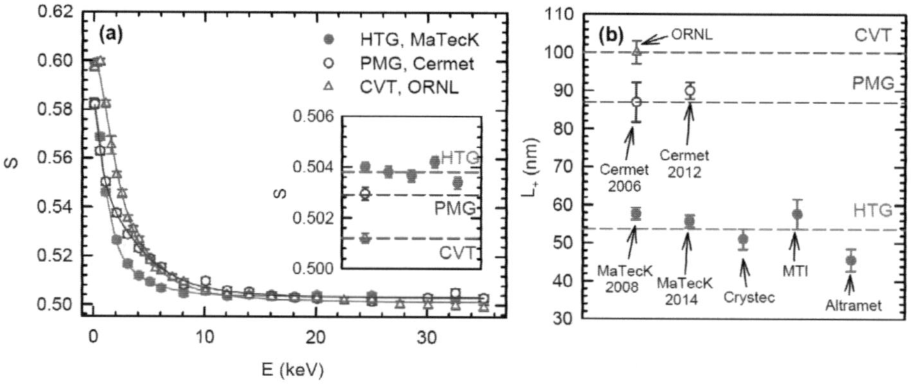

Figure 3. Results of positron back-diffusion measurements of as-grown ZnO crystals: (a) examples of $S(E)$ curves; solid lines are model curves calculated by VEPFIT. The inset shows the S parameters for ZnO bulk obtained from fitting; (b) the mean positron diffusion lengths L_+ obtained from fitting.

Ab-initio theoretical calculations of positron lifetimes were employed in order to identify defects in ZnO crystals [10]. Calculated positron lifetimes depend on the approximation of electron-positron correlation used. Hence, Table 1 shows not only the calculated lifetimes but also their ratios to the bulk lifetime τ_B calculated with the same approach that depend only slightly on the approximation used. Our calculations are in a satisfactory agreement with τ / τ_B ratios calculated in Ref. [14], see Table 1. The oxygen vacancy (V_O) is only a shallow trap incapable of positron confinement. On the other hand, zinc vacancy (V_{Zn}) is a deep positron trap (positron binding energy > 1 eV) characterized by positron lifetime exceeding 200 ps. Zinc and oxygen di-vacancy ($V_{Zn}V_O$) is a deep trap as well.

Table 1. Lifetimes τ for various positron states in ZnO obtained by *ab-initio* theoretical calculations and their ratios to the bulk lifetime τ/τ_B. Detailed description of the calculations is given in Ref. [10].

Positron state	bulk	V_O	V_{Zn}	$V_{Zn}V_O$	V_{Zn+H}
Lifetime τ (ps)	154	154	207	253	179
τ/τ_B	1	1	1.34	1.64	1.16
Ref. [14]	1	1	1.34	1.53	

From comparison of experimental data with theoretical calculations one can conclude that the lifetime of 165 ps measured for PMG, BG and CVT crystals is ≈10 ps longer than the calculated bulk lifetime τ_B = 154 ps. This indicates that PMG, BG and CVT crystals exhibit relatively low density of defects and the majority of positrons is annihilated in the free state. Experimental lifetimes slightly higher than the calculated τ_B might be caused by lattice expansion due to absorbed hydrogen. It supports analysis of CVT crystal which revealed hydrogen concentration as high as 1.03 wt.%. HTG crystals, however, exhibit lifetime which is on one hand remarkably higher than the bulk ZnO lifetime but on the other hand smaller than the lifetime calculated for positrons trapped at V_{Zn}. Hence, HTG crystals contain defects with open volume smaller than V_{Zn} and the concentration of these defects is so high that virtually all positrons are annihilated in the trapped state (saturated positron trapping).

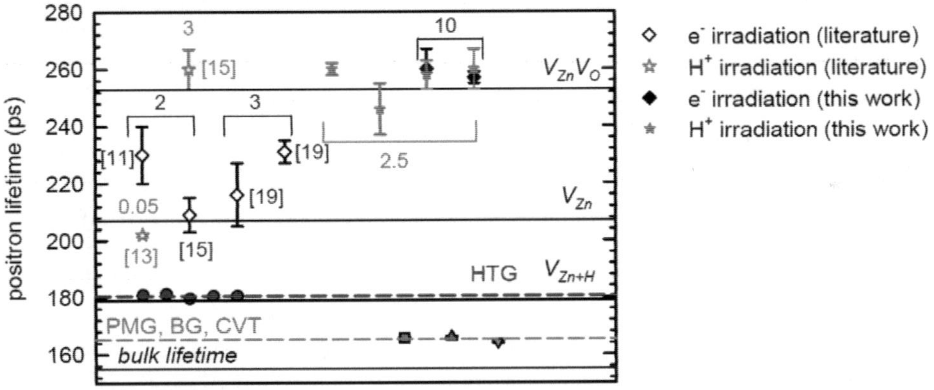

Figure 4. Positron lifetimes for radiation-induced defects in electron and proton irradiated ZnO crystals. Results measured in the present work and data taken from literature are plotted by full and open symbols, respectively. The labels in the figure denote energies of bombarding electrons or protons (in MeV). Solid lines indicate calculated lifetimes for various positron states, see Table 1.

The most probable candidates for defects in HTG ZnO crystals are V_{Zn} associated with some impurities. It has been demonstrated that hydrogen is the most important impurity in ZnO crystals because its concentration is roughly two orders of magnitude higher than the concentration of any other impurity [10]. Thus, it is natural to consider V_{Zn} associated with hydrogen (V_{Zn+H}). Theoretical calculations revealed that there is attractive interaction between H and V_{Zn} and the calculated lifetime of positrons trapped at V_{Zn+H} was found to be 179 ps [10]. This value is in excellent agreement with the lifetime of ≈181 ps measured in HTG crystals.

The concentration of V_{Zn+H} can be estimated from positron back-diffusion measurement using the formula $[V_{Zn+H}] = \nu^{-1}\tau_B^{-1}\left(L_{+,B}^2/L_+^2 - 1\right)$, where τ_B is the bulk ZnO lifetime and $\nu \approx 10^{15}$ s^{-1} is the specific positron trapping rate typical for negatively charged vacancies in semiconductors [17]. Assuming that positron diffusion length determined for CVT crystal is the value corresponding to a perfect ZnO lattice, i.e. $L_{+,B} \approx 100$ nm, one gets $[V_{Zn+H}] \approx 2\times10^{-5}$ at.$^{-1}$ (10^{18} cm^{-3}) with $L_+ = 54$ nm. Using the two-state simple trapping model [17] one can easily calculate that such a concentration of defects corresponds to the free positron component with lifetime of ≈20 ps and intensity of ≈3% only. It is too weak to be resolved in positron lifetime spectrum. Thus, the estimated $[V_{Zn+H}]$ in HTG crystals is consistent with saturated positron trapping. Note that hydrogen content in HTG crystals determined by nuclear reaction analysis [10] is roughly 20 times higher than $[V_{Zn+H}]$. Hence, all V_{Zn} available in HTG crystals were coupled with H and remaining hydrogen occupies interstitial sites in ZnO lattice.

Recently it has been proposed that positrons in HTG crystals are trapped by negatively charged substitutional Li impurities [18]. Although this might be a plausible explanation, the concentration of Li in some of our HTG crystals is lower than $[V_{Zn+H}]$. More importantly there are strong variations of Li content in HTG crystals (e.g. MaTecK sample exhibits [Li] = 3×10^{16} cm^{-3} while crystal from Altramet has [Li] = 6×10^{17} cm^{-3}) but no variations of positron lifetimes were observed. Thus, we propose that in HTG crystals positrons are trapped at V_{Zn+H}.

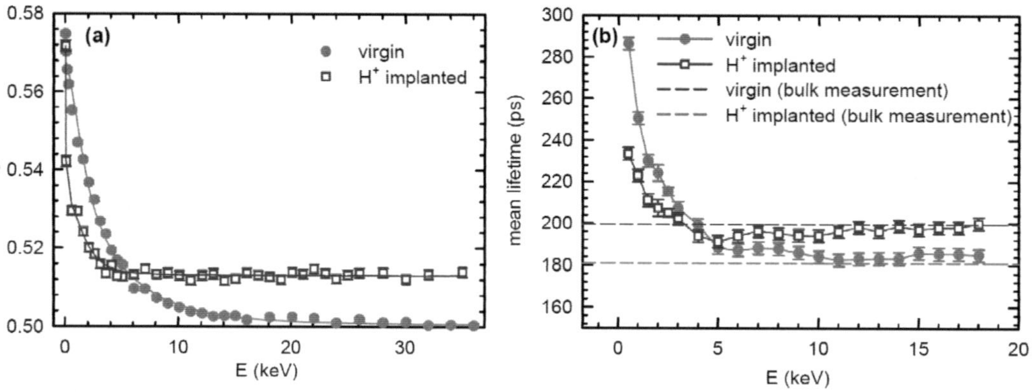

Figure 5. Results of positron annihilation studies of HTG ZnO crystals supplied by MaTecK in the as-grown state and after irradiation by 2.5 MeV protons: (a) $S(E)$ curves, solid lines are model curves calculated by VEPFIT; (b) the dependence of the mean positron lifetime on the energy of incident positrons. Dashed lines indicate the mean lifetimes obtained by bulk positron lifetime spectroscopy.

3.2. Irradiated crystals

Additional knowledge about defects in ZnO crystals can be obtained from irradiation experiments. Irradiated ZnO crystals exhibit typically two-component positron lifetime spectra. The longer component represents a contribution of positrons trapped at defects created by irradiation. Figure 4 shows the lifetime of radiation-induced defects for ZnO crystals irradiated by electrons (diamonds) and protons (stars). Lifetimes for irradiation-induced defects reported in literature are plotted in the figure as well (open symbols). The energies of bombarding electrons or protons (in MeV) are indicated by labels in the figure. The points for the same energies are ordered according to increasing fluence. The investigations of irradiated samples supported the calculated positron lifetimes for V_{Zn} of 207 ps and for $V_{Zn}V_O$ of 253 ps. One can conclude that electron irradiation introduced a mixture of V_{Zn} and $V_{Zn}V_O$. The fraction of $V_{Zn}V_O$ increases with increasing electron energy and increasing fluence. Irradiation by protons with energy exceeding 1 MeV introduced always $V_{Zn}V_O$ as dominating defects. Recently it has been reported [20] that lifetime of positrons trapped at $V_{Zn}V_O$ is practically the same as that for V_{Zn}. This, however, contradicts not only our calculations but also the calculations in Ref. [14].

Figure 5a shows a comparison of $S(E)$ curves for virgin and 2.5 MeV proton irradiated HTG crystals. Proton irradiation led to shortening of L_+ and increase of the S parameter inside ZnO due to $V_{Zn}V_O$ created by irradiation. Figure 5b shows a comparison of the mean positron lifetime measured on a pulsed slow positron beam in a HTG crystal in the virgin state and after irradiation by 2.5 MeV protons. The mean lifetime gradually decreases from the surface value at very low energies down to a value corresponding to the situation when virtually all positrons are annihilated inside the ZnO crystal. This value is in good agreement with the lifetime around 181 ps obtained in the bulk positron lifetime measurements as indicated by the red dashed line in the figure. The sample irradiated by 2.5 MeV protons exhibits an enhanced lifetime inside the crystal due to defects created by irradiation. Again this value is in good agreement with the mean lifetime obtained in the bulk positron lifetime measurement. This testifies that depth concentration profile of radiation-induced defects is rather flat.

4. Conclusions

Defects in ZnO single crystals grown by various methods were characterized by positron annihilation spectroscopy. ZnO crystals prepared by melt growth, Bridgman growth and chemically assisted vapor phase transport are characterized by positron lifetime of 165 ps which is close to the calculated bulk ZnO lifetime of 154 ps. These crystals exhibit a low concentration of defects and the majority of positrons are annihilated in the free state. Hydrothermally grown ZnO crystals exhibit saturated positron trapping at defects characterized by the lifetime of 181 ps. We propose that these defects are V_{Zn} associated with H. Irradiation experiments revealed that V_{Zn} and $V_{Zn}V_O$ are characterized by lifetimes of $\approx$ 210 ps and $\approx$ 255 ps. These values are in good agreement with theoretical calculations.

Acknowledgments

This work received funding from the People Program (Marie Curie Actions) under REA grant agreement number 609642, the Ministry of Education, Youths and Sports of the Czech Republic (project 7AMB15DE003) and the Deutscher Akademischer Austauschdienst (project-ID 57155407).

5. References

[1] Avrutin V, Silversmith D J and Morkoç H 2010 *Proc. IEEE* **98** 1269
[2] Triboulet R 2014 *Prog. Mater. Sci.* **60** 1
[3] Look D C, Reynolds D C, Sizelove J R, Jones R L, Litton C W, Cantwell G and Harsch W C 1998 *Solid State Commun.* **105** 399
[4] Nause J and Nemeth B 2005 *Semiconductor Science and Technology* **20** S45
[5] Schulz D, Ganschow S, Klimm D and Struve K 2008 *J. Cryst. Growth* **310** 1832
[6] Ehrentraut D, Sato H, Kagamitami Y, Sato H, Yoshikawa A and Fukuda T 2006 *Prog. Cryst. Growth Charact. Mater.* **52** 280
[7] Bečvář F, Čížek J, Procházka I and Janotová J 2005 *Nucl. Instrum. Methods A* **539** 372
[8] Anwand W, Brauer G, Butterling M, Kissenger H and Wagner A 2012 *Def. Diff. Forum* **331** 25
Sperr P, Egger W, Kögel G, Dollinger G, Hugenschmidt Ch, Repper R and Piochacz C 2008 *Appl. Surf. Sci.* **255** 35
Brauer, G, Anwand W, Grambole D, Grenzer J, Skorupa W, Čížek J, Kuriplach J, Procházka I, Ling C C, So C K, Schulz D and Klimm D 2009 *Phys. Rev. B* **79** 115212
[9] Tuomisto F and Look D C 2007 *Proc.SPIE* **6474** 647413
[10] Chen Z Q, Kawasuso A, Xu Y, Naramoto H, Yuan X L, Sekiguchi T, Suzuki R and Ohdaira T 2005 *Phys. Rev. B* **71** 115213
[11] Chen Z Q, Yamamoto S, Maekawa M, Kawasuso A, Yuan X L, and Sekiguchi T 2003 *J. Appl. Phys.* **94** 4807
[12] Tuomisto F, Ranki V, Saarinen K and Look D C 2003 *Phys. Rev. Lett.* 91 205502
[13] Brunner S, Puff W, Balogh A and Macher P 2001 *Mater. Sci. Forum* **363-365** 141
[14] van Veen A, Schut H, Clement M, de Nijs J M M, Kruseman A and Ijpma M R 1995 *Appl. Surf. Sci.* **85** 216
[15] Krause-Rehberg R and Leipner H 1999 *Positron Annihilation in Semiconductors* (Berlin: Springer)
[16] Johansen K M, Zubiaga A, Makkonen I, Tuomisto F, Neuvonen P T, Knutsen K E, Monakhov E V, Kuznetsov A. Yu and Svensson B G 2011, *Phys. Rev. B* **83** 245208
[17] Chen Z Q, Wang S J, Maekawa M, Kawasuso A, Naramoto H, Yuan X L and Seiguchi T 2007 *Phys. Rev. B* **75** 245206
[18] Makkonen I, Korhonen E, Prozheeva V, Tuomisto F 2016 *J. Phys.: Condens. Matter* **28** 224002

Positron annihilation study of cavities in black Au films

O Melikhova[1], J Čížek[1], P Hruška[1], M Vlček[1], I Procházka[1], W Anwand[2], M Novotný[3] and J Bulíř[3]

[1]Faculty of Mathematics and Physics, Charles University,
V Holešovičkách 2, 18000, Praha 8, Czech Republic
[2]Institute of Radiation Physics, Helmholtz-Zentrum Dresden-Rossendorf
Bautzner Landstr. 400, 01328 Dresden, Germany
[3]Institute of Physics, Academy of Sciences of the Czech Republic,
Na Slovance 2, 182 21 Prague 8, Czech Republic

E-mail address: oksivmel@yahoo.com

Abstract. Defects in a black Au film were studied using variable energy positron annihilation spectroscopy. Black Au films exhibit porous morphology similar to cauliflower. This type of structure enhances the optical absorption due to a multiple reflections in the micro-cavities. A nanostructured black Au film was compared with conventional smooth Au films with high reflectivity. The black Au film exhibited a remarkably enhanced *S*-parameter in sub-surface region. This is caused by a narrow para-Positronium contribution to the annihilation peak.

1. Introduction

Light absorbers are needed in optical applications for preventing crosstalk between optical connections and for thermal light-emitting sources. A complete absorption of light from the visible into the infrared spectral region over the whole range of incident angles remains still a challenge [1]. This goal could be achieved in nanostructured metal surfaces containing a broad size distribution of cavities that sustain localized optical excitations [2]. Metals evaporated in the presence of a gas with pressure of about 100 Pa deposit as porous layers called black metals [3]. Their porous structure with percolated micro-cavities leads to high absorbance in the visible to infrared spectral region. Black metals, in particular black Au, are therefore highly suitable as absorbers [4]. The surface morphology and the structure of cavities in sub-surface region are crucial for performance of black metals absorbers. In the present work variable energy positron annihilation spectroscopy (VEPAS) was employed for characterization of cavities in black Au film. Microstructures of black Au and a conventional smooth Au film were compared. Note that a conventional smooth Au films have been already studied by VEPAS in Ref. [5].

2. Experimental details

Black Au films with thickness of $\approx$ 900 nm were prepared by thermal deposition on glass substrates in nitrogen atmosphere at the pressure of 200 Pa. Conventional smooth Au films were deposited in vacuum also on the glass substrate. Figure 1 shows Au films after deposition. The smooth Au film deposited in vacuum and shown in Figure 1a exhibits a high reflectivity. On the other hand the nanostructured Au film deposited at 200 Pa appears black, see Figure 1b, because incident light undergoes multiple reflections in cavities and is almost completely absorbed. Note that light absorption occurs on the nanostructured surface only. This is demonstrated in Figure 1c which shows the back side of the black Au film, i.e. a view through the glass substrate. Obviously the back side of the black Au film is smooth and exhibit typical metallic appearance with high reflectivity.

Bulk Au (99.9%) annealed at 1000°C for 1h in vacuum (10^{-3} mbar) was used as a reference. The reference sample was characterized by conventional positron lifetime (LT) spectroscopy using a 1 MBq ^{22}Na source deposited on 2 μm Mylar foil. The LT measurements were performed on a digital spectrometer [6] with time resolution of 145 ps. The source contribution consisting of two components with lifetimes of ≈368 ps, ≈1.5 ns and intensities of ≈10%, ≈1% was always subtracted from spectra.

VEPAS studies were performed on a slow positron beam SPONSOR [7] with positron energy adjustable from 0.03 to 36 keV. Doppler broadening of the annihilation line was measured by a HPGe detector with energy resolution of (1.09 ± 0.01) keV at 511 keV and evaluated using the S and W parameters that were normalized to bulk values S_0, W_0 measured in the annealed bulk Au reference for the energy of incident positrons E = 35 keV. The dependence of the S parameter on the positron energy was analyzed using the VEPFIT code [8] assuming Makhovian positron implantation profile with parameters $m = 1.9$, $n = 1.6$ and A = 20.73 Å keV$^{-1.6}$. The F parameter which is proportional to the fraction of positrons forming positronium (Ps), was calculated as $F = (R - R_{min})/(R_{max} - R_{min})$, where R is the ratio of annihilation events recorded below the annihilation peak (energy range 430-500 keV where 3-γ annihilation of ortho-Ps contributes) to those in the 511 keV peak. R_{min} and R_{max} denote the minimum and the maximum R measured in the experiment. R_{min} corresponds to situation when no Ps is formed (E > 15 keV) while R_{max} corresponds to maximum Ps yield on the surface (E = 30 eV).

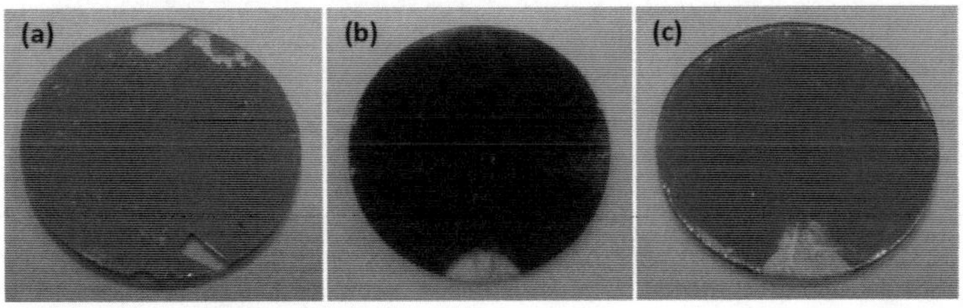

Figure 1. Appearance of Au films deposited on glass substrate: (a) smooth Au film; (b) black Au film - front side; (c) black Au film - back side (look through the glass substrate).

3. Results and discussion

3.1 bulk Au samples

LT investigations of annealed bulk Au revealed single component spectrum (except of the source contribution) with lifetime of (118.8 ± 0.2) ps, which agrees with calculated Au bulk lifetime [9] Hence, the sample exhibits very low defect density and almost all positrons annihilate in the free state.

Figure 2a shows the $S(E)$ curve for the bulk Au annealed at 1000°C. At very low energies positrons are annihilated at the surface and S attains the surface value. At $E \approx 1$ keV the $S(E)$ curve exhibits a local maximum followed by a continuous decrease to the bulk value. The local maximum is likely caused by positrons annihilated in a thin surface oxide layer formed during annealing at 1000°C. Clearly the vacuum in the furnace was not good enough to prevent oxidation completely. This hypothesis was verified by etching the annealed sample in aqua regia (a mixture of HNO_3 and HCl in the molar ratio 1:3) which removed the oxide layer. Figure 2a shows that the $S(E)$ curve for the annealed bulk Au sample after etching does not contain the local maximum at 1 keV anymore and S exhibits a monotonic decrease from the surface value down to the bulk value. The S-W plot for annealed and etched bulk Au presented in Figure 2b is a straight line connecting the point corresponding to positron annihilation on the surface with the point corresponding to annihilation in Au bulk. This testifies that there are only two positron states in the annealed and subsequently etched sample: (i) positrons trapped in the surface state and (ii) positrons annihilating in Au bulk. With increasing energy positrons penetrate deeper and deeper into the sample and the fraction of positrons diffusing back to the surface gradually decreases while the fraction of positrons annihilated in bulk grows. The S-W plot for the annealed sample measured prior to etching exhibits first a shift to higher S and lower W due to positrons annihilated in oxide layer. It is followed by a gradual transition to the point for Au bulk. Solid lines in Figure 2a show model curves calculated by VEPFIT assuming a two layer model for annealed bulk Au (thin oxide layer and Au bulk) and a single layer model for the annealed and etched sample. One can see in Figure 2a that the model curves describe the experimental points accurately. Fitting yielded a positron diffusion length in Au $L_{+,B}$ = (134 ± 1) nm for both samples. This value falls into the range 100 – 200 nm expected for positron diffusion length in defect-free metals. The thickness of the oxide layer of (40 ± 1) nm was determined by fitting for the annealed bulk Au sample.

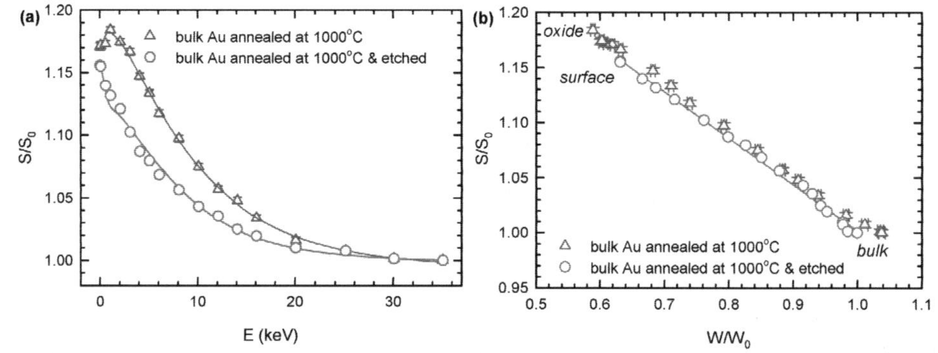

Figure 2. Results of VEPAS studies of bulk Au samples: (a) *S(E)* curves; (b) *S-W* plot.

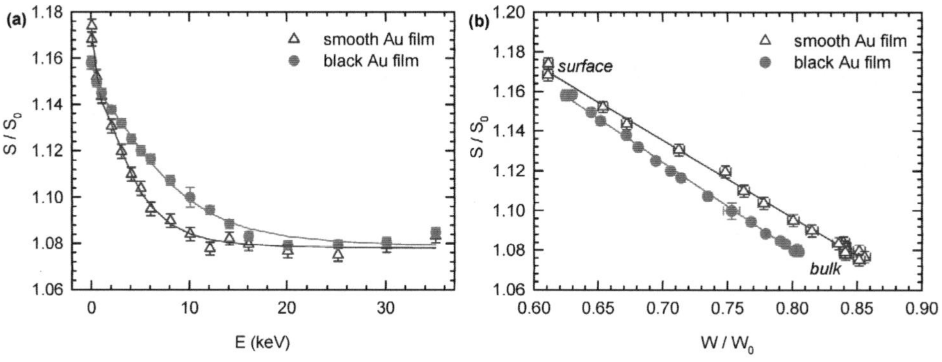

Figure 3. Results of VEPAS studies of Au films: (a) *S(E)* curves; (b) *S-W* plot.

3.2 Au films

Figure 3a shows *S(E)* curves for the smooth and the black Au film. For both films *S* decreases with increasing positron energy from a surface value down to a value when virtually all positrons are annihilated inside the film. This value $S/S_0 \approx 1.08$ is significantly higher than that for annealed bulk Au, cf. Figure 2a. The mean positron diffusion length L_+ = (24 ± 1) nm obtained for the smooth Au film by VEPFIT is remarkably shorter than $L_{+,B}$ determined for the bulk annealed Au. Higher *S* parameter and shorter positron diffusion length testify that Au film contains considerable concentration of defects. The *S-W* plot for the smooth Au film shown in Figure 3b is a straight line connecting the surface point with the point corresponding to annihilations inside the Au film. The *S(E)* curve for the black Au film is at high energies similar to that for the smooth Au film testifying to similar concentration of defects in both films. However, at lower energies (E < 15 keV) the black Au sample exhibits higher *S* parameters than the smooth Au film indicating that its structure differs in the sub-surface region. Proper description of the *S(E)* curve for the black Au film by VEPFIT requires a two-layer model consisting of (i) sub-surface region where the microstructure differs from that for smooth Au film; (ii) deeper region with microstructure similar to that for smooth Au film. The *S(E)* curve for black Au film can be well fitted assuming that positron diffusion length in the region (ii) is the same as that in the smooth Au film. One can see in Figure 1c that the black Au film appears metallic when looking at its back side through the transparent substrate. Hence, film layer attached directly to the substrate is smooth while the region close to the surface is porous. The sub-surface region with an enhanced S in the black Au films obviously corresponds to a nanostructured region containing microscopic cavities (diameter ≈200 nm as determined by AFM [2]) and likely also smaller nano-cavities. The thickness of the sub-surface region obtained from fitting is (79±3) nm and it is characterized by the positron diffusion length L_+ = (62±4) nm. Hence the sub-surface region exhibits a higher S parameter compared to the smooth Au film and the positron diffusion length is higher as well. This can be explained by formation of Ps in micro- and nano-cavities inside the sub-surface layer (i). In metals Ps does not form since any bound state of positron and electron is quickly destroyed by the screening of conduction electrons. However, a thermalized positron escaping from Au through inner surface into a cavity may form Ps by picking an electron on the surface. Narrow para-Ps contribution to the annihilation photo-peak [10] likely caused the observed enhancement of S in the sub-surface region. Enhanced L_+ can be explained by Ps diffusion through percolated network of cavities in the sub-surface region. However, cavities in the sub-surface region may not necessarily be interconnected. Isolated cavities with the mean size and/or concentration gradually decreasing with depth from the surface can also explain enhanced L_+. Formation of Ps in the sub-surface region (i) was confirmed by behavior of the F parameter plotted in Figure 4. At very low energies (E< 0.5 keV) Ps is always formed on metal surfaces. Ps yield on the surface of black Au is lower compared to the smooth Au due to its higher roughness and consequently higher probability of pick-off annihilation. At higher energies F quickly drops to zero in smooth Au while in black Au F is enhanced also in the sub-surface region due to Ps formed in cavities.

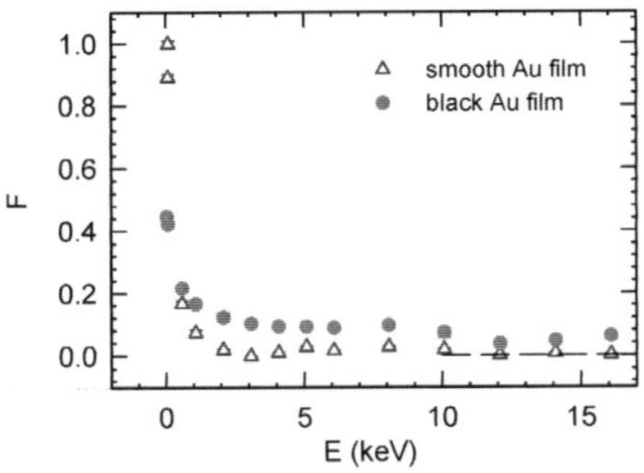

Figure 4. Dependence of the F parameter on the positron energy for Au films

4. Conclusions

The microstructure of nanostructured black Au and smooth Au film with high reflectivity was compared. At the depth higher than 80 nm both films contain comparable concentration of defects which is considerable higher than that in well annealed bulk Au reference. Black Au film exhibits enhanced S and F parameters and higher positron diffusion length in sub-surface region. This is likely due to Ps formation in nanostructured sub-surface layer containing micro- and nano-cavities.

Acknowledgements
This work has been supported by the Czech Science Foundation (project P108/12/G043).

References

[1] Teperik T V, García de Abajo F J, Borisov A G, Abdelsalam M, Bartlett PN, Sugawara Y and Baumberg JJ 2008 *Nature Photonics* **2** 299
[2] Novotný M, Fitl P, Krasilnikova Sytchkova A, Bulíř J, Lančok J and Pokorný P 2009 *Cent. Eur. J. Phys.* 7 327
[3] Pfund A H 1930 *Rev. Sci. Instrum.* **1** 397
[4] Advena D J, Bly V T and Cox J T 1993 *Appl. Optics* **32** 1136
[5] Reiner M, Pikart P and Hugenschmidt C 2014 *J. Alloys Comp.* **587** 515
[6] Bečvář F, Čížek J, Procházka I and Janotová J 2005 *Nucl. Instrum. Methods A* **539** 372
[7] Anwand W et al. 2012 *Def. Diff. Forum* **331** 25
[8] van Veen A, Schut H, Clement M, de Nijs J M M, Kruseman A and Ijpma M R 1995 *Appl. Surf. Sci.* **85** 216
[9] Campillo Robles M, Ogando E and Plazaola F 2007 *J. Phys.: Condens. Matter* **19** 176222
[10] Van Veen A, Hakvoort R, Schut H and Mijnarends P 1995 *Journal de Physique IV* **5** Cl-37

Investigation of point defects in HfO_2 using positron annihilation spectroscopy: internal electric fields impact

M Alemany[1,2,4], A Chabli[2], E Oudot[1,3,5], F Pierre[3], P Desgardin[4], F Bertin[3], M Gros-Jean[1] and M F Barthe[4]

[1] STMicroelectronics, 850 rue Jean Monnet, 38926 Crolles, France.
[2] Univ. Grenoble Alpes, INES, F-73375 Le Bourget du Lac, France, CEA, LITEN, Department of Solar Technologies, F-73375 Le Bourget du Lac, France
[3] Univ. Grenoble Alpes, F-38000 Grenoble, France, CEA, LETI, MINATEC Campus, F-38054 Grenoble, France.
[4] CNRS, CEMHTI UPR3079, Univ. Orléans, F-45071 Orléans, France.
[5] Univ. Grenoble Alpes, Lab. LTM (CEA-LETI/Minatec), 38000 Grenoble, France

Email : mathias.alemany@st.com

Abstract.
In this work, we report on the PAS characterization of sintered HfO_2 bulk ceramic and HfO_2 layers deposited with various methods on a silicon substrate with a layer thickness ranging from 25 to 100 nm. PAS measurements are sensitive to the deposition process type and the post-deposition annealing. Chemical and structural characterisations have been performed on the same samples. The PAS results are discussed regarding to the material defects of the different layers. In addition, a built-in electrical field induced by charged defects located at the HfO_2/Si interface as well as in the HfO_2 layer must be taken into account in the PAS data fitting. Both non-contact internal electrical field measurements and internal electrical field simulations support the PAS finding.

1. Introduction

Due to aggressive scaling for MOS devices, the transistors gate length drop requires new materials introduction. For example, high-k dielectrics (HfO_2) and metals (TiN) are used in transistor generations embedding High-k Metal Gate (HKMG) stacks, raising new issues such as shifts in transistor threshold voltages [1]. Oxygen vacancies in both HfO_2 and SiO_2 are usually invoked to explain this shift [2]. The activation spike annealing causes charged oxygen vacancies creation in high-k material, resulting in the formation of a dipole at the high-k/metal interface [3]. This dipole causes the Fermi level pinning phenomenon, responsible for the threshold voltage shift in HKMG devices [3, 4]. In addition, this shift is enhanced by oxygen vacancies creation in SiO_2 interfacial layer [5]. To asses these mechanisms, techniques capable of characterizing oxygen vacancy density in the HKMG stacks for 14 and 10 nm nodes are required. Today, two techniques are envisioned to reach these requirements: Positron Annihilation Spectroscopy (PAS) known as the most sensitive characterization method of vacancy nature and concentration in solids [6] and Electron Energy Loss

Spectroscopy (EELS) that is potentially sensitive to oxygen vacancies with a spatial resolution compatible with the dimensions of the HKMG structures [7].

Today, PAS is intensively used to characterize vacancies in diverse materials but only few studies are dedicated to high-k dielectrics [8]. Nevertheless, Uedono et al. worked on thin 3 nm HfO_2 and HfSiON layers and highlighted the existence of electric fields in the silicon substrate, near interface with HfO_2. They suggested the presence of negative charges in HfO_2 before deposition of TiN, and positive charges in HfO_2 after deposition of metal alloy and they proposed that TiN could act as a catalyst for oxygen vacancies formation. However, Uedono et al. don't observe oxygen vacancies, and suggest existence of void defects wider than vacancies.

In this paper, a slow positron beam coupled with a Doppler broadening spectrometer (DBS) has been used to study vacancy defects in HfO_2 layers on silicon. First results on PAS capabilities in this configuration are presented and discussed in terms of identification of charged defects.

2. Experimental details

2.1. Samples elaboration and characterisation

In order to assess the annihilation characteristics of the HfO_2 crystalline structure and to identify those due to oxygen vacancy, different deposition techniques have been used. HfO_2 layers with thickness ranging from 10 to 100 nm, as measured by spectroscopic ellipsometry and by SEM imaging, have been deposited on p-type 100-oriented Si substrates (B doped at ≈ 2.10^{15} cm^{-3}). An interfacial layer (IL) of native silicon oxide is located between HfO_2 layers and substrate. This IL presents a typical thickness ranging from 0.5 to 2 nm. For thick layers, Physical Vapor Deposition (PVD) technics have been used. Sputtering of an Hf target within an Ar/O_2 plasma has been used to deposit layers on 300°C heated substrates with thicknesses ranging from 25 to 100 nm. In addition, Atomic Layer Deposition (ALD) films of 10, 25 and 50 nm thickness, closer to the thickness used in microelectronic devices, were grown at 300°C using $HfCl_4$ and H_2O precursors. After deposition, spike annealing under low N_2 pressure at 900°C has been used to induce either layer densification or oxygen vacancies creation.

2.2. PAS: Doppler broadening spectroscopy (DBS)

In this work, the DBS system is a standard gamma-spectroscopy system equipped with a high purity germanium detector that offers a high energy resolution (<1.3 keV at 511 keV) and a high efficiency (>25% at 1.33 MeV). The positron emitted from a ^{22}Na radioactive source are first converted in a mono-energetic beam using a 5 µm thick tungsten foil and then accelerated at a kinetic energy ranging from 0.2 to 25 keV. The flux is ~10^5 cm^{-2} s^{-1}. The momentum distribution of the annihilated (e^+, e^-) pairs is measured at 300 K by recording the Doppler broadening of the 511 keV annihilation line. It is characterized by the low S and the high W momentum annihilation fractions in, respectively, the momentum windows (0–|2.80|$\times 10^{-3} m_0c$) and (|10.61|–|26.35| $\times 10^{-3} m_0c$), where m_0 is the rest mass of the electron and c the speed of light. The S fraction corresponds essentially to annihilations with low momentum electrons, thus more predominantly with valence electrons. The W fraction corresponds to annihilations with high momentum electrons, thus it is essentially related to positron annihilations with core electrons. The S and W fractions are extracted from the DBS measurements acquired at energies ranging from 0.2 to 25 keV with step increasing from 0.2 to 1 keV. The positron implantation depth profile can be calculated (p. 33 in [9]), and broadens with positron energy. For the 0.2-25 keV energy range, the first 8 µm in bulk silicon can be probed. The probing depth resolution depends on the positron energy and on its mobility in the sample after kinetic energy loss, mobility influenced by trap concentration and/or by a possible electric field. Thus each point of the $S(E)$ curves results from the positron interaction on a depth section of the sample that can be larger than the HfO_2 layer thickness. Therefore fitting with VEPFIT program [10] is used to extract from experimental curves the S and W values as a function of depth. The VEPFIT program considers that the sample is made of one or several homogenous layers of material with specific annihilation characteristics. It takes into account the implantation and diffusion properties of positron in the different layers. VEPFIT supplies information about *S*, *W*, thickness, the positron effective diffusion length L^+ and the electric field of each layer.

3. Results and discussion

Figure 1.a, shows measured *S*(*E*) curves for different PVD layers of HfO_2 and for a hot sintered bulk ceramic hafnia. For each films *S* first decreases quickly up to approximately 1.5 or 3 keV. Then it increases monotonically towards values very close to the ones reported for annihilation characteristics in the Si lattice [11] when the positron energy becomes higher than 1.5 keV or 4 keV according to the thickness of the HfO_2 layer. This variation is due to the increasing of annihilation probability in the Si substrate as the positron beam energy increases. As the thickness of the layers decreases the substrate characteristic values are reached for lower positron energy. For the 100 nm thick film, *S* parameter remains nearly constant in the 1-4 keV energy range. Compared to bulk hafnia curve, this *S* values are close to the annihilation characteristics of a bulk HfO_2. For the 25 nm thick film, the steady area on *S* curve disappears in favor of a "valley" feature for 1-2 keV energy range. This result is consistent with the corresponding implantation profiles that show a 1-3 keV range governed by the DBS characteristics of HfO_2 but this imply for thickness lower of 25 nm, the HfO_2 contribution is embedded in surface and Si contribution.

The *S-W* plots (figure 1.b) show for all PVD HfO_2/Si structures, a monotone and linear decrease between (*S*,*W*) values of Si and values of HfO_2 called $D_{Si/HfO2}$ and deviation from this line for SW points measured in the steady state area for the 50 and 100 nm layers, suggesting that vacancy type defects are detected in the HfO_2 layers.

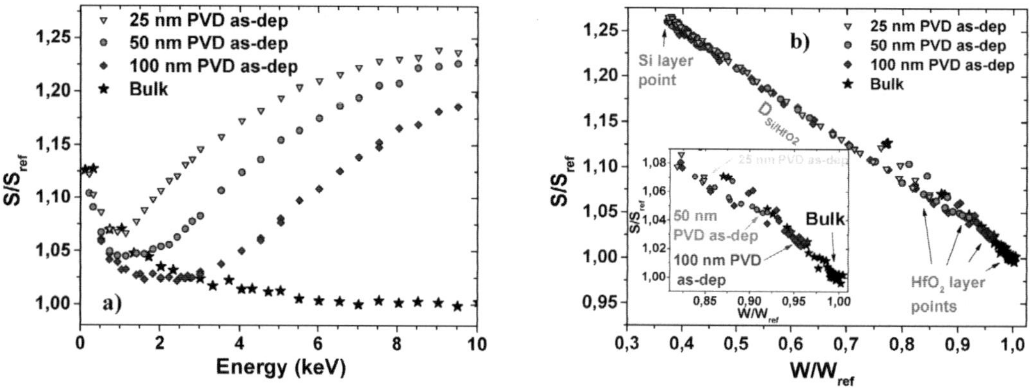

Figure 1. *S* parameter as function of incident positron energy (a) and *S-W* plot, with a zoomed region (b) in as-deposited PVD HfO_2/Si structures with HfO_2 thickness ranging from 25 to 100 nm compared to bulk HfO_2.We use S and W of bulk hafnia as S_{ref} and W_{ref}, with $S_{ref} = 0.3606$ and $W_{ref} = 0.0900$.

In Figure 2.a, *S* curves are plotted for the as-deposited ALD 50 nm thick layer compared to PVD one. The *S* steady state energy range is the same for both type of layers and the *S* parameter value is higher for ALD layer than for the PVD one. In the *S-W* plot (fig. 2.b) it has to be noticed that the ALD HfO_2/Si structure points don't follow the $D_{Si/HfO2(i)}$ line suggesting a higher initial defect concentration in ALD than in PVD layers. An explanation could be that the ALD process is usually optimized for 1-10 nm range thicknesses.

After annealing, 50 nm ALD layer shows an increase of the *S* value in the lower energy range as indicated in Figure 2.a. Moreover, the overall shape of the curves is significantly modified and very steady *S* and *W* values appear in the 1-5 keV energy range, with a slower evolution towards the Si substrate *S* characteristics. It's important to clarify that there is no modification in film thickness. According to the implantation depth profile simulation, we can assess the existence of an electric field at the ALD HfO_2/Si interface or in the HfO_2 layer that induces a shift of the DBS curves towards the high energy range. This suggests the existence of charged defects in HfO_2, or at the HfO_2/Si interface created during annealing [3].

We compare a 10 nm as-deposited ALD thin film with 50 nm ALD ones (see fig. 2.a and 2.b). According to positron implantation profile calculation, a 10 nm thick film should be probed for energy below 1 keV. In this energy range the fraction of annihilation at the surface is high. As observed for 50 nm ALD layer (ii), the SW points measured in the as-deposited 10 nm layer are aligned on a new straight line $D_{Si/HfO2(iv\text{-}a)}$ with a lower slope than the $D_{Si/HfO2(i)}$ line. This shift suggests a non-negligible influence of silica interfacial layer between HfO_2 film and silicon substrate due to the presence of an electric field in HfO_2 or in depletion area of the substrate, as mentioned by Uedono et al [8]. This hypothesis seems supported by the behavior shown in Figure 2 for annealed 10 nm thick ALD layer where the *S-W* slope ($D_{Si/HfO2(v)}$) is similar to the $D_{Si/HfO2(iii)}$ line. This behavior suggests an electrical field modification in the stack due to charge creation induced by annealing.

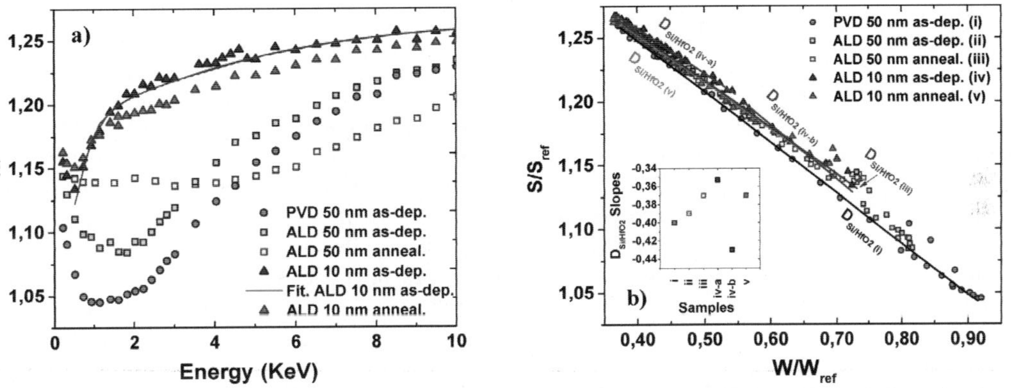

Figure 2. *S* parameter as function of incident positron energy (a) and *S-W* plot (b) in as-deposited and annealed ALD HfO_2/Si structures with 10 and 50 nm HfO_2layer thickness compared to as-deposited 50 nm PVD HfO_2/Si, all samples are numbered from (i) to (v) on legend. All $D_{Si/HfO2}$ slopes are represented as function of samples numbers on the left of figure 2 (b).

To evidence this charge creation or modification during annealing, electrical measurements have been performed using Corona Oxide Characterization of Semi-Conductors (COCOS) [12], a non-contact method, on 10 nm samples. Negatives charges are detected in the as-deposited sample, with a density of 2×10^{11} cm^{-2}. After annealing, COCOS measurements reveal positive charges with a larger density (×3) as observed in some studies [1-3]. As COCOS measurement has no depth resolution, we assume that charges are created near the interfaces in accordance with previous works [1, 3, 4]. The electric field induced by the measured charge density in the 10 nm HfO_2/Si structure has been calculated using UTOX, a Poisson solver developed at STMicroelectronics [13]. The simulation, with parameters indicated in Figure 3, results in a positive (directed towards the HfO_2/Si interface) and constant electric field, E_{HfO2}, of ≈ 3 kV/cm in HfO_2, a negative electric field E_{IL} in SiO_2 interfacial layer of ≈ 0.7 kV/cm, and a weaker negative electric field E_{Si} in the Si substrate. The E_{Si} module vanishes over a length of ≈ 200 nm. In the following we have approximated E_{Si} by a constant value of -100 V/cm over 200 nm in the Si substrate. For the annealed sample case, positive charges are imposed in hafnia (not shown), the simulation results in an inverted electric fields configuration with in a negative E_{HfO2} and a positive E_{Si}.

Injecting this value of the electric field in VEPFIT, the best fit of *S* curves found for the 10 nm as-deposited ALD sample is shown in Figure 2.a. It is obtained for an E_{HfO2} value of about 400 V/cm and the 50 V/cm for E_{IL} in SiO_2 interfacial layer, which is an order of magnitude smaller than the 3 kV/cm for E_{HfO2}, and the 0.7 kV/cm for E_{IL} calculated using UTOX. Nevertheless, the developed model is in agreement with COCOS measurements for as-deposited sample, and with Uedono et al. results for the electric field in Si [8].

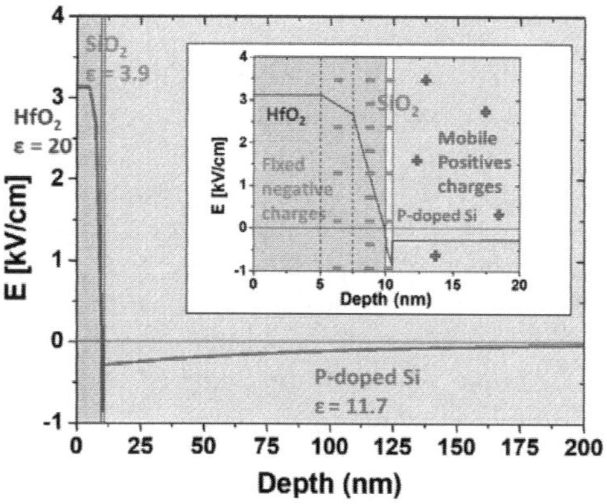

Figure 3. UTOX electric field simulation on 10 nm as-deposited layer assuming that the negative charges detected using the COCOS technique are located in HfO_2, close to the interface. Mobile positive charges in Si are attracted by fixed negative charges in HfO_2.

For the annealed sample, data fitting by VEPFIT does not allow to obtain realistic S and W values regardless of free or imposed electric fields parameters reflecting COCOS measures. Theses unrealistic S and W found during data fitting probably can highlight VEPFIT limitation concerning negative E_{HfO2} or positive E_{Si} presence. On the other hand, alternative electric fields configurations reflecting positive charges in hafnia than simulated previously, can exist. In the future more realistic charge distributions in HfO_2 will be tested in order to obtain a better agreement for the as-deposited 10 nm specimen. It is necessary to investigate on hypothetical VEPFIT limitation and to rebuild alternative charge distribution and electric field configuration model for annealed 10 nm sample.

4. Conclusion

Doppler broadening measurements performed with a slow positron beam have shown an effective sensitivity to the properties of HfO_2 layers down to a thickness of 10 nm. The qualitative analysis of the DBS curves show that the void defect concentration depends on the deposition process of the layer and it changes after annealing. The as-deposited ALD material is revealed to include more defects than the PVD one due to the out-of-range thickness for the ALD process. In addition, for ALD layers the defects concentration is higher after annealing than in the as-deposited state. The role of a built-in electrical field related to charged defects at the HfO_2/Si interface or in the HfO_2 layer is highlighted through comparison with electrical measurement by COCOS and electric field simulation. However, a quantitative analysis based on data reduction using complete DBS simulations is still limited by the lack of knowledge of the annihilation characteristics in the HfO_2 material. Significant information about defect parameters could be brought by DFT simulation. Also, to fulfill the nano-electronic specifications, coupling PAS with EELS-TEM and cathodoluminescence analyses is mandatory.

Acknowledgment

This work was supported by the National Research Agency (ANR) through the French "Recherche Technologique de Base" Program. The experiments were performed in the frame of the joint development program with STMicroelectronics and the Nanocharacterisation platform (PFNC) at MINATEC. We warmly thank A. Roule, H. Grampeix from LETI for providing ALD and PVD deposition.

References

[1] Shiraishi K, Yamada K,Torii K, Akasaka Y, Nakajima K, Konno M, Chikyow T, Kitajima H, Arikado T and Nara.Y 2006 *Thin Solid Films* **508** 305
[2] Guha S and Narayanan V 2007 *Physical Review Letters* **98** 196101
[3] Kechichian A, Barboux P, and Gros-Jean 2013 M *ECS Transactions* **58** 325
[4] Robertson J, Sharia O, and Demkov A A 2007 *Applied Physics Letters* **91** 132912

[5] Bersuker G, Park C S, Wen H C, Choi K, Price J, Lysaght P, Tseng H H, Sharia O, Demkov A, Ryan J T and Lenahan P 2010 *IEEE Transaction on Electron Devices* **57** 2047

[6] Uedono A, Naito T, Otsuka T, Ito K, Shiraishi K, Yamabe K, Miyazaki S, Watanabe H, Umezawa N, Chikyow T, Ohdaira T, Suzuki R, Akasaka Y, Kamiyama S, Nara Y and Yamada K 2007 *Japanese Journal of Applied Physics* **46** 3214

[7] Calka P, Martinez E, Delaye V, Lafond D, Audoit G, Mariolle D, Chevalier N, Granpeix H, Cagli C, Jousseaume V and Guedj C 2013 *Nanotechnology* **24** 085706

[8] Uedono, Naito T, Otsuka T, Ito K, Shiraishi K, Yamabe K, Miyazaki S, Watanabe H, Umezawa N, Chikyow T, Akasaka Y, Kamiyama S, Nara Y, Yamada K 2006 *Journal of Applied Physics* **100** 034509

[9] Krause-Rehberg R and Leipner H S 1999 *Positron Annihilation in Semiconductors: Defect Studies*, (Berlin: Springer-Verlag) p 33

[10] Van Veen A, Schut H, De Vries J, Hakvoort R A and Ijpma M R 1991 *SLOPOS 4 (Ontario) AIP Conf. Proc.* **218** 171

[11] Kauppinen H, Corbel C, Liszkay L, Laine T, Oila J, Saarinen K, Hautojärvi P, Barthe M F and Blondiaux G 1997 *Journal of Physics: Condensed Matter* **9** 10595

[12] Wilson M, Lagowski J, Jastrzebski L, Savtchouk A and Faifer V 2001 *Conf.Characterization and Metrology for ULSI Technology 2000 (Gaithersburg) AIP Conf. Proc.* **550** 220

[13] Garetto D, Rideau D, Dornel E, Tavernier C, Leblebici Y, Schmid A, Jaouen H 2011. *Proc. NSTI Nanotech 2011 (Boston)*, vol 2 607

IOP Conf. Series: Journal of Physics: Conf. Series **791** (2017) 012020 doi:10.1088/1742-6596/791/1/012020

Slow positron beam studies of zirconium exposed to dry sliding

Jerzy Dryzek[1] and Paweł Horodek[1,2]

[1]Institute of Nuclear Physics Polish Academy of Sciences, ul. Radzikowskiego 152, PL-31342 Kraków, Poland

[2]Joint Institute for Nuclear Research, Joliot-Curie 6, 141980 Dubna, Moscow region, Russia

E-mail: jerzy.dryzek@ifj.edu.pl

Abstract. Slow positron beam studies of Zr samples exposed to dry sliding are reported. The observed increase of the annihilation line shape parameter indicates generation of vacancy clusters although an increase of the positron diffusion length suggests recovery process indicated by sliding. The increase of the number of positronium created on the worn surface in comparison to the non damaged surface was noticed.

1. Introduction

Technological treatments of surface by ion implantation [1], laser treatment [2], blasting [3], polishing [2], cutting [4], sliding, and etc. induce elastic and plastic deformation which stands behind the crystalline defects generated in the subsurface region. These defects are distributed like strain in a certain volume of the samples adjoining the surface and finally affect physical properties like hardness and performance of products. Thus the knowledge about defect distribution can be significant from practical point of view. Sliding contact accompanies many technological processes. The defect depth distribution induced in such a case is not well recognized especially at the depth close to the worn surface. One can expect that for instance a flow of heat accompanying friction can significantly affect defectstructure causing recovery the zone adjoining the surface. Another process is the mass transport at the worn surface between contra parts in sliding contact.

The experimental technique which is suitable for studies of defects close to the worn surface is the slow positron beam technique. The application of slow positron beam to study of subsurface defects and other phenomena is a natural extension of conventional bulk spectroscopy employing beta decay positrons. In this case the low energetic positrons are injected at the depth of about 1 μm and after thermalization they annihilate with electrons. Open value defects like, vacancies, their clusters and dislocation with jogs can localize positrons. The annihilation with low energy electrons which occupy such defects is reflected in the annihilation radiation parameters like Doppler broadening of annihilation line shape parameter or mean positron lifetime. This is a convenient method for observation of defects and changes of their concentration.

The constant defect concentration at the depth less than 350 nm in gold exposed to dry sliding was observed [5]. Additionally the concentration of vacancies increases with the increase of the applied load. This indicates that the zone at such a depth is uniform. However, for copper a maximum at a depth between 300 and 500 nm in defect concentration was resolved using the VEPFIT code [6]. In other cases the defect concentration at this depth remains constant. This indicates that sliding causes close to the surface only the deep plastic deformation uniformly distributed. To confirm this we intend to perform similar studies for pure Zr. This metal has poor tribological properties thus one can expect new features which are not present for other studied metals commonly applied in tribojunctions.

Zr has only niche applications. Its excellent resistance for corrosion allows its using as an alloying element in materials that are exposed to aggressive environments, for instance alloys for surgical instruments, chemical and biomedical industries. Because of low neutron-capture cross section of Zr nuclei, zircaloys are used for cladding of nuclear reactor fuels.

In the studies the variable energy positron beam (VEP) is applied to studies of Zr exposed to dry sliding.

2. Sample preparation

In our studies we used Zr purity of 99.2% purchased from Goodfellow. Samples had a shape of discs of 10 mm in diameter and 3 mm thick each, their surfaces were polished. Before treatments they were annealed in the flow of N_2 gas at a temperature of 550 °C for 1 hr, and then slowly cooled to room temperature. After annealing, the samples were etched in the etching paste TS-K 2000 delivered by Pelox to reduce their thickness by 10 μm and clean their surface. After this procedure only residual defects were present. In fact, for virgin samples the measured positron lifetime spectra contained only one lifetime component equal to 162±1 ps. This corresponds well with the data reported in the literature as the bulk lifetime for Zr [7] .

One of a virgin sample of Zr was located in a tribotester and the base surface of the sample was sliding against the rotating disc with the load of 25 N. This disc of diameter of 50 mm was made of the martensitic steel (steel SW18 hardness about 670 HV0.1). The speed of the disc relative to the surface of the sample was about 5 cm/s. The treatment was performed in air during a certain time, no oxidation was observed.

3. Measurements details

Doppler broadening (DB) spectra of studied samples were measured at room temperature using VEP at JINR. Positrons formed in the 5 mm diameter beam with intensity of 10^5 e+ /s were implanted with incident energy range between 50 eV and 36 keV. Annihilation gamma quanta were registered with high purity germanium (HPGe) detector whose energy resolution was equal to 1.20 keV at 511 keV.

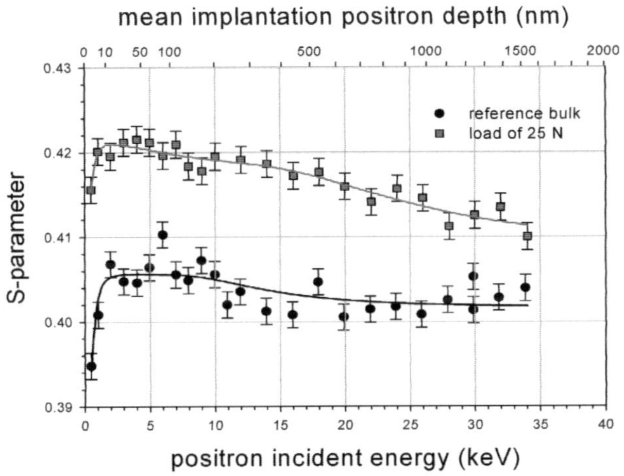

Figure 1. The dependency of the S-parameter as the function of the positron incident energy in VEP experiment measured for reference (closed circles) and exposed to the dry sliding (closed squares) samples. The solid lines represents the best fits using VEPFIT code.

4. Results and discussion

Obtained dependencies of the S-parameter as a function of the positron incident energy are depicted in Figure 1. This parameter is defined as the integral of the central part normalized to the total area of the measured annihilation line. It is well proved experimentally that this parameter is extremely sensitive to the presence of open volume defects. The values of the S-parameter for the sample exposed to dry sliding (closed squares) are placed above the results for the reference sample, (closed circles). However, the dependencies do not like results obtained for other metals like steel [8] or copper [6] where the continuous decrease of the S-parameter values takes place. Increase of the S-parameter is caused by the presence of a very thin, only few atoms layer, mainly oxides on the surface, which reduces the value of the S-parameter on the surface, as it observed for silicon [9].

For description of the dependencies VEPFIT code was used [10]. The Makhovian function was used as the positron implantation profile with the following parameters: m=1.720, n=1.667 and A=2.84 $\mu g/cm^2 keV^n$, which corresponds to the values for copper [11]. The mean positron implantation depth is depicted in the top axis in Figure 1 The both samples exhibit an increase of the S-parameter at low energies less than 4 keV, i.e., at the depth of c.a.10 nm, followed by a gradual decrease for higher energies. In the air due to the passivation process the surface of the Zr is covered by thin ZrO_2 monolayer which shield the interior. The S-parameter in this layer is lower than in the metal. The striking feature of the obtained results clearly visible in Figure 1 is the increase of the value of the S-parameter at the surface, from 0.4058(10) to 0.4215(6) caused by sliding. Thus sliding significantly changes the properties at the surface. However, the best description of the obtained dependencies requires assumption of the presence of two layers. The value of the S-parameter in the adjoined surface layer also increases, from 0.4055(5) to 0.4180(5) for the samples exposed to dry sliding. The thickness of this layer is about 400 nm for the reference sample and 1060 nm for the second sample and positron diffusion length obtained from the fitting procedure are as follows: 51(26) nm and 130(40)nm, respectively. This is a surprising results. On one point the rise of value of the S-parameter indicates the increases of the defects concentration but this is not accompanied of reduction of the positron diffusion length, as one should expect. This can be explained by recovery process induced by sliding, what can cause the generation of vacancy clusters for which the S-parameter has a large value. The value of the S-parameter deeper than 400 nm for reference sample and 1060 nm for the sample after sliding are about 0.4016(4) and 0.4085(8), and corresponding positron diffusion length are as follows: 3 nm and 44 nm, respectively. This indicates presence of open volume defects due to the significant deformation in the latter sample.

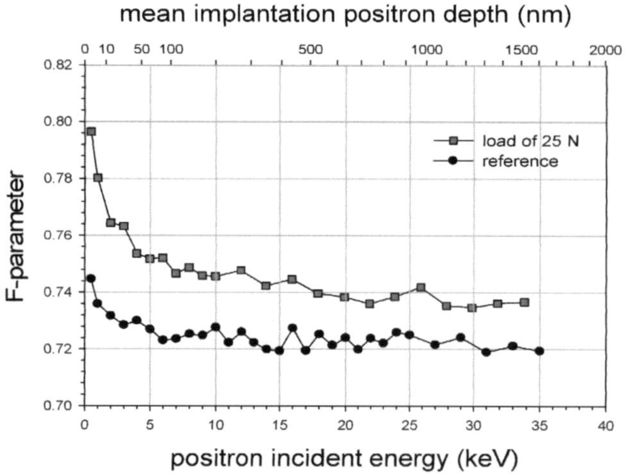

Figure 2. The dependency of the F-parameter as the function of the positron incident energy for the reference sample of Zr and the sample exposed to dry sliding.

It is well known that at the surface of metals positronium, i.e., the bound state between electron and positron can be created. Three quanta annihilation at of this state is reflected in the increase of the so called F-parameter, called "valley to peak" extracted from the energy spectrum of the measured photons. Energy spectrum of quanta from triplet annihilation is continuous from zero to about 511 keV, thus they contribute to the "valley" region in the front of 511 keV peak. This parameter is defined as number of counts in the energy range between 340 and 504 keV to the number of counts in the range from 504 to 518 keV. In Figure 2 the dependencies of this parameter as a function of the positron incident energy are depicted. One can notice the increase of this parameter for the sample exposed to dry sliding in comparison to the reference sample. This indicates that the positronium is created at the worn surface with higher probability than at the surface of the reference sample. This can be caused by the presence of surface defects and/or oxides. The reason can be that the worn surface is more rough after sliding affecting the amount of positronium creation.

5. Conclusions

Slow positron beam studies of pure Zr exposed to the dry sliding were performed. The increase of number of vacancy clusters in the region adjoin the worn surface was detected. The increase of number of positronium at the worn surface was also observed.

References

[1] Horodek P, Dryzek J and Skuratov V A 2016 *Rad. Phys. Chem.* **122** 60
[2] Chai L, Chen B, Wang S, Zhou Z and Huang W 2015 *Materials Characterization* **110** 25
[3] Horodek P and Dryzek J 2016 *Acta Phys. Polon B Proceedings Supplement* **9** 1001
[4] Horodek P, Dryzek J, and Wróbel M 2013 *Tribol. Lett.* **45** 341
[5] Dryzek J, Dryzek E and Cleff B, 1997 *Appl. Surf. Sci.* **116** 236
[6] Dryzek J, Dryzek E, Börner F and Krause-Rehberg R 2001 *Tribol. Lett.* **11** 29
[7] Campillo R J M, Ogando E and Plazaola F J 2007 *J. Phys.: Condens. Matter* **19** 176222
[8] Dryzek J, Dryzek E, Krause-Rehberg R and Redman F 2001 *Tribol. Lett.* **11** 121
[9] Eichler S, Gebauer J, Börner F, Polity A, Krause-Rehberg R, Wendler E, Weber B, Wesch W and Börner H 1997 *Phys. Rev. B* **56** 1393
[10] van Veen A, Schut H, de Vries J, Hakvoort R A and Ijmpa W R, *Positron beam for solids and surfaces*, in: Schultz J, Massoumi G R, Simpson P J (Eds.), AIP Conf. Proc., vol.218, AIP Press, New York, 1990, p. 171
[11] Dryzek J and Horodek P 2008 *Nucl. Instrum. Methods in Phys. Res. B* **266** 4000

IOP Conf. Series: Journal of Physics: Conf. Series **791** (2017) 012021 doi:10.1088/1742-6596/791/1/012021

New insights into the nanostructure of innovative thin film solar cells gained by positron annihilation spectroscopy

S W H Eijt[1], W Shi[1], A Mannheim[1], M Butterling[1], H Schut[1], W Egger[2], M Dickmann[2], C Hugenschmidt[3], B Shakeri[4], R W Meulenberg[5], V Callewaert[6], R Saniz[6], B Partoens[6], B Barbiellini[7], A Bansil[7], J Melskens[8], M Zeman[8], A H M Smets[8], M Kulbak[9], G Hodes[9], D Cahen[9] and E Brück[1]

[1] Department of Radiation Science and Technology, Delft University of Technology, NL-2629 JB Delft, Netherlands
[2] Institut für Angewandte Physik und Messtechnik, Universität der Bundeswehr München, D-85579 Neubiberg, Germany
[3] Physik-Department & Heinz Maier-Leibnitz Zentrum (MLZ), Technische Universität München, D-85748 Garching, Germany
[4] Department of Chemistry, University of Maine, Orono, ME 04469, USA
[5] Department of Physics and Astronomy and the Laboratory for Surface Science and Technology, University of Maine, Orono ME 04469, USA
[6] Department of Physics, University of Antwerp, B-2020 Antwerp, Belgium
[7] Department of Physics, Northeastern University, Boston, MA 02115, USA
[8] Photovoltaic Materials and Devices, Faculty of Electrical Engineering, Mathematics and Computer Science, Delft University of Technology, NL-2628 CD, Netherlands
[9] Department of Materials & Interfaces, Weizmann Institute of Science, Rehovot, 76100, Israel

E-mail: s.w.h.eijt@tudelft.nl

Abstract. Recent studies showed that positron annihilation methods can provide key insights into the nanostructure and electronic structure of thin film solar cells. In this study, positron annihilation lifetime spectroscopy (PALS) is applied to investigate CdSe quantum dot (QD) light absorbing layers, providing evidence of positron trapping at the surfaces of the QDs. This enables one to monitor their surface composition and electronic structure. Further, 2D-Angular Correlation of Annihilation Radiation (2D-ACAR) is used to investigate the nanostructure of divacancies in photovoltaic-high-quality a-Si:H films. The collected momentum distributions were converted by Fourier transformation to the direct space representation of the electron-positron autocorrelation function. The evolution of the size of the divacancies as a function of hydrogen dilution during deposition of a-Si:H thin films was examined. Finally, we present a first positron Doppler Broadening of Annihilation Radiation (DBAR) study of the emerging class of highly efficient thin film solar cells based on perovskites.

1. Introduction

Solar cell technology is rapidly developing into one of the main sources of electrical energy. The efficiencies of photovoltaic (PV) devices have reached high values for many types of solar cells [1, 2]; for example, record values of 25.6% and 21.7% for solar cells based on c-Si wafers and $Cu(In,Ga)Se_2$

thin films, respectively, were reported recently [2]. Solar cells based on a-Si:H and mc-Si:H thin films, which have a high market potential for applications requiring flexible solar cell modules, reached efficiencies of up to 14.0% in triple-junction configurations [3]. Also, new concepts, such as PV devices based on PbS quantum dot (QD) films have currently reached already respectable efficiencies of 10.7% [4]. Furthermore, recent years showed a rather spectacular increase in efficiency for the emerging class of hybrid organic-inorganic perovskite solar cells, with a record of 21.0% reached for a mixture of methyl ammonium lead bromide ($MAPbBr_3$) and formamidinium lead iodide ($FAPbI_3$) [5].

In this field, the application of positron annihilation methods for sophisticated thin film analysis is promising, as new insights can be obtained into the presence and types of open volume defects, the local nanostructure and the electronic structure of the various functional layers of thin film PV devices [6-12], and in-situ studies can be envisioned. Moreover, the depth range accessible to low-energy positron beam systems, typically between ~10 nm and several μms below the surface, matches very well with the thicknesses of a wide range of technologically relevant functional solar cell layers.

In our previous studies, we found that positrons implanted in thin films of PbSe or CdSe QDs have a very high probability to trap at the surfaces of the colloidal QDs. The observed positron-electron momentum distributions can be used to extract information on the composition of their surfaces, gaining new insights into ligand-surface interactions and surface oxidation processes [8, 13]. Furthermore, they provide information about the electronic structure and electronic coupling of CdSe and PbSe QDs [7-10, 13]. Positron annihilation was further applied as one of the main techniques to demonstrate that the nanostructure of hydrogenated amorphous silicon (a-Si:H) is more complex than the conventional model of a Continuous Random Network (CRN) [11, 12]. Instead, open volume deficiencies such as divacancies are present in a CRN-like environment and form a key ingredient to understand the nanostructure and functional properties of a-Si:H solar cell absorber layers. Finally, new insights into the evolution of vacancies in ZnO:Al transparent conductive oxide (TCO) electrodes were obtained upon both fast laser annealing [14] and accelerated environmental degradation [15].

The present paper reports on recent results on advanced characterization of thin film solar cells with positrons. We first provide evidence of trapping of positrons at the surfaces of CdSe QDs using positron annihilation lifetime spectroscopy (PALS). We then apply the 2D-ACAR method to study the nanostructure of divacancies in a-Si:H. The collected momentum distributions are converted by Fourier transformation to a direct space representation given by the electron-positron autocorrelation function $B^{2\gamma}(r)$. The evolution of the nanostructure as a function of hydrogen dilution used in the deposition of a-Si:H was determined, since hydrogen dilution is an important factor influencing the stability against light-induced-degradation. Finally, we present a first positron Doppler broadening study of the emerging class of highly efficient thin film solar cells based on perovskites.

2. Experimental

CdSe QDs with a size of ~4 nm were synthesized according to the methods described in [16, 17] capped with stearic acid and trioctylphosphine oxide ligands. The number density of ligands on the surface of the QD can be systemically tuned by control of the QD cleaning process [17]. The CdSe QDs were drop casted as a thick layer (> 4 μm) onto a $10\times10\,\mathrm{mm}^2$ sapphire substrate for the PALS study at the PLEPS spectrometer of the NEPOMUC facility. Lifetime spectra consisting of $4\cdot10^6$ counts were collected at positron implantation energies in the range of 1-18 keV. The lifetime analysis was performed using both the LT and POSWIN software packages in order to check consistency.

A series of a-Si:H thin films with a thickness of ~450 nm was deposited by Plasma Enhanced Chemical Vapor Deposition on Corning Eagle XG glass substrates, at various hydrogen dilution flow ratios $R=[H_2]/[SiH_4]$ in the range of R=0 to R=10 [18]. The samples were characterized by Doppler depth-profiling in the range of 0.1-25 keV using the VEP facility. Positron 2D-ACAR distributions were collected using the POSH-ACAR facility at a positron implantation energy of 3.7 keV.

Two perovskite $CsPbBr_3$ thin films were spin-coated on In-doped tin oxide (ITO) coated glass substrates using the synthesis method described in [19]. Positron Doppler broadening S and W

parameter depth-profiles [11] were collected on the as-deposited films and on the two samples after 3 and 5 months of storage under ambient conditions in the dark. The depth-profiles collected in the range of 0.1-25 keV using the VEP facility were analyzed using the VEPFIT software package [11].

3. Trapping of positrons at the surfaces of CdSe QDs: evidence from lifetime spectroscopy

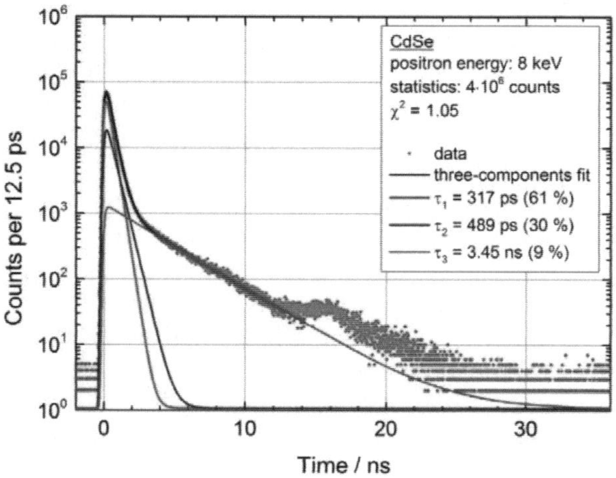

Figure 1. Positron annihilation lifetime spectrum of a thick layer of CdSe QDs deposited on a sapphire substrate, collected at a positron implantation energy of 8 keV. The full lines are the fit curves obtained from a three-lifetime-component analysis.

Previous positron 2D-ACAR studies of CdSe QDs indicated that positrons trap at the local open space present near the surfaces of the QDs, based on the information contained in electron-positron momentum distributions [7]. Here, we present evidence based on positron lifetime spectroscopy that a large fraction of positrons implanted into a thick layer of CdSe QDs indeed annihilate from a trapped state at the surfaces of CdSe QDs, similar to what was previously revealed for PbSe QDs [8].

Figure 1 shows the positron annihilation lifetime spectrum of the CdSe QD layer collected at a positron implantation energy of 8 keV. A three-component lifetime analysis leads to a satisfactory fit. The small bump in the spectrum at ~16 ns is due to back-scattering of positrons, and does not affect the three-component analysis. In the range of 4-16 keV, the three-lifetime-component analysis revealed a short lifetime component τ_1 in the range of $\sim 320-350$ ps with a high intensity of $\sim 60\%-85\%$, and a second (short) lifetime τ_2 in the range of $\sim 460-550$ ps with an intensity of $\sim 30\%-10\%$. The lifetime of both components likely corresponds to annihilation from positrons trapped at the surfaces of CdSe QDs, where the second lifetime component could be associated with a surface state with less overlap of the positron wave function with electron orbitals. Notably, the average lifetime of the first two components was found to be constant throughout the positron implantation range of 4 keV to 18 keV, with $\tau_{mean} \approx 379 \pm 5$ ps and an intensity of $92 \pm 1\%$, *i.e.* quite close to the values obtained in our previous positron studies on PbSe QD layers [8]. The positron lifetime is much larger than that of defect-free CdSe (275 ps [9]) and, combined with its high intensity, provides strong support for trapping and annihilation of positrons at the surfaces of the CdSe QDs, consistent with previous 2D-ACAR studies [7, 10]. The preliminary outcome of our ab-initio calculations on CdSe slabs using the Local Density Approximation and a Corrugated-Image-Potential indicates that the positron density peaks just below the CdSe surface, while preliminary calculations using the Weighted Density Approximation point to the presence of a true positron surface state.

The third, long lifetime component with $\tau_3 \sim 3.5 \pm 0.1$ ns and $I_3 = 8 \pm 1\%$ corresponds to pick-off annihilation of o-Ps formed in the available open space in between the ligands at the surfaces of the CdSe QDs. The lifetime of this o-Ps component is larger than the values of ~2.0-2.7 ns for PbSe QDs capped with oleic acid ligands [8]. This finding infers a somewhat larger open space of ~0.8 nm in diameter for the case of CdSe QDs with stearic acid and TOPO ligands compared to ~0.65 nm for PbSe QDs with oleic acid ligands, as deduced from the Tao-Eldrup model [20].

4. Size evolution of small vacancy clusters in hydrogenated amorphous silicon: 2D-ACAR study

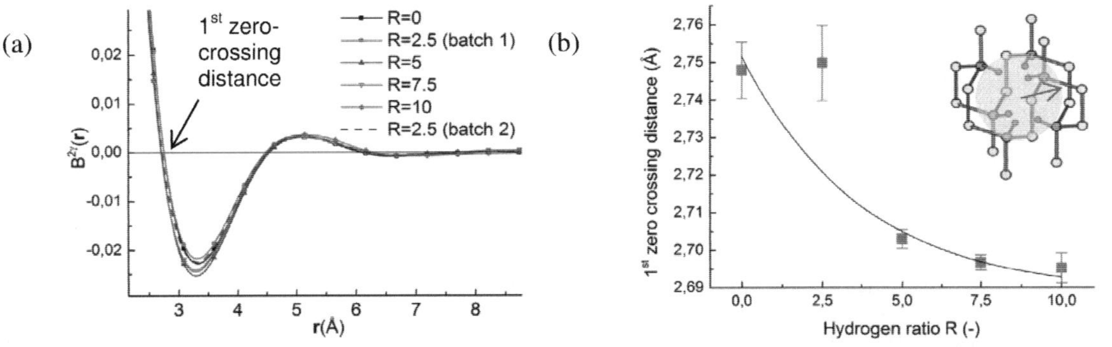

Figure 2. (a) Electron-positron autocorrelation function $B^{2\gamma}(r)$for as-deposited a-Si:H as a function of hydrogen dilution ratio *R*. (b) First zero-crossing distance as a function of *R*. The line is a guide-to-the-eye. The inset is a schematic representation of a Si-divacancy decorated with 6 hydrogen atoms.

In Figure 2(a), we present the electron-positron autocorrelation function $B^{2\gamma}(r)$ as a function of hydrogen dilution ratio *R*, in order to examine the evolution in the size of open volume deficiencies in a-Si:H films on hydrogen concentration. For each sample, the autocorrelation function was extracted from the collected isotropic 2D-ACAR momentum distribution, using a Fourier transformation of the corresponding 1D-ACAR distribution $N(p)$ [21, 22]. $B^{2\gamma}(r)$ represents the real space autocorrelation function of the electron orbitals modulated by the wave function of the positron, and can thus provide insights into the local spatial environment of the positron annihilation site. In particular, the first zero-crossing in the autocorrelation function provides an indication of the shortest average distance, in any direction, to the site which cannot be occupied by a positron [23]. This can be understood since this distance corresponds to the distance of the center of the positron annihilation site to a nearest neighbour atomic site where the positron wave function $\psi_+(\vec{r})$ necessarily vanishes.

In Figure 2(b), we present the extracted first zero-crossing distances as a function of hydrogen dilution ratio *R*. The observed first zero-crossing distances of ~2.75 Å provide additional support to the picture that the dominant open volume deficiencies in high-quality a-Si:H films are small vacancies (in particular divacancies), as previously proposed based on Doppler broadening *S* parameter measurements and Fourier transform infrared spectroscopy [11, 12]. Namely, the distance of the center of the positron wave function in an unrelaxed divacancy (in c-Si) to one of the six nearest neighbour Si positions is 2.96 Å, while the corresponding distance of 2.35 Å for an unrelaxed monovacancy is much shorter. The observed first zero-crossing distance of ~2.75 Å indicates a local inward relaxation of the Si atom positions around the divacancy, as could be expected [24]. Furthermore, the presence of up to six hydrogen atoms decorating the nearest neighbour Si atoms will also reduce the locally available open space. With increased hydrogen dilution ratio, the first zero crossing is seen to move towards smaller distances of ~2.70 Å, correlating with the observed reduction in *S* parameter for the same set of films. The decrease in the first zero-crossing distance corresponds to ~5% reduction in open volume, and can be caused by additional hydrogen locally present near the boundary of the divacancy. Figure 2(a) moreover shows that the first minimum in $B^{2\gamma}(r)$ at ~3.25 Å becomes more pronounced with an increase in *R*, as the smaller open space induces a larger overlap between the positron wave function and electron orbitals, and thus, a stronger spatial correlation of the positron and electrons. In short, the 2D-ACAR studies provide additional support for the presence of small open volume deficiencies (in particular divacancies) in a CRN environment in a-Si:H films. With increased hydrogen dilution, more hydrogen is present in the close vicinity of the divacancy, effectively reducing the open space available to the positron.

IOP Conf. Series: Journal of Physics: Conf. Series **791** (2017) 012021 doi:10.1088/1742-6596/791/1/012021

5. Doppler broadening depth profiling study of $CsPbBr_3$ perovskite solar cell films

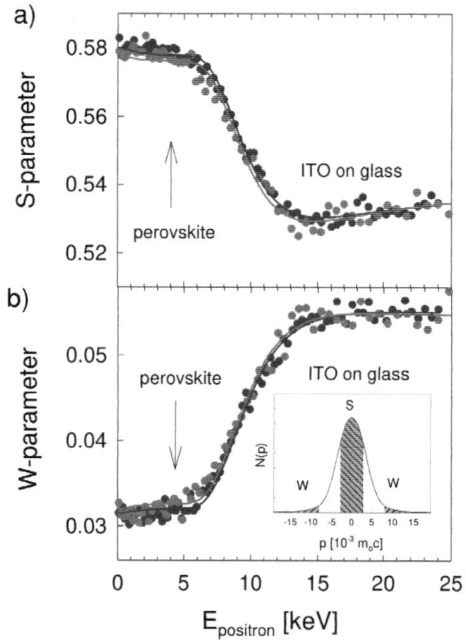

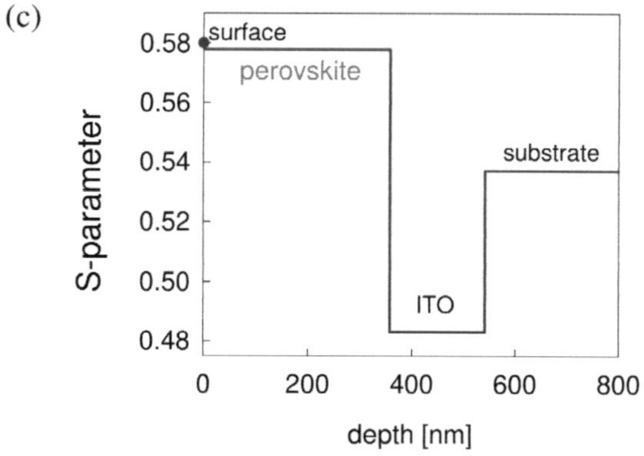

Figure 3. Positron (a) *S* and (b) *W* parameter depth profiles of a $CsPbBr_3$ perovskite thin film on an ITO coated glass substrate in the as-received state (•) and after 5 months under ambient conditions in the dark (•). Full lines are VEPFIT curves. (c) Two layer VEPFIT model used in the analysis of the as-received sample.

Finally, $CsPbBr_3$ perovskite thin films were studied, since the role of point defects in the degradation mechanism of the emerging class of perovskite solar cells is a topic of debate [25], and, furthermore, vacancies may form unwanted recombination centres for charge carriers [2], limiting the open circuit voltages and solar cell efficiencies reached in practice.

Figure 3 shows the Doppler depth-profiles of one of the $CsPbBr_3$ thin films on an ITO coated substrate in the as-received state and in the final state after 5 months of exposure to ambient conditions in the dark. The changes in depth-profiles are very small for both samples, as the change in the fit values for *S* and *W* for the perovskite layer is only $\Delta S / S_0 \sim -0.3 \pm 0.1\%$ and $\Delta W / W \sim 1 \pm 1\%$, respectively. It is instructive to compare these modest changes to typical values for the shift in *S* and *W* upon formation of cation monovacancies. For example, saturation trapping in Zn monovacancies in c-ZnO leads to substantially larger shifts of $\Delta S / S_{c-ZnO} \sim +5.5\%$ and $\Delta W / W_{c-ZnO} \sim -19\%$ [26, 27]. Furthermore, the sign of the small change in *S* for $CsPbBr_3$ is opposite. Doppler broadening spectroscopy thus indicates that (negatively charged or neutral) V_{Cs} or V_{Pb} cation vacancies are not visibly generated during prolonged exposure of $CsPbBr_3$ to ambient conditions in the dark. $CsPbBr_3$ perovskite solar cells are substantially more stable against environmental degradation than the methyl ammonium lead bromide ($MAPbBr_3$) based solar cells [19]. Therefore, positron depth profiling studies of $MAPbBr_3$, including also degradation under AM1.5 solar illumination conditions [19], seem promising in order to unravel whether or not the formation of cation vacancies is a key element in the degradation of perovskite-based solar cells [25]. It should be noted that, since V_{Br} monovacancies mostly are positively charged, positrons are not expected to trap in the anion vacancies, and are largely insensitive to the presence of such monovacancies. Therefore, the formation of V_{Br} monovacancies cannot be excluded for the $CsPbBr_3$ films investigated in this study.

6. Conclusions

PALS has revealed that a high fraction of positrons implanted in CdSe QD films trap at the surfaces of the QDs. Depth-profiling positron methods show high potential as a characterization tool for the surface composition and electronic structure of colloidal semiconductor QDs embedded as the functional layer of innovative PV devices. Positron methods can aid the development of such QD-based PV devices, since present limitations in their solar cell efficiency involve surface recombination of generated charge carriers and inefficient transport of charge carriers involving hopping conductivity, requiring optimization of surface passivation and electronic coupling between neighbouring QDs as well as a reduction of transport barriers. These factors form key contributions to the relatively high loss in open circuit voltage in PV devices, and tailoring of the surface properties and electronic coupling of QDs may lead to further large improvements in solar cell efficiencies.

Further, 2D-ACAR was used to examine the evolution of the size of divacancies in a-Si:H films as a function of hydrogen dilution during synthesis. Such open volume deficiencies form a key ingredient to understand and tailor the nanostructure of a-Si:H films, which may aid the development of synthesis procedures to stabilize a-Si:H absorber layers against light induced degradation.

Finally, positron Doppler depth-profiling was applied to examine the relatively high stability of $CsPbBr_3$ perovskite solar cells. Exposure to ambient conditions in the dark for 5 months did not lead to visible formation of cation vacancies.

Acknowledgments

The work at Delft University of Technology was supported by the China Scholarship Council (CSC) grant of W.S., by ADEM, A green Deal in Energy Materials of the Ministry of Economic Affairs of The Netherlands (www.adem-innovationlab.nl), and the STW Vidi grant of A.S., Grant No. 10782. The PALS study is based upon experiments performed at the PLEPS instrument of the NEPOMUC facility at the Heinz Maier-Leibnitz Zentrum (MLZ), Garching, Germany, and was supported by the European Commission under the 7th Framework Programme, Key Action: Strengthening the European Research Area, Research Infrastructures, Contract No. 226507, NMI3. The work at University of Maine was supported by the National Science Foundation under Grant No. DMR-1206940. Research at the University of Antwerp was supported by FWO grants G022414N and G015013. The work at Northeastern University was supported by the US Department of Energy (DOE), Office of Science, Basic Energy Sciences grant number DE-FG02-07ER46352 (core research), and benefited from Northeastern University's Advanced Scientific Computation Center (ASCC), the NERSC supercomputing center through DOE grant number DE-AC02-05CH11231, and support (applications to layered materials) from the DOE EFRC: Center for the Computational Design of Functional Layered Materials (CCDM) under DE-SC0012575. The work at the Weizmann Institute was supported by the Sidney E. Frank Foundation through the Israel Science Foundation, by the Israel Ministry of Science, and the Israel National Nano-Initiative. D.C. holds the Sylvia and Rowland Schaefer Chair in Energy Research.

References

[1] Nayak P K, Bisquert J and Cahen D 2011 *Adv. Mater.* **23** 2870
[2] Polman A, Knight M, Garnett E C, Ehrler B and Sinke W 2016 *Science* **352** aad4424
[3] Sai H, Matsui T and Matsubara K 2016 *Appl. Phys. Lett.* **109** 183506
[4] Kim G-H *et al.* 2015 *Nano Lett.* **15** 7691
[5] Brittman S, Adhyaksa G W P, and Garnett E C 2015 *MRS Communications* **5** 7
[6] Tuomisto F and Makkonen I 2013 *Rev. Mod. Phys.* **85** 1583
[7] Eijt S W H, van Veen A, Schut H, Mijnarends P E, Denison A B, Barbiellini B, Bansil A 2006 *Nature Mater.* **5** 23
[8] Chai L *et al.* 2013 *APL Materials* **1** 022111
[9] Weber M H, Lynn K G, Barbiellini B, Sterne P A and Denison A B 2002 *Phys. Rev. B* **66** 041305(R)
[10] Eijt S W H, Mijnarends P E, van Schaarenburg L C, Houtepen A J, Vanmaekelbergh D, Barbiellini B and Bansil A 2009 *Appl. Phys. Lett.* **94** 091908
[11] Melskens J, Smets A H M, Schouten M, Eijt S W H, Schut H and Zeman M 2013 *IEEE J. of Photovoltaics* **3** 65
[12] Melskens J *et al.* 2015 *Phys. Rev. B* **91** 245207
[13] Shi W, Eijt S W H, Suchand Sandeep C S, Siebbeles L D A, Houtepen A J, Kinge S, Brück E, Barbiellini B and Bansil A 2016 *Appl. Phys. Lett.* **108** 081602
[14] Scorticati D *et al.* 2015 *Acta Mater.* **98**, 327
[15] Shi W *et al.*, in preparation
[16] Lee J R I, Whitley H D, Meulenberg R W *et al.*, 2012 *Nano Lett.* **12** 2763
[17] Shakeri B and Meulenberg R W 2015 *Langmuir* **31** 13433
[18] Melskens J, PhD thesis, Delft University of Technology (2015)
[19] Kulbak M, Gupta S, Kedem N, Levine I, Bendikov T, Hodes G and Cahen D 2016 *J. Phys. Chem. Lett.* **7** 167
[20] Gidley D W, Peng H-G and Vallery R S 2006 *Annu. Rev. Mater.* **36** 49
[21] Ho K F, Ching H M, Beling C D, Fung S, Ng K P, Biasini M, Ferro G and Gong M 2003 *J. Appl. Phys.* **94** 5549
[22] Kobayashi T 2001 *Mater. Sci. Tech.* **363-365** 689
[23] Britton D T, Minani E, Knoesen D, Schut H, Eijt S W H, Furlan F, Giles C, Härting M 2006 *Appl. Surf. Sci.* **252** 3194
[24] Makkonen I and Puska M J 2007 *Phys. Rev. B* **76** 054119
[25] Miller J L 2014 *Physics Today* **67** 13, and Refs. therein
[26] Kahn E H, Weber M H and McCluskey M D 2013 *Phys. Rev. Lett.* **111** 017401
[27] Makkonen I, Korhonen E, Prozheeva V and Tuomisto F 2016 *J. Phys.: Condens. Matter.* **28** 224002

IOP Conf. Series: Journal of Physics: Conf. Series **791** (2017) 012022 doi:10.1088/1742-6596/791/1/012022

Structure of nitride layer formed on titanium alloy surface by N_2-gas exposure at high temperatures

Y Takeda[1,4], K Iida[2], S Sato[2,3], T Matsuo[3], Y Nagashima[2], N Okubo[4], K Kondo[4] and T Hirade[1,4]

[1] Graduate School of Science and Engineering, Ibaraki University, 2-2-1, Bunkyo, Mito, Ibaraki, 310-8512, Japan

[2] Department of Physics, Tokyo University of Science, 1-3, Kagurazaka, Shinjuku, Tokyo, 162-8601, Japan

[3] Fukushima College, National Institute of Technology, 30 Nagao, Kamiarakawa, Taira, Iwaki, Fukushima, 970-8034, Japan

[4] Nuclear Science and Engineering Center, Japan Atomic Energy Agency, 2-4, Shirakata, Tokai, Ibaraki, 319-1195, Japan

E-mail: 2210080@alumni.tus.ac.jp

Abstract. Nitride treatment is used for hardening of titanium alloy surfaces. In this study, we prepare samples under two different conditions, (1) 810°C, for 600 min, and (2) 850°C, for 720 min. A depth-profile analysis of the surfaces of the samples is conducted through Doppler broadening measurements of positron annihilation gamma rays using a slow positron beam. Lager S-parameters are obtained near the surface for all samples. These parameters can be explained by the positron annihilation at the grain boundaries between the nano-grains confirmed by transmission electron microscope images. Furthermore, the variation in the W-parameters near the surface is independent of the variation in S–parameters, thus indicating a depth dependence of the W-parameters on the chemical composition of the precipitate at the grain boundaries, also confirmed by energy dispersive X-ray spectroscopy measurements. Variations in the W-parameters was observed to occur in the deeper regions, indicating that the nitrated layer is thicker in the sample prepared under condition (2) and the increased hardness of the sample surface measured using Vickers Hardness method is likely caused by this thicker nitrated layer.

1. Introduction

Titanium alloy is widely used in manufacturing of different objects, such as golf club heads and structural materials for aircrafts, because of its high strength, low weight and high corrosion resistance. The alloy surface can be significantly hardened through a nitride treatment of the surface, but there are some difficulties in the use of titanium nitride such as exfoliation and cracking of the surface layer because of increased stress. A previous study reported that these properties depend on the treatment temperature and time [1]. Two different nitride layers were reported to have formed on the surface, a TiN layer near the surface and a Ti_2N layer behind the TiN layer, and higher interfacial fracture toughness was observed near the thicker nitride layer [1]. Therefore, it is necessary to further study the structure of the nitride layer more to understand this interfacial fracture mechanism. One of our authors (S Sato) has already proven that samples prepared at higher temperatures for a longer period of time, i.e.

samples C and D in the present study, exhibit higher surface hardness, as shown in Table 1. In this study, a depth-profile analysis was conducted for the surfaces of the samples prepared through Doppler broadening (DB) of positron annihilation gamma-ray measurements [2]. The surface layers showing higher S values in samples C and D, i.e. the samples prepared at higher temperatures and for longer durations, were thinner than those of samples A and B, as shown in Figure 1. Hence, it is necessary to study the surface structures to determine the higher S-parameters. Transmission electron microscopy (TEM) and Energy dispersive X-ray spectrometry (EDS) were used to understand the structure near the surface in detail.

2. Experiment

DB measurements of positron annihilation gamma rays were conducted using a slow-positron beam line at the Tokyo University of Science [3]. The annihilation gamma rays were detected using a Ge semiconductor detector coupled with a NaI(Tl) scintillation detector. Measurements were performed at positron energies in the range of 0.1–30 keV and an energy interval of 1 keV. The measurement time and total count for the annihilation peak at each energy interval were 500 s and approximately 10000, respectively. The same procedures were performed several times for each sample.

The samples analysed were of titanium alloy (6 wt% aluminium, 4 wt% vanadium). A nitride layer was formed on the surface of each sample by heating under a nitrogen-gas atmospheric pressure of 1.3. Four samples were prepared under two different conditions: samples A and B were heated at 810°C for 600 min, and samples C and D were heated at 850°C for 720 min. Samples A and C were placed in the lower space and samples B and D in the upper space of the electrical furnace. The samples for TEM and EDS observation were prepared using focused ion beams (FIB) of gallium.

3. Results and Discussion

The TEM images for the four samples (A, B, C and D) are shown in Figure 2, indicating that the surfaces of these samples are covered with fine grains. The size of the grains were observed to be slightly different; however, they were found to be very small in general; i.e. less than 10 nm in size. The positron trap site in the nano-grained structures is formed mainly of grain boundaries as thermalized positrons can diffuse easily in grains and reach the grain boundaries [4]. The thicknesses of the nano-grain layers are approximately 200–300 nm for samples A and B and approximately 100–150 nm for samples C and D. Therefore, we conclude that the higher S parameters near the surface (Figure 1) are caused by the positron annihilation at the grain boundaries.

The depth dependences of S and W parameters are shown in Figure 2. Movements on the S–W plane are not on a diagonal line; hence, W-parameter variations provide qualitative information regarding

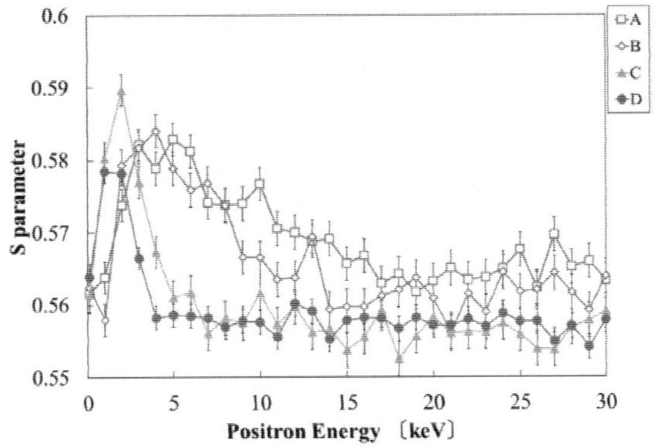

Figure 1.

Positron beam-energy dependence of the S-parameters of the samples prepared at 810°C for 600 min (A, B) and 850°C for 720 min (C, D).

Table 1. Vickers Hardness (VH) with 100 mN before and after nitriding treatment.

Virgin	(1) 810°C 600min		(2) 850°C 720min	
	A	B	C	D
354.1	724.1	729.0	1044.4	1036.2

the chemical elements present at the positron annihilation sites. The patterns in the plots, as shown in Figure 2, of the four samples correspond to the depth from the surface of the four TEM images. Larger S-values near the surface are explained by the positron annihilation at the grain boundaries [4], as mentioned above. At the depth where the S parameters change, the W parameters change independently, indicating that there is a depth dependence of the chemical composition of the precipitate at the grain boundaries.

The EDS results for sample C are shown in Figure 3. High nitrogen (N) concentration is observed up to a depth of 100 nm from the surface and it decreases gradually to almost half until the depth of 400 nm. This indicates that the layer showing higher N concentration (up to a depth of approximately 100 nm) contains TiN and the lower concentration layer contains Ti_2N. The thickness of the TiN layer

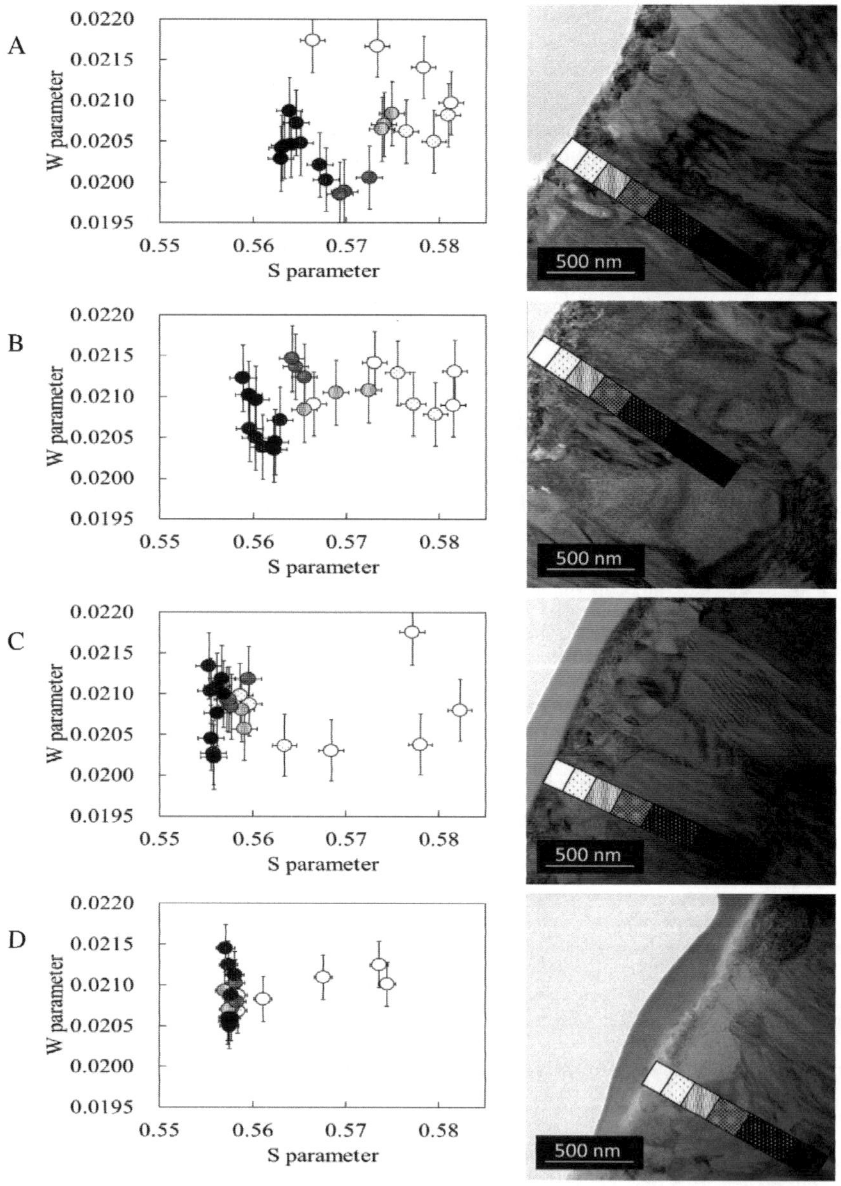

Figure 2. S–W plots and TEM images for samples A, B (810°C 600 min) and samples C, D (850°C 720 min). The grey shading in the plots corresponds to the shades on the bars in the TEM images indicating depth from the surface.

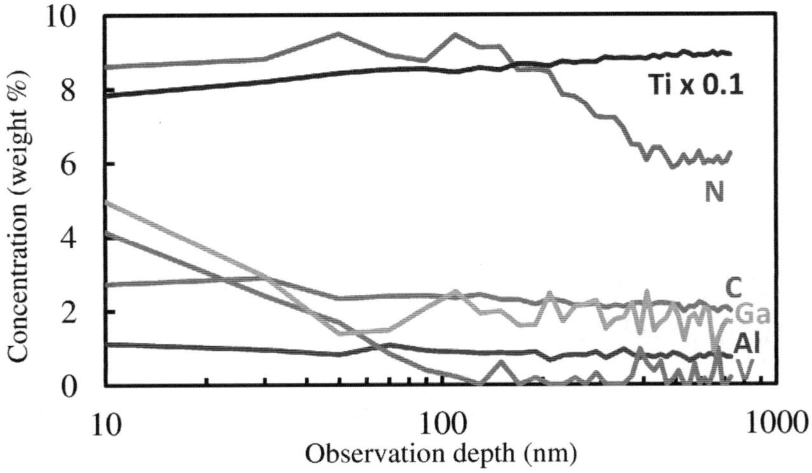

Figure 3.
Depth dependence of concentrations of elements measured by EDS. Carbon was coated on the sample surface when TEM and EDS samples were prepared using FIB. Implanted ions by FIB in the sample were observed to cause gallium contamination.

corresponds to the thickness of the nano-grain layer. In sample C, W parameter variations could be observed up to a depth of approximately 100 nm (Figure 2). According to the EDS measurement shown in Figure 3, density changes for Vanadium occur at the same depth region. Vanadium is likely to be the chemical element of the precipitate present at the grain boundaries.

The thicknesses of the nitrated layers with TiN and Ti_2N can be determined from the depths giving the lowest W values for the four samples. Hence, samples C and D have thicker nitrated layers than samples A and B. The thicker nitrated layers are probably the reason for the higher hardness in samples C and D.

The location of the occurrence of exfoliation near the surface is not known; however, a fracture along the grain boundaries can be a possible mechanism. W parameter variations appear near the surface and they are likely caused by the change in the chemical composition of the precipitate at the grain boundaries. DB measurements by slow-positron beams can be used a tool in the future to non-destructively estimate nitrated layers.

4. Conclusion

Higher S-parameters observed near the alloy surface can be explained by the positron annihilation at the grain boundaries in the nano-grained structures, as indicated by the TEM images. Variations of W-parameters that are independent of the changes in S-parameters indicate a depth dependence on the chemical composition of the precipitate at the grain boundaries. On the basis of the W-parameter variations, the nitride layers of samples C and D were found to be thicker than those of samples A and B. These thick nitride layers can be considered a reason for the increased hardness.

References

[1] Yan X, Kato M, Nakasa K and Morita K 2001 *J. Soc. Mat. Sci. Japan*, **50**, 764
[2] Takeda Y and Iida K 2015 *Bachelor's degree thesis, Tokyo University of Science "Study of surface layer on Titanium alloy by Doppler Broadening of positron annihilation gamma-rays measurement."*
[3] Nagashima Y, Hakodate T, Miyamoto A, Michishio K and Terabe H 2009 *J. Phys. Conf. Ser.* **194** 012039
[4] Hidalgo C, de Diego N and Plazaola F 1985 *Phys. Rev. B* **31** 6941

IOP Conf. Series: Journal of Physics: Conf. Series **791** (2017) 012023 doi:10.1088/1742-6596/791/1/012023

Prediction of positron-annihilation parameters for vacancy-type defects in ternary alloy semiconductors by data-scientific approach

Shoji Ishibashi[1,2], Hiori Kino[2], Akira Uedono[3], Takashi Miyake[1,2] and Kiyoyuki Terakura[2]

[1] Research Center for Computational Design of Advanced Functional Materials (CD-FMat), AIST, Tsukuba, Ibaraki 305-8568, Japan

[2] Center for Materials research by Information Integration (CMI[2]), NIMS, Tsukuba, Ibaraki 305-0047, Japan

[3] Division of Applied Physics, University of Tsukuba, Tsukuba, Ibaraki 305-8573, Japan

E-mail: `shoji.ishibashi@aist.go.jp`

Abstract. We calculated positron annihilation parameters for mono- and di-vacancies in ternary semiconductors $Al_{0.5}Ga_{0.5}N$ and $In_{0.5}Ga_{0.5}N$. It has been found that the obtained annihilation parameters are well correlated with structural parameters. By constructing multiple linear regression models using selected (about 1/4 of the total) datasets as training sets in order to reduce computational cost, we could predict annihilation parameters for the rest.

1. Introduction

Group-III nitrides are important materials for optoelectronic and power devices. Since vacancy-type defects often play a critical role in device performances, we have been investigating such defects by the positron annihilation technique [1, 2, 3]. As is well known, the positron annihilation technique is a powerful tool to study vacancy-type defects in various solids [4, 5]. Alloying is one of the key technologies for semiconductor devices by controlling the band gap and/or the lattice parameters. In alloy semiconductors, atomic sites are not equivalent to one another because of randomness. In our previous studies, we calculated positron annihilation S and W parameters as well as positron lifetimes τ for 64 types of cation mono-vacancies and those for 104 types of cation-nitrogen di-vacancies among 256 possibilities in $In_{0.5}Ga_{0.5}N$ [2, 3]. In the present study, we have completed calculations for the rest of the di-vacancies and calculated positron annihilation parameters for mono- and di-vacancies in $Al_{0.5}Ga_{0.5}N$ also.

2. Computational details

For calculations of positron states and annihilation parameters, we used our computational code QMAS (Quantum MAterials Simulator), which is based on the plane-wave basis and the projector augmented-wave (PAW) method [6]. Ternary semiconductors were modeled as random alloys by the special quasirandom structure (SQS) approach [7]. Their lattice parameters and high-frequency dielectric constants were evaluated by averaging experimental values for binary systems. Further details are described in Ref. [2]. For data-scientific analyses, we used the **lm()**

function of R [8] to perform multiple linear regression. With the help of the **step()** function, we selected descriptors in regression models.

3. Results and discussion

3.1. Multiple linear regression with optimized structural parameters

In Fig. 1, an atomic configuration around a cation mono-vacancy is shown together with two types of configurations for di-vacancies. For cation mono-vacancies, we used the following 3 descriptors: x_{Ga}, Ga concentration among neighboring 12 cations, d_N, average distance of 4 nitrogen atoms from the center of tetrahedron which 4 nitrogen atoms form, d_C, average distance of 12 cations from the center of the tetrahedron. In Fig. 2, predicted positron annihilation parameters are plotted as a function of originally-calculated values, which were obtained using QMAS with atomic positions computationally-optimized. Root-mean-square-error (RMSE) values in percentage for S, W and τ are 0.037 %, 0.61 % and 0.061 % for $Al_{0.5}Ga_{0.5}N$, 0.16 %, 0.56 %, 0.42 % for $In_{0.5}Ga_{0.5}N$, respectively. Multiple linear regression works very well for these cases. Here, we do not present regression coefficients because their values do not have essential meanings when there is significant correlation among descriptors.

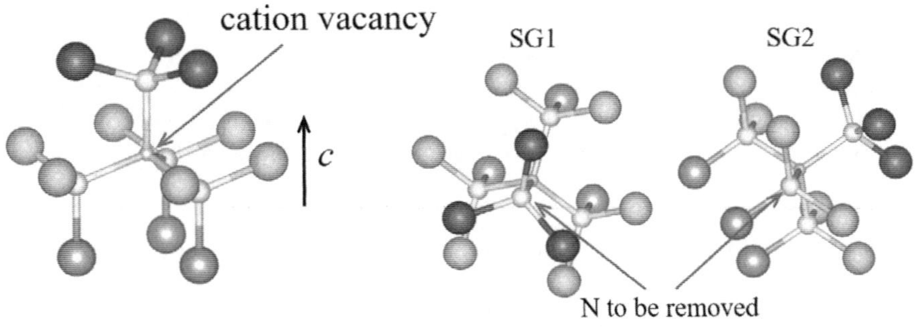

Figure 1. Atomic configuration around cation mono-vacancy (left) and two types of configurations for di-vacancy (middle and right). Yellow small balls represent nitrogen atoms while larger balls represent cations. In case of an ideal wurtzite structure, balls with the same color are crystallographically equivalent to each other. d_N and d_C in the text are defined as average distances from the center of the nitrogen tetrahedron of 4 nitrogen atoms and 12 cations, respectively. SG1 and SG2 correspond to two subgroups of di-vacancies described in the text.

Among 4 nitrogen atoms around a cation mono-vacancy, one nitrogen shows a different geometry from those for the rest because of the wurtzite structure. We therefore divided 256 patterns of di-vacancies into two subgroups consisting of 64 and 192 patters (denoted as SG1 and SG2), respectively. Surrounding 12 cations should be classified into two categories depending on the positron of the vacant nitrogen site V_N (a part of di-vacancy). For the di-vacancy case, the following 5 descriptors were chosen: $x1_{Ga}$, Ga concentration among 3 cations adjacent to V_N, $x2_{Ga}$, that among the other 9 cations, d_N, average distance of 3 nitrogen atoms from the center of tetrahedron formed by 3 nitrogen atoms and the center of 3 cations, $d1_C$, average distance of 3 cations from the center of the tetrahedron, $d2_C$, that of the other 9 cations. In Fig. 3, predicted positron annihilation parameters are plotted as a function of originally-calculated values. RMSE values for the SG1 are 0.12 %, 1.04 % and 0.28 % for $Al_{0.5}Ga_{0.5}N$, 0.24 %, 0.88 %, 0.54 % for $In_{0.5}Ga_{0.5}N$, and those for the SG2 are 0.15 %, 1.20 % and 0.33 % for $Al_{0.5}Ga_{0.5}N$, 0.30 %, 0.85

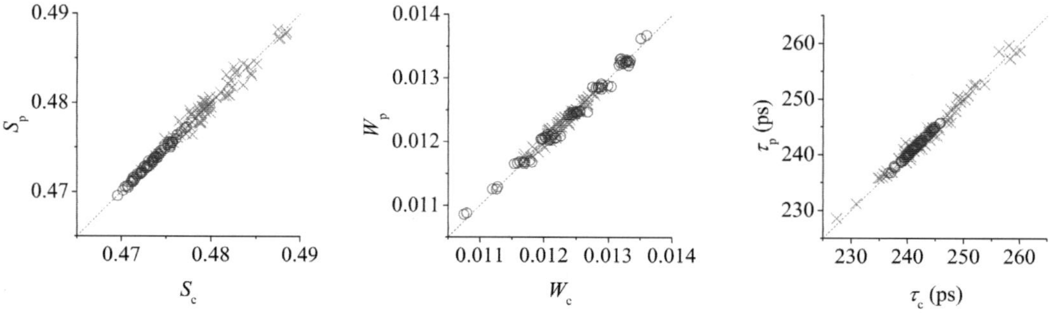

Figure 2. Positron annihilation parameters predicted by linear regression models (S_p,W_p,τ_p) for cation mono-vacancies in $Al_{0.5}Ga_{0.5}N$ (open circles) and $In_{0.5}Ga_{0.5}N$ (crosses) plotted as a function of originally-calculated values (S_c,W_c,τ_c).

%, 0.66 % for $In_{0.5}Ga_{0.5}N$. Although these are slightly worse than those for the mono-vacancy case, multiple linear regression is still working well.

Apart from the regression issue, differences in the calculated parameters between two materials should be mentioned. Whether for mono- or di-vacancy, S_c and τ_c show wider distributions while W_c shows a narrower one in the case of $In_{0.5}Ga_{0.5}N$ as shown in Figs. 2 and 3. It is thought to be closely related to the difference in ionic radii of cations and to the presence or absence of *d* electrons. This problem will be discussed elsewhere.

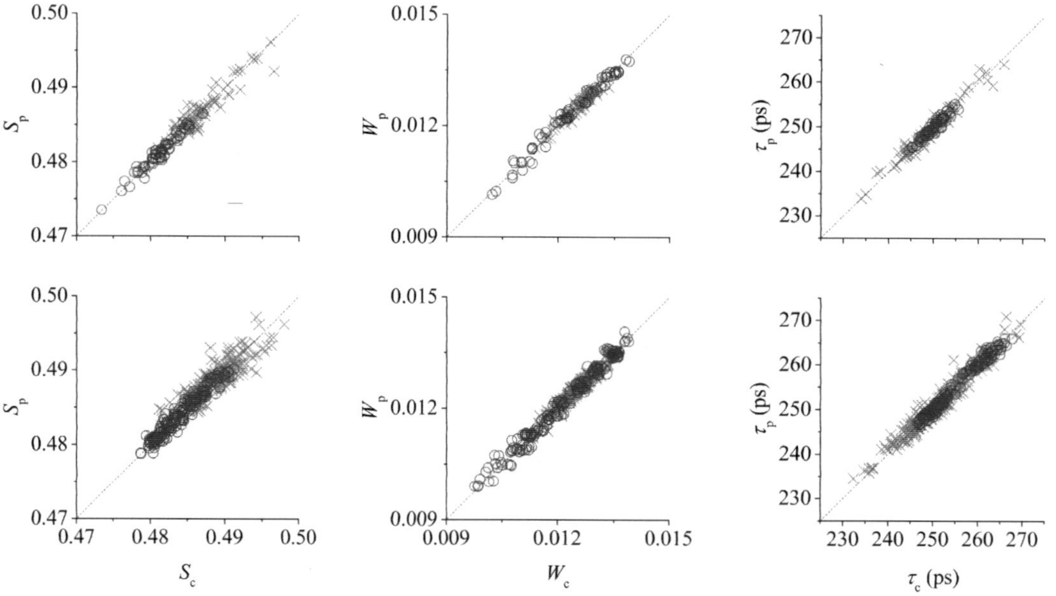

Figure 3. Positron annihilation parameters predicted by linear regression models (S_p,W_p,τ_p) for di-vacancies in $Al_{0.5}Ga_{0.5}N$ (open circles) and $In_{0.5}Ga_{0.5}N$ (crosses) plotted as a function of originally-calculated values (S_c,W_c,τ_c). The top and bottom panels are for the SG1 and SG2.

3.2. Multiple linear regression with pre-optimized structural parameters

To obtain descriptors d_N and d_C, structural optimization is required and its computational cost is significant. We tried to predict positron annihilation parameters using pre-optimized structures, which are the SQS-modeled vacancy-free structures with internal atomic positions fully-relaxed. As a first attempt, we replaced d_N and d_C with values evaluated on the vacancy-free structures but, as a matter of course, it was not successful. The degree of structural relaxation accompanied with the vacancy creation is crucial for the prediction of positron annihilation parameters. It is not determined solely by the distances from the vacancy site. Considering that the structural optimization is performed using force values acting on the constituent atoms, we added two more descriptors f_N and f_C, which are force components of 4 nitrogen atoms and 12 cations toward the vacancy just after removal of the cation. One-shot calculations of the force components require much less computational time than full optimization. Thus, the following 5 descriptors were chosen: x_{Ga} (exactly the same as the above), d^b_N, average distance of 4 nitrogen atoms from the cation to be removed to create a mono-vacancy on the vacancy-free structure, d^b_C, average distance of 12 cations from the cation, f_N, average force component of 4 nitrogen atoms toward the vacancy just after removal of the cation, f_C, average force component of 12 cations toward the vacancy.

To construct regression models, we must know several correct answers for target parameters (S,W,τ). Full structural optimization is required to obtain them. We call the selected datasets "training set". To assess the minimal number of datasets for constructing regression models, we evaluated RMSE values as a function of the number of datasets in the training set n. As training sets, in order of d^b_N, the 1st. $k+1$st, $2k+1$st, $3k+1$st,... ($k = 2, 3, 4, 5, 6, 7, 9$) and 64th datasets among the 64 datasets are chosen. Figure 4 represents training-set size dependence of RMSE values of positron annihilation parameters for mono-vacancies. The results at $n = 64$, where all the datasets were used to construct the regression models, are not much different from those obtained using fully-optimized structures. Down to $n = 17$, RMSE values are similar to those at $n = 64$. Even with the pre-optimized structures, the regression models were successfully constructed using about 1/4 of the total datasets.

For the di-vacancy case, similar replacements of descriptors to those in the mono-vacancy case were made. The resultant descriptors are $x1_{Ga}$ (exactly the same as the above), $x2_{Ga}$ (exactly the same as the above), d^b_N, average distance of 3 nitrogen atoms from the cation to be removed on the vacancy-free structure, $d1^b_C$, average distance of 3 cations from the cation, $d2^b_C$, that of the other 9 cations, f_N, average force component of 3 nitrogen atoms toward the cation vacancy just after removal of the cation-nitrogen pair $f1_C$, average force component of 3 cations, $f2_C$, that of the other 9 cations. In Fig. 5, training-set size dependence of RMSE values is shown. Even at $n = 64$ for the SG1 and at $n = 192$ for the SG2, RMSE values are significantly worse than those for predictions from the optimized structures especially for W. It shows room for improvement in choosing descriptors. Despite this situation, at most 3 % errors are often acceptable depending on purposes. Again, utilization of about 1/4 of the total datasets brings similar results to those obtained using all the datasets.

4. Summary and concluding remarks

Close relationships between positron annihilation parameters and structural parameters are confirmed for mono- and di-vacancies in $Al_{0.5}Ga_{0.5}N$ and $In_{0.5}Ga_{0.5}N$. It is shown that positron annihilation parameters are predictable by constructing multiple linear regression models with descriptors from the pre-optimized structures and that about 1/4 of the total datasets are sufficient to prepare the training set though there is room for improvement in choosing descriptors in the di-vacancy case. In future, the present study will be extended utilizing nonlinear regression techniques such as the Gaussian process.

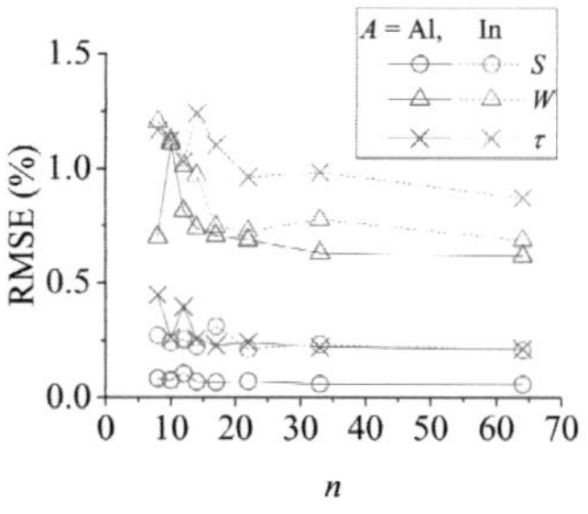

Figure 4. Training-set size dependence of RMSE values of positron annihilation parameters for mono-vacancies.

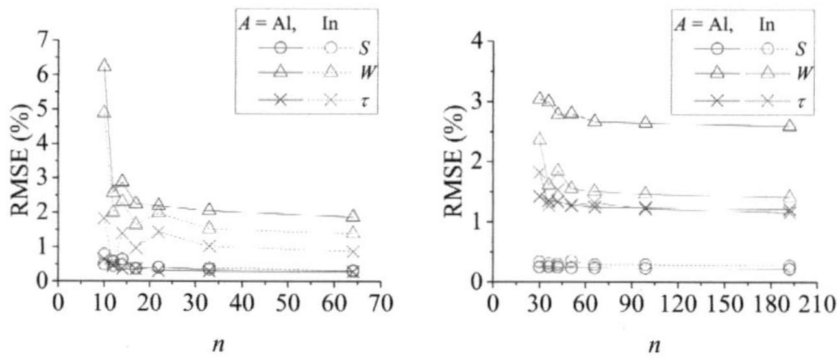

Figure 5. Training-set size dependence of RMSE values of positron annihilation parameters for di-vacancies. The left panel is for the SG1 and the right panel is for the SG2.

Acknowledgments

This work was supported in part by "Materials research by Information Integration" Initiative (MI^{2}I) project of the Support Program for Starting Up Innovation Hub from JST.

References

[1] Uedono A, Ishibashi S, Oshima N and Suzuki R 2013 *Jpn. J. Appl. Phys.* **52** 08JJ02
[2] Ishibashi S and Uedono A 2014 *J. Phys.: Conf. Ser.* **505** 012010
[3] Ishibashi S and Uedono A 2016 *J. Phys.: Conf. Ser.* **674** 012020
[4] Krause-Rehberg R and Leipner H S 1999 *Positron Annihilation in Semiconductors* (Berlin: Springer)
[5] Tuomisto F and Makkonen I 2013 *Rev. Mod. Phys.* **85** 1583
[6] Blöchl P E 1994 *Phys. Rev.* B **50** 17953
[7] Zunger A, Wei S-H, Ferreira L G and Bernard J E 1990 *Phys. Rev. Lett.* **65** 353
[8] R Core Team 2015 *R: A language and environment for statistical computing* (Vienna: R Foundation for Statistical Computing)

Positron beam studies of saponite film prepared by precipitation method

K Numata[1,2], K Sato[1], K Fujimoto[1] and K Ito[3]

[1]Department of Environmental Sciences, Tokyo Gakugei University, 4-1-1 Nukuikita, Koganei, Tokyo 184-8501, Japan.
[2]Tokyo Electric Power Company (TEPCO), 16-46 Kashiwazaki, Niigata 945-0186, Japan.
[3]National Institute of Advanced Industrial Science and Technology (AIST), Tsukuba, Ibaraki 305-8565, Japan.

E-mail: koichiro@u-gakugei.ac.jp

Abstract. Micrometer-sized saponite particles were isolated by means of gravitational sedimentation method applying Stokes' law. The sample prepared with this method was film-like state with a thickness of ~ 30 μm and was semitransparent, indicating that the particles are homogeneously stacked all together. Positron lifetime spectroscopy using a slow positron beam yielded two components for *ortho*-positronium with lifetimes of 1.5 ns and 8.4 ns with the respective relative intensities of 13 % and 10 %. The results suggest that the self-assembly of the larger particles toward structural densification is suppressed for the present film in comparison with the smaller nanoparticles.

1. Introduction

Saponite inorganic layered compound is composed of two-dimensional (2D) nanosheets with angstrom-scale interlayer spaces [1]. The minimum structural unit of 2D nanosheets spontaneously agglomerate with well-defined local structures through their mutual interactions in the time scale more than 10 days [2-3]. This agglomeration process toward structural densification, so-called self-assembly of 2D nanosheets, is increasingly of importance in the field of environmental science [4-6] as well as materials science [1]. Recently, we found two kinds of local molecular structures responsible for self-assembly on the basis of the data of positronium (Ps) lifetime spectroscopy combined with molecular dynamics (MD) simulation [2]. One is the local molecular structure, in which two nanosheets are inserted into the interlayer spaces forming open space with a size of ~ 0.9 nm. There are particular molecular sites such as nanosheet edge and wedge-shaped part inside the open space originating from two-nanosheet insertion, which becomes adsorption sites for Cs and CO_2 molecules [4-5]. The local molecular structure of the two-nanosheet insertion type is however a stable structure exclusively existing at dehydrated state and gradually annihilates along with self-assembly due to hydration. One of possible approaches to preserve the local molecular structure of the two-nanosheet insertion type is suppression of self-assembly. In the present study, saponite particles larger than 2 μm in diameter were isolated by means of gravitational sedimentation method applying Stokes' law for the purpose of suppressing the self-assembly process. Well-assembled saponite particles available after complete hydration were studied by a slow positron beam together with field-emission type scanning electron microscopy (FE-SEM).

2. Experiments

2.1. Sample preparation

A Na-type saponite powder (54.71 % SO_2, 5.02 % Al_2O_3, 0.03 % Fe_2O_3, 30.74 % MgO, 2.15 % Na_2O, 0.07 % CaO, 0.67 % SO_3, 6.64 % H_2O) produced by Kunimine Industries Co. Ltd., Japan was employed in this study. The average particle size is approximately 20 nm in diameter and the powder is apparently white. For isolating saponite particles larger than 2 μm in diameter from the bear powder, gravitational sedimentation method was conducted, the principle of which is detailed below. In fluid, two main forces: gravity and drag force are in principle applied for a particle. Immediately after dispersing into fluid, the particle is approximated to be at rest and no drag force is applied, by which the particle begins to be accelerated only by gravity. With increasing acceleration, the contribution of the drag force in the direction opposite to the particle's motion increases retarding further acceleration. Finally gravity and the drag force are balanced each other yielding the constant velocity, which is known as the terminal velocity under the steady state condition. Assuming the spherical shape of particle, terminal velocity v of the particle in fluid under the steady state condition is described by Stokes' low as

$$v = \frac{g(\rho_s - \rho_w)d^2}{18\mu}, \quad (1)$$

where g, ρ_s, ρ_w, d, and μ are gravity, density of solid, density of water, particle diameter, and molecular viscosity, respectively. It is obvious from equation (1) that the particles with smaller size take long to sediment. For example, the terminal velocity v of the 2-μm particle is calculated to be 1.7×10^{-2} cm/min by putting $g = 980$ cm/s^2, $\rho_s = 2.3$ g/cm^3, $\rho_w = 1.0$ g/cm^3, $d = 2.0 \times 10^{-4}$ cm, and $\mu = 1.0 \times 10^{-2}$ g/cms into equation (1). The present experiments were conducted with the sedimentation distance of 20 cm in the graduated cylinder, which takes 20 h for complete sedimentation. The supernatant solution containing the particles smaller than 2 μm was discarded after 20 h. This process was repeated 5 times so as to enhance the homogeneity of particle size. The residual particles were dried in the thermostatic chamber for a half year and then 1-year hydration was conducted under the atmospheric condition to obtain fully hydrated state.

The sample was film-like state with a thickness of ~ 30 μm and semitransparent unlike the bear powder. The semitransparent nature indicates that the diffuse reflection of visible light from the surface of the particle interface is suppressed. It is thus inferred that the saponite particles with the size larger than 2 μm are more homogeneously stacked all together than those of the bear powder.

2.2. Sample characterization

A series of positron annihilation experiments using slow positron beams were conducted at AIST [7]. Positron annihilation γ-ray spectra were measured by using a ^{22}Na-source-based magnetically-guided beam at various positron incident energies from 80 eV to 20 keV. The γ-ray spectra measured at each positron incident energy were analyzed by taking the S parameter, so that the S parameters as a function of positron incident energy (S-E plot) are obtained. Positron lifetime measurements were carried out by utilizing a pulsed-positron beam. A multi-exponential analysis was applied to the recorded positron lifetime data to deduce *ortho*-Ps (*o*-Ps) lifetimes [8].

3. Results and Discussions

Figure 1 shows the S parameters plotted as a function of positron incident energy for the film sample. The S parameter rapidly increases as positron incident energy increases and becomes saturated above 1 keV, where the information on the bulk of the film sample is dominantly available. The positron lifetime measurements were thus conducted at the positron incident energy of 1 keV. Two components of *o*-Ps lifetimes τ_3 and τ_4 are evaluated on the basis of multi-exponential analysis. Table 1 lists the *o*-Ps lifetimes τ_3 and τ_4 and their relative intensities I_3 and I_4 together with those of saponite nanoparticles [2]. The values of τ_3 and τ_4 are ~ 1.5 ns and ~ 8.4 ns and the relative intensities I_3 and I_4 are ~ 13 % and ~ 10 %. The sizes of open spaces derived from Tao-Eldrup model [9-10] are ~ 2 Å and ~ 6 Å in radius, which are found to be similar to those of nanoparticle one (~ 3 Å and ~ 6 Å). It is thus expected that two species of open spaces in the film sample probed by the present slow positron beam are essentially identical to those in the nanoparticle sample.

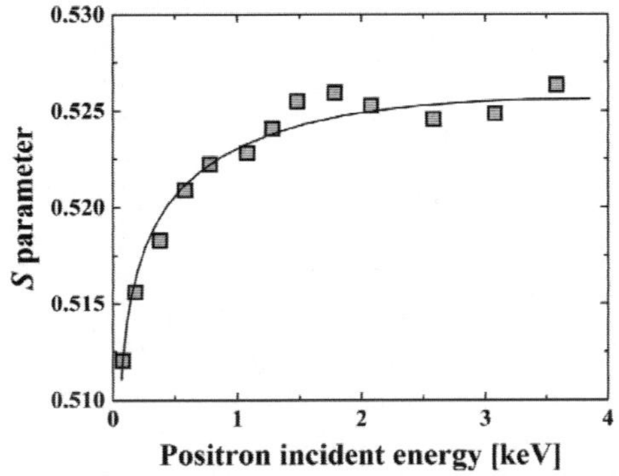

Figure 1. *S* parameters plotted as a function of positron incident energy for the film sample. The solid line are drawn for guiding the eye.

The local molecular structures with open spaces attributable to τ_3 (2.5 ns) and τ_4 (9.1 ns) in saponite nanoparticles have been already revealed based on the results of MD calculations [2]. According to the rheological model with the 2D nanosheets established recently, saponite nanoparticles possess two kinds of local molecular structures of one- and two-nanosheet insertion types before self-assembly. H_2O molecules adsorbed at the Na^+ cations in the interlayer spaces act as a lubricant and are the driving force for the rheological motion of nanosheets parallel to the layer direction. One of the two nanosheets inserted into the interlayer spaces is thus released away. The local molecular structures with the larger open spaces in the two-nanosheet insertion type are gradually altered to those with the smaller open spaces of one-nanosheet insertion type which become dominant for self-assembled saponite nanoparticles as evident in Table 1.

Table 1. The *o*-Ps lifetimes τ_3 and τ_4 and their relative intensities I_3 and I_4. The results of saponite nanoparticles [2] are presented for comparison.

sample	τ_3 [ns]	I_3 [%]	τ_4 [ns]	I_4 [%]
film	1.5	13	8.4	10
nanoparticle	2.5	15	9.1	1

In light of the fact that two kinds of hole structures are obtained for both the film and nanoparticles, they could correspond to the open spaces in the local molecular structures of one- and two-nanosheet insertion types. It is noted here that the relative intensity I_4 for the film sample is much higher than that of the nanoparticle sample, although I_3 for both the sample are comparable to each other. This indicates that the local molecular structure of two-nanosheet insertion type does not disappear after complete hydration (even after 1-year hydration). As mentioned above, the micrometer-sized saponite particles are more homogeneously stacked all together for the present film sample than those of nanoparticle one. It is thus expected that the larger particles are well in-plane ordered in the film sample and highly stable. In such a rigid agglomeration of the large particles, the open spaces in the one- and two-nanosheet insertion type could slightly shrink as suggested from the *o*-Ps lifetimes τ_3 and τ_4 shorter than those of nanoparticle one (see Table 1). In addition to that, the self-assembly toward structural densification could be suppressed compared with that in the nanoparticle one. The local molecular structure of two-nanosheet insertion type has been found to gradually disappear along with hydration in spite of usefulness for adsorbing both Cs and CO_2 molecules [5]. It is thus beneficial to demonstrate that the local molecular structure of two-nanosheet insertion type is preserved after complete hydration for the present film sample.

4. Conclusion

Micrometer-sized saponite particles were isolated by means of gravitational sedimentation method applying Stokes' law. The sample prepared with this method was film-like state with a thickness of ~ 30 μm and semitransparent, indicating that the particles are homogeneously stacked all together. Positron lifetime spectroscopy using a slow positron beam yielded two components for *o*-Ps with lifetimes of 1.5 ns and 8.4 ns with the intensities of 13 % and 10 %, respectively. Considering the similarity of open spaces in size to those of nanoparticle samples, they could correspond to the open spaces in the local molecular structures of one- and two-nanosheet insertion types. The relative intensity I_4 for the film sample was found to be much higher than that of nanoparticles, demonstrating that the local molecular structure of two-nanosheet insertion type does not disappear after complete hydration. The results suggest that the self-assembly of the larger particles toward structural densification is suppressed for the present film in comparison with the smaller nanoparticles.

Acknowledgements

This work was partially supported by a Grant-in-Aid of the Japanese Ministry of Education, Science, Sports and Culture (Grant Nos. 15K05308 and 16K05394).

References

[1] Sato K, Numata K, Dai W and Hunger M 2014 *Appl. Phys. Lett.* **104** 131901
[2] Sato K, Fujimoto K, Kawamura K, Dai W and Hunger M 2012 *J. Phys. Chem. C* **116** 22954
[3] Sato K, Numata K, Dai W and Hunger M 2014 *Phys. Chem. Chem. Phys.* **16** 10959
[4] Sato K, Fujimoto K, Dai W and Hunger M 2013 *J. Phys. Chem. C* **117** 14075
[5] Sato K, Fujimoto K, Dai W and Hunger M 2016 *J. Phys. Chem. C* **120** 1270
[6] Numata K, Sato K and Fujimoto K 2012 *Inter. J. Nanosci.* **11**, 1240034
[7] Kobayashi Y, Kojima I, Hishita S, Suzuki T, Asari E and Kitajima M 1995 *Phys. Rev. B* **52** 823
[8] Kirkegaard P and Eldrup M 1974 *Comput. Phys. Commun.* **7** 401
[9] Tao S J 1972 *J. Chem. Phys.* **56** 5499
[10] Eldrup M, Lightbody D, Sherwood J N 1981 *Chem. Phys.* **63** 51

Antifouling enhancement of polysulfone/TiO_2 nanocomposite separation membrane by plasma etching

Z Chen[1], C Yin[1], S Wang[1], K Ito[2], Q M Fu[1], Q R Deng[1], P Fu[1], Z D Lin[1] and Y Zhang[1]

[1]Hubei Key Laboratory of Plasma Chemical and Advanced Materials & Department of Material Science and Engineering, Wuhan Institute of Technology, Wuhan, 430073, China

[2] National Institute of Advanced Industrial Science and Technology (AIST), Tsukuba, Ibaraki 305–8565, Japan

E-email: zhezchen@gmail.com

Abstract. A polysulfone/TiO_2 nanocomposite membrane was prepared via casting method, followed by the plasma etching of the membrane surface. Doppler broadened energy spectra vs. positron incident energy were employed to elucidate depth profiles of the nanostructure for the as-prepared and treated membranes. The results confirmed that the near-surface of the membrane was modified by the plasma treatment. The antifouling characteristics for the membranes, evaluated using the degradation of Rhodamin B, indicated that the plasma treatment enhances the photo catalytic ability of the membrane, suggesting that more TiO_2 nanoparticles are exposed at the membrane surface after the plasma treatment as supported by the positron result.

1. Introduction

Separation membranes are widely applied to the water filtration. During the filtration process for organic wastewater treatments, a fouling on the membrane surface due to the adsorption of organic pollutants is a major issue, which reduces the water flux, increases the energy consumption, and shortens the service lifetime of the membrane. According to the conventional filtration blocking theory [1], the fouling mechanism includes the following models: (1) standard blocking (gradual pore constriction), (2) complete blocking (sealing of pore entrances), (3) intermediate blocking (sealing and accumulation on membrane surface), and (4) cake filtration. All the above models indicate the membrane fouling is related to the accumulation of foulants on the membrane.

Introducing photo-active nanoparticles, capable to transfer the organic foulant to CO_2 and H_2O, is a promising way to provide a photo catalytic ability to the membranes. Yang, et al. have reported that polysulfone/TiO_2 composite membranes exhibit excellent water permeability, hydrophilicity and good antifouling ability [2]. For the prepared membrane, most TiO_2 nanoparticles, responsible to the photo-catalytic antifouling ability, are wrapped in the bulk. Thus, the antifouling ability of the membrane may be enhanced by exposing more TiO_2 nanoparticles at the membrane surface.

In this work, a polysulfone/TiO_2 nanocomposite membrane was prepared via casting method, followed by the plasma etching of the membrane surface. The difference in photocatalytic activity between the two membranes was tested in order to examine the effect of the plasma treatment on the antifouling ability. Positron annihilation γ-ray spectroscopy was employed to probe depth profiles of the nanostructure for the as-prepared and modified membranes. The obtained results are discussed in comparison with the antifouling property of the membranes.

2. Experimental

TiO_2 nanoparticles with a nominal average size of ~30 nm were purchased from Degussa. The nanoparticles were dispersed into N-methyl-2-pyrrolidinone (NMP) to obtain a 1-wt% suspension, followed by dissolving a proper amount of polysulfone in the suspension with stirring and ultrasonic vibration to obtain a homogeneous suspension with 19-wt% polysulfone and 1-wt% TiO_2. The prepared suspension was then poured at room temperature onto a glass plate with a 800-μm-thick blade. After exposing to air for 30 s, the suspension on the glass was immersed in a water bath at ~25 °C for 1 day to remove the remaining solvent from the membrane.

A plasma treatment was performed in a RF-plasma reactor (YZD08-2A) purchased from Tangshan Yanzhao technical institute. The RF discharge was excited with plain parallel electrodes placed in about 50 mm distance. The treatment with placing the membrane on one electrode was carried out for 5 min with an RF power of 80 W and an operating argon pressure of ~ 10 Pa.

Positron annihilation γ-ray spectra for the membranes were measured with a ^{22}Na-based, magnetically-guided positron beam at AIST. The line-shape *S* parameter was determined as the ratio of the counts appearing in the central region (510.3 keV–511.7 keV) to the total counts of the 511 keV annihilation photo peak (506.8 keV–515.2 keV) for each spectrum recorded with a high-purity Ge detector. The mean positron implantation depth *z* [nm] is estimated by the following formula [4]

$$z = \frac{40}{\rho} E^{1.6} \qquad (1)$$

where *E* is the incident positron energy in keV and ρ is the density of the measured membranes in g*cm^{-3}.

The photocatalytic activities of the present membranes were evaluated by monitoring the degradation due to the irradiation of a 300-W Xe light to a rhodamin B (RhB) solution which was obtained by immersing a 2-cm-2-cm-sq. membrane sheet into 500-ml RhB solution. The distance between the Xe lamp and the RhB solution was about 15 cm. The reaction apparatus for the irradiation was put in a water bath at ~30 °C. After stirring in darkness for 30 min to attain the adsorption-desorption equilibrium, the Xe lamp was turned on. A 2-ml solution was sampled every 30 min, followed by quantifying the remained RhB in the solution using a UV-vis spectrometer. A blank experiment was carried out without the photocatalyst, as a result, the photolysis effect for RhB was confirmed to be negligible, which ensures the stability of RhB.

3. Results and Discussion

3.1. Positron annihilation γ-ray spectroscopy

Figure 1 shows the variation of the *S* parameter for the membranes before and after plasma treatment as a function of positron incident energy *E*. For both the membranes, the *S*(*E*) curves exhibit similar shape, namely *S* increases with increasing *E* from 80 eV to 2 keV, and it reaches a flat stage at $E > 2$ keV. In comparison with the untreated membrane, the *S*(*E*) curve for the treated one exhibits a distinct decrease in the *S* values at the respective *E* less than 1 keV, corresponding to a mean implantation depth around 40 nm.

For polyolefin systems, *S* is generally related to the Ps formation probability and the momentum of the electrons belonging to the constituents of the polymer chains. Both of these two factors are significantly influenced by the chemical structure of the material. According to the spur reaction model [5], the Ps formation is partially inhibited by positron trapping and annihilation with polar groups in the matrix reducing the S parameter. In light of this, the lower S at the near-surface region for the present modified membrane suggests inhibition of the Ps formation due to polar groups or more metal oxide nanoparticles at the near-surface region. Moreover, the difference in S between the treated and untreated membranes becomes significant at $E < 1$ keV, corresponding to a mean implantation depth of < 40 nm. This indicates that the chemical structure of the treated membrane at the near-surface region is different from that for the untreated one.

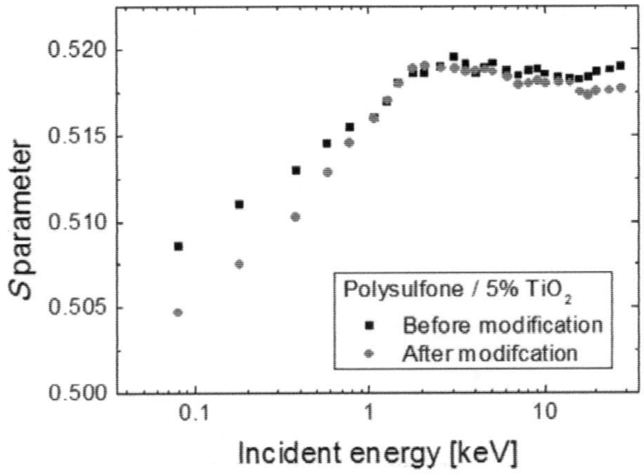

Figure 1. Variation of S parameter for the samples (membrane as obtained and membrane after modification) as a function of positron incident energy.

3.2. Photo catalytic ability for aqueous RhB

Figure 2 shows the variation of the relative RhB concentration C/C_0 in the immersing solutions for the membranes before and after the plasma treatment as a function of Xe irradiation time. C/C_0 for both the membranes monotonously decreases with increasing the irradiation time, in which the reduction rate of C/C_0 for the treated membrane is significantly higher than that for the untreated membrane, This indicates that the plasma treatment effectively enhances the photocatalytic activity, implying that more TiO_2 nanoparticles are indeed exposed on the membrane surface due to the plasma etching, as expected by the result from the S parameter measurements.

The cycling performance of the treated membrane was further investigated to evaluate the photocatalysis stability. Figure 3 shows the repetition test of the photocatalytic activity for the treated membrane. As shown in the figure, even after several cycles for the photo degradation test of RhB, the treated membrane does not exhibit any significant loss of photo catalytic activity, indicating that the polysulfone/TiO_2 membrane after the plasma treatment is stable enough for the photocatalytic oxidation of the pollutant molecules.

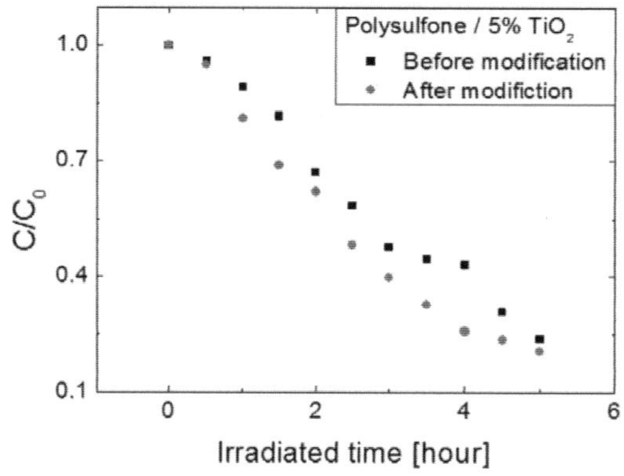

Figure 2. Variation of the relative RhB concentration in the immersing solutions for the membranes before and after the plasma treatment as a function of Xe irradiation time.

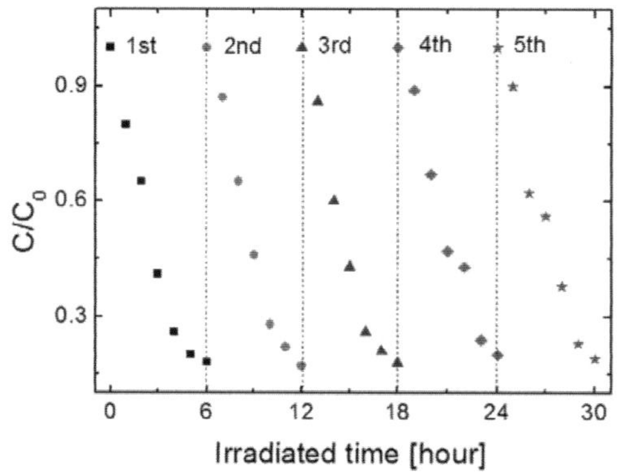

Figure 3. Repetition test of the photocatalytic activity for the treated membrane.

4. Summary

TiO_2 nanoparticles were introduced into a polysulfone membrane to attain the photocatalytic activity for antifouling, and the membrane near-surface was etched by the plasma treatment. A comparison of the photo degradation of RhB between after and before the plasma treatment showed that the photocatalytic activity is effectively enhanced, implying that more TiO_2 nanoparticles are exposed on the membrane surface, as expected by the result from the *S* parameter measurements.

Acknowledgment
This work was supported by NSFC11205118, 11404248 and 11504277.

References

[1] Mondal S and De Sir S 2010 *Sep. Purif. Technol.* **75** 222
[2] Yang Y, Zhang H, et. al. 2007 *J. Membrane Sci.* **288** 231
[3] Chen Z, Wang Z, et.al. 2014 *Jpn. J. Appl. Phys. Conf. Proc.* **2** 011202
[4] Kobayashi Y, Ito K, et al. 2008 *Appl. Surf. Sci.* **255** 174
[5] Mogensen O E 1974 *J. Chem. Phy.* **60** 998

Degradation of electron-irradiated polyethylene studied by positron annihilation lifetime spectroscopy

K Onodera[1], T Oka[1,2], Y Kino[1] and T Sekine[1,2]

[1] Department of Chemistry, Tohoku University, Sendai, Miyagi 980-8578, Japan
[2] Institute for Excellence in Higher Education, Tohoku University, Sendai, Miyagi 980-8576, Japan

E-mail: karin.onodera.t3@dc.tohoku.ac.jp

Abstract. Degradation of electron beam irradiated high-density polyethylene was studied by positron annihilation lifetime spectroscopy (PALS), micro-FT-IR, and gel fraction measurements. The obtained results indicated that *ortho*-positronium intensity is influenced not only by the irradiation but also the post oxidation, which illustrates that PALS may be a promising tool to monitor/evaluate the degradation of polyethylene induced by irradiation and long-term storage.

1. Introduction

Polyethylene (PE) is widely used for insulators and packaging materials in various radiation environments such as nuclear power plants and accelerator facilities [1]. During long-term usage, PE which is subjected to a severe radiation environment is degraded mainly through oxidation, cross-linking, and scission. The degradation by radiation has been studied because it has to be properly evaluated for ensuring the safety of the relevant facilities. In this work, we investigated the structural change of electron beam irradiated PE during irradiation and after long-term storage, and discuss the usability of positron annihilation lifetime spectroscopy (PALS) for the degradation assessment.

2. Experimental

Additive free high-density polyethylene (HDPE), having a density of 0.951 g cm^{-3} and a thickness of 2 mm, were examined. The samples were irradiated by an electron beam with up to 1000 kGy under vacuum at room temperature at Takasaki Advanced Radiation Research Institute, National Institutes for Quantum and Radiological Science and Technology (QST). The dose rate, the acceleration voltage, and the beam current were 10 kGy $pass^{-1}$ (2.23 m min^{-1}), 2 MV, and 2 mA, respectively. The irradiated samples were stored under vacuum at room temperature, followed by exposure to air just before the first measurement.

The irradiated samples were immersed in *p*-xylene at ~140 °C for 24 h and were dried in air so that the sol fraction in the samples is removed. The gel fraction was calculated by the equation,

$$\text{gel fraction (\%)} = (W/W_0) \times 100,$$

where W and W_0 are the sample weights after and before the above treatment, respectively.

The amount of carbonyl groups induced by the irradiation was evaluated by micro-FT-IR spectroscopy. The details of measurement are described elsewhere [2].

Positron lifetime spectra were recorded with a digital storage oscilloscope by determining the time interval between the detection of the 1.27-MeV γ-ray from the β^+ decay of the ^{22}Na positron source and one of the 0.511-MeV annihilation γ-rays [3]. In the measurements, the positron source, sealed with 7.5-μm-thick Kapton foils, was sandwiched between two samples taken from identical preparation lots. The obtained spectra were analyzed assuming three exponential components to deduce the lifetime τ_3 and relative intensity I_3 of the longest-lived *ortho*-positronium (*o*-Ps) component by the PALSfit program [4]. The time resolution of the system was ~200 ps full width at a half maximum [3]. Correction was made for the positrons annihilating in the source material. The validity of our system was confirmed with certified reference materials NMIJ CRM 5601-a and NMIJ CRM 5602-a provided by National Metrology Institute of Japan, National Institute of Advanced Industrial Science and Technology (AIST) [5,6].

3. Results and discussion

The variations of the gel fraction for the HDPE samples taken just after the irradiation and stored for 1 year as a function of dose are shown in Figure 1. For both the samples the gel fraction increases with increasing dose, possibly ascribed to the increased amount of the produced radicals along with dose. As for the sample after 1-year storage, the gel fraction with a dose lower than 500 kGy significantly reduced from each value for the sample taken just after the irradiation. This implies that the subsequent chain-scission process in those samples with doses lower than 500 kGy has a considerable effect on the resultant molecular weight as compared to the samples with higher doses. Irradiation of HDPE produces two types of radicals, that is, alkyl and allyl radicals [7]. The alkyl radicals are unstable at room temperature, therefore most of them instantly disappear, followed by the production of cross-linking or carbonyl groups with oxygen remaining in the polymer matrix during the irradiation. On the other hand, the allyl radicals are relatively stable under vacuum at room temperature. Hence, they are expected to survive until the samples were exposed to air so that they induce to form hydroperoxides, otherwise they also induce to form carbonyl groups during the irradiation [7,8].

Figure 2 shows the depth distribution of the carbonyl absorbance at 1716 cm^{-1} for 500-kGy-irradiated HDPE. After storage beyond half a year, the absorbance near both surfaces of the sample drastically increased, indicative of an enhanced amount of carbonyl groups. With further storage over 1 year the absorbance around 1.0 mm in depth was further increased. Through thermolytic and/or photolytic

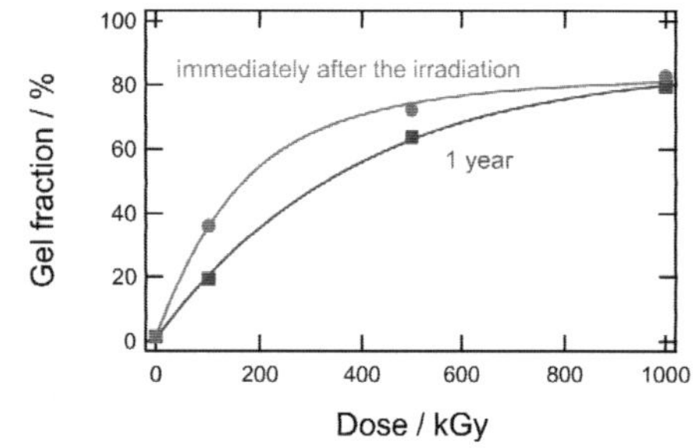

Figure 1. Relationship between the gel fraction and dose for HDPE. The circle and square plots denote the gel fraction measured immediately after the irradiation and 1 year from the irradiation, respectively. Solid lines are drawn to guide the eyes.

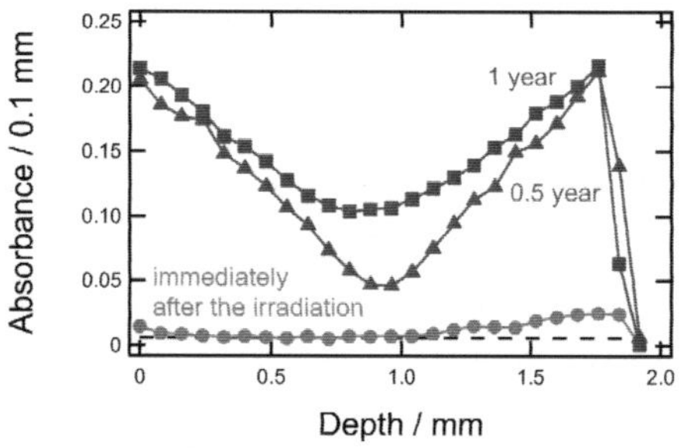

Figure 2. Depth distribution of the carbonyl groups for 500-kGy-irradiated HDPE. The dashed line denotes the absorbance of unirradiated HDPE. The circle, triangle, and square symbols denote the absorbance measured immediately after the irradiation, 0.5 year from the irradiation, and 1 year from the irradiation, respectively.

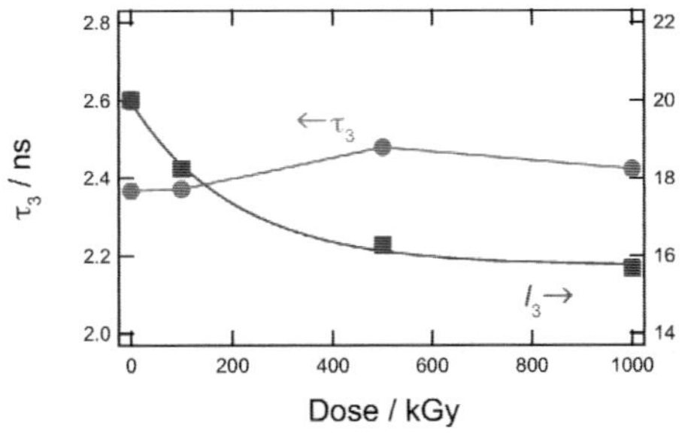

Figure 3. Plots of the *o*-Ps lifetime τ_3 (circle symbols, left-hand axis) and the *o*-Ps intensity I_3 (square symbols, right-hand axis) for HDPE vs dose. Data are taken immediately after the irradiation. Solid lines are drawn to guide the eyes.

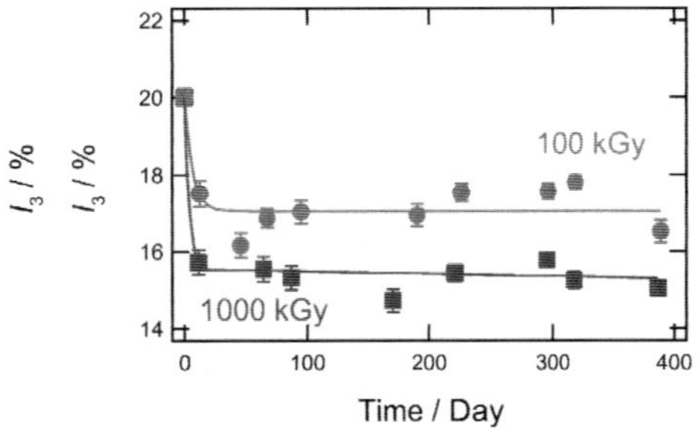

Figure 4. Variations of I_3 during the long-term storage for HDPE. The circle and square symbols denote 100-kGy- and 1000-kGy-irradiated samples, respectively. Solid lines are drawn to guide the eyes.

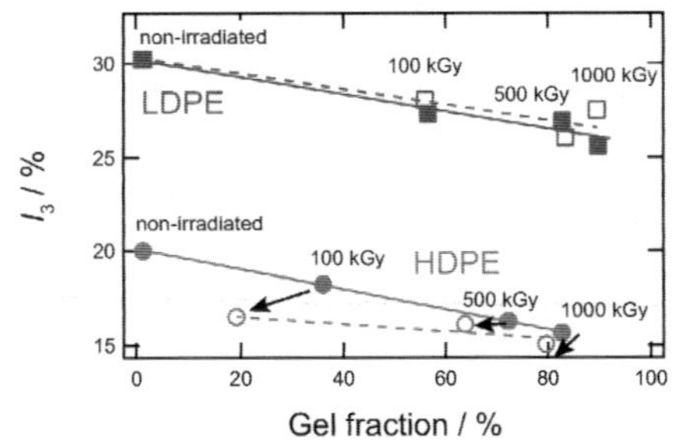

Figure 5. Relationship between I_3 and the gel fraction measured immediately after the irradiation (filled symbols) and 1 year from the irradiation (open symbols). The circle and square symbols denote HDPE and LDPE, respectively. Solid and dotted lines are drawn to guide the eyes. The arrows represent the respective directions toward long-term storage for 1 year.

reactions hydroperoxide produces carbonyl groups as well as alternative radicals [7–9]. These radicals lead to a cyclic oxidation process during long-term storage, gradually increasing the amount of carbonyl groups. This cyclic oxidation process induces polymer-main-chains scission, resulting in reduced molecular weight, that is, a decrease in gel fraction as evident in Figure 1.

Figure 3 shows the variation of the *o*-Ps lifetime τ_3 and its relative intensity I_3 measured immediately after the irradiation as a function of dose. τ_3 was 2.38 ± 0.02 ns and the free-volume hole size, estimated based on the Tao-Eldrup model [10–12], was ~0.64 nm in diameter. I_3 significantly decreases with increasing dose, whereas τ_3 or the free-volume hole size does not. Figure 4 shows the time dependence of I_3 for 100-kGy- and 1000-kGy-irradiated HDPE samples. While no time dependence for τ_3 was observed, I_3 for 100-kGy-irradiated HDPE levels off around 17 % whereas I_3 for 1000-kGy-irradiated HDPE slightly decreases from 16 % to 15 % with increasing time.

According to the spur model, Ps forms in the terminal spur as a result of positron-electron recombination [13]. Ito *et al.* reported that the thermalized positron can diffuse over a distance of an order of 10^2 nm in PE before it annihilates into 2γ-rays [14]. This diffusion length is long enough for the positron to be trapped by a low concentration of oxygen-containing polar groups formed by thermal oxidation. Thus part of the positrons may not reach an electron to form Ps provided that a fraction of them are trapped by polar groups. For the present system, the produced radical could also be a positron trapping site [15, 16], in addition to the polar carbonyl groups involved with the cyclic oxidation process from the hydroperoxides. Thus these trapping sites of the polar groups and radicals possibly cause a reduction in the Ps formation probability, that is, I_3.

Finally, we discuss the relationship between Ps formation and the gel fraction. Figure 5 shows a plot of I_3 vs the gel fraction, obtained for the samples just after the irradiation and stored for 1 year from the irradiation, in which the result for low-density polyethylene (LDPE) with a density of 0.920 g cm^{-3} is also included. In LDPE alkyl and allyl radicals are produced by irradiation as in HDPE, but their lifetimes are much shorter than in HDPE [7,8]. Hence almost of them are eliminated during the irradiation, so that the hydroperoxides and subsequent carbonyl-group formation in LDPE should be much less than that in HDPE. Interestingly, I_3 for both LDPE and HDPE linearly decreased with increasing gel fraction. The reduction in I_3 for the samples taken just after irradiation is probably ascribed to radicals that survive until the measurement. After long-term storage, the relationship for HDPE deviates from the solid line to the dashed line. In light of the FT-IR and the gel fraction results, this shift is due to the reduction of I_3 and main chain scission along with long-term oxidation. On the other hand, no deviation was observed for LDPE because there is only a small increase in the carbonyl groups along with the chain scission. Note that I_3 may be influenced by both polar groups and radicals. It is necessary to consider these effects independently for better understanding of the irradiation effect and the long-term degradation of polyethylene.

4. Conclusion

Structural changes of electron beam irradiated HDPE during irradiation and long-term storage were investigated by micro-FT-IR, gel fraction measurement, and PALS. Significant changes of τ_3 were not observed whereas I_3 decreased not only with dose but also long-term storage. Carbonyl groups suppress the formation of Ps, therefore I_3 decreases with the formation of carbonyl groups. Furthermore, I_3 seems to correlate with the gel fraction for both HDPE and LDPE immediately after irradiation. Although further study is required to understand the relationship between I_3 and the gel fraction, the obtained results demonstrated that I_3 measured by PALS can be useful as an indicator of the structural change induced by irradiation and long-term storage in PE.

Acknowledgments

We thank Drs. N. Nagasawa, S. Saiki (QST), and T. Hirade (Japan Atomic Energy Agency) for the electron beam irradiation and their valuable comments and suggestions. The authors also thank Dr. K. Ito (AIST) for providing the data accumulation/analysis software for PALS and for useful discussions and expert advice on the interpretation of the data.

References

[1] Ohki Y, Yamada T and Hirai N 2013 *IEEE Trns. Dielectr. Electr. Insul.* **20** 2099
[2] Hama Y, Oka T, Kudoh H, Sugimoto M and Seguchi T 2003 *Nucl. Instrum. Methods. Phys. Res. B* **208** 123
[3] Oka T, Sano Y, Kino Y and Sekine T 2014 *Eur. Phys. J. D.* **68** 156
[4] Kirkegaard P and Eldrup M 1972 *Compt. Phys. Commun.* **3** 240
[5] Ito K *et al.* 2008 *J. Appl. Phys.* **104** 3
[6] Ito K *et al.* 2008 *Mater. Sci. Forum* **607** 248
[7] Dole M 1972 The Radiation Chemistry of Macromolecules Volume 1, ed M Dole (New York: Academic Press) chapter 14 pp 335–347
[8] Hama Y, Oka T, Uchiyama J, Kanbe H, Nabeta K and Yatagai F 2001 *Radiat. Phys. Chem.* **62** 133
[9] Oka T *et al.* 2007 *Polymer Durability and Radiation Effects*, ed M C Celina and R A Assink (Washington, DC: Oxford University Press)
[10] Tao S J 1972 *J. Chem. Phys.* **56** 5499
[11] Eldrup M, Lightbody D and Sherwood J N 1981 *Chem. Phys.* **63** 51
[12] Nakanishi H and Jean Y C 1988 *Positron and Positronium Chemistry*, ed D M Schrader and Y C Jean (New York: Elsevier) chapter 5 pp 159–192
[13] Mogensen O E 1995 *Positron Annihilation in Chemistry* (Berlin: Springer)
[14] Ito K, Kobayashi Y and Nanasawa A 2003 *Appl. Phys Lett.* **82** 654
[15] Kobayashi Y, Zheng W, Hirata K and Suzuki T 1997 *Radiat. Phys. Chem.* **50** 589
[16] Sato K, Shanai D, Hotani Y, Ougizawa T, Ito K, Hirata K and Kobayashi Y 2006 *Rhys. Rev. Lett.* **96** 228302

Nanopore structure relevant to the D_2O permeation into silica thin films as studied by secondary ion mass spectrometry, ellipsometric porosimetory and positron annihilation

S Yoshimoto[1,3], K Ito[2], H Hosomi[1], T Nakamura[1] and M Takeda[1]

[1]Toray Research Center, Inc., Otsu, Shiga 520–8567, Japan
[2]National Institute of Advanced Industrial Science and Technology, Tsukuba, Ibaraki 305–8565, Japan
[3]Graduate School of Engineering, Hiroshima University, Higashi-Hiroshima, Hiroshima 739–8527, Japan

E-mail: Shigeru_Yoshimoto@trc.toray.co.jp

Abstract. Subnanoscopic pore structure relevant to the water permeation for two types of silica thin films, fabricated by plasma-enhanced chemical vapor deposition (PECVD) and thermal oxidization (TO), were examined by means of dynamic secondary ion mass spectrometry (D-SIMS), vapor-adsorption ellipsometric porosimetry (EP), and low-energy positron annihilation spectroscopy (PALS). The D^- secondary ion intensity for the PECVD film, observed using D-SIMS, was much higher than for the TO film, indicative of the higher permeance of D_2O molecules in the matrix of the PECVD film. The results from PALS and EP suggest that this is ascribed to the larger pores and the higher open porosity of the PECVD film as compared to the TO film.

1. Introduction

Silica thin film has been widely utilized as the top coating layer of various materials in the nanotechnology fields, such as the barrier layer of the display for electronic devices and the active layer of separation membranes. Better understanding of molecular permeation in silica films is necessary to develop innovative materials. In this work we examine the pore structure relevant to the water permeation for two types of silica thin films by means of dynamic secondary ion mass spectrometry (D-SIMS), vapor-adsorption ellipsometric porosimetry (EP), and low-energy positron annihilation spectroscopy (PALS).

2. Experimental

2.1. Sample preparation

Two types of silicon oxide films were prepared through a plasma-enhanced chemical vapor deposition (PECVD) and a thermal oxidation (TO), respectively. The PECVD film was fabricated on a silicon wafer at 400 °C from a mixture of tetraethyl orthosilicate (TEOS) and O_3 diluted in a helium gas. The TO film was grown on a silicon wafer at 950 °C in a mixture of H_2 and O_2 gas sources. The thickness, the refractive index and the density for the obtained films are listed in table 1. FT-IR spectra for both films are shown in figure 1, and the enhanced IR absorption over 3000 cm^{-1} to 3800 cm^{-1} for the PECVD film is attributable to the stretching vibration of OH groups [1], indicating that hydrophilic terminal silanols are introduced more into the silicon oxide network of the PECVD film. For depth profiling of D^- ions

Table 1. Thickness L, refractive index n_{630} at a wavelength of 630 nm and density ρ for the present films.

	L^a [nm]	n^a_{630}	ρ^b [g/cm^3]
PECVD film	90.0	1.427	2.10
TO film	99.4	1.456	2.29

[a]Determined by spectroscopic ellipsometry.
[b]Determined by x-ray reflectometry.

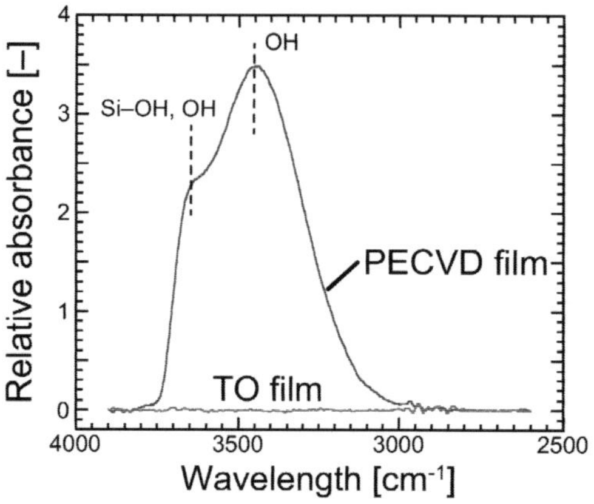

Figure 1. FT-IR spectra for the present films.

due to D_2O permeated from the film surface, both the films were exposed to deuterium oxide (D_2O: 99.9 %, Wako Japan) vapor at 70 °C for 3 hours in a thermostat chamber.

2.2. Dynamic secondary ion mass spectrometry (D-SIMS)
D-SIMS was conducted at 25 °C in a ultra high vacuum chamber (2.7×10^{-8} Pa) using a 50 nA, 3 keV Cs^+ primary ion beam incident at 60° with respect to the sample normal, and the actual sputtered area was 300×600 μm^2. In the measurement, negative secondary ions including D^- were detected for the present films exposed to D_2O vapor. Charge neutralization was induced through 0.5 keV electron beam irradiation to the vicinity of the sputtered area. The depth scale was obtained by calibrating the retention time at the film-substrate interface to the respective film thicknesses.

2.3. Ellipsometric porosimetry
Adsorption isotherms of water and methanol (MeOH) at 26 °C for the present films were elucidated using the specially-designed flow-type ellipsometric porosimeter at AIST [2]. Prior to the adsorption measurements, the samples were heated at 300 °C under a nitrogen gas flow to remove impurities adsorbed on the film surface. Ellipsometric parameters (Δ, Ψ) with various wavelengths λ were measured with varying the flow rate ratio f_r, proportional to the adsorbate concentration. The flow rate ratio f_r is defined by $f_s/(f_d + f_s)$, where f_s and f_d represent the flow rates for a nitrogen gas saturated by the adsorbate vapor and a dry nitrogen gas, respectively. The observed parameters (Δ, Ψ) were analyzed based on an ambient-film-substrate three-layers model under the assumption of the Cauchy equation [3], describing the refractive index n as a function of λ. In order to obtain the isotherms, the overall refractive index n_o at $\lambda = 630$ nm upon successive adsorption of the adsorbates were elucidated as a function of f_r. In the case of water, the actual relative humidity was adapted for the parameter instead of f_r.

Based on the Lorentz–Lorenz equation, $A\rho = (n^2 - 1)/(n^2 + 2)$ (A is a constant and ρ is the density), the specific amounts of the adsorbate in a film are assumed to be proportional to the volume fraction occupied by the adsorbate molecules V, quantified by the following equation

$$V = \left[\left(\frac{n_o^2 - 1}{n_o^2 + 2}\right) - \left(\frac{n_f^2 - 1}{n_f^2 + 2}\right)\right] \Big/ \left(\frac{n_a^2 - 1}{n_a^2 + 2}\right) \quad . \tag{1}$$

Here, n_f and n_a were fixed to the refractive indices for the corresponding films with $f_r = 0$ and for the bulk adsorbates (water: 1.331 and MeOH: 1.328). By using this equation with the measured n_o, the adsorption isotherms for the present films were obtained as the variation of V as a function of f_r.

2.4. Low-energy positron annihilation lifetime spectroscopy

Positron lifetime measurements were carried out by utilizing a ^{22}Na-based pulsed-positron beam system (Fuji-Imvac, Japan). The measurements were performed at room temperature in vacuum. For each measurement 5 million counts of the annihilation events were accumulated. A multi-exponential analysis was applied to the recorded lifetime data to deduce the long-lived *ortho*-positronium (*o*-Ps) lifetime τ. The pore radius r [nm] was estimated from the obtained τ [ns] based on the following equation [4, 5],

$$\tau = 0.5\left[1 - \frac{r}{r+0.166} + \frac{1}{2\pi}\sin\left(\frac{2\pi r}{r+0.166}\right)\right]^{-1} . \quad (2)$$

3. Results and discussion

Figure 2 shows the depth profile of the D^- intensities for the PECVD and TO films, exposed to D_2O vapor. The D^- intensity for the PECVD film is almost constant throughout the film, while that for the TO film slightly decreases from the near surface to shallow depths, and becomes constant deeper than ~30 nm with a value which is about 10^3 times lower than for the PECVD film. This indicates that in comparison with the TO film a larger amount of D_2O molecules are homogeneously distributed inside the film after the exposure to D_2O vapor, and D atoms remaining even in a high vacuum were detected by the D-SIMS measurements. On the other hand, the gradual decrease in D^- intensity for the TO film suggests a transitional state of D_2O permeation from the film surface into bulk. Based on the profile for the TO film, the diffusion coefficient is estimated to be of the order of 10^{-20} m^2/s, comparable to the value of water in the bulk silica at 60 °C (3.1×10^{-20} m^2/s) [6].

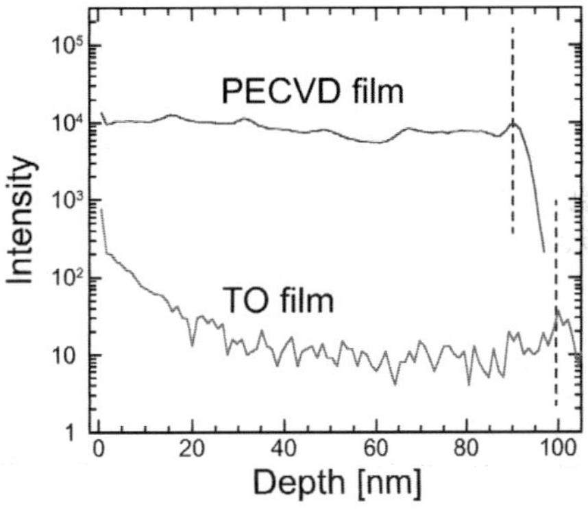

Figure 2. Depth profile of the D^- intensities for the PECVD and TO films obtained by D-SIMS. The vertical broken lines represent the respective boundaries between the film and the substrate.

Figure 3 shows the adsorption isotherms of water and MeOH for the two films. Whereas both the isotherms for the TO film show little amount of the adsorption, those for the PECVD film increase with increasing relative humidity and flow rate ratio, where V at the higher concentration of water and MeOH approach, respectively, 8.0 % and 1.6 %, defined as the total adsorption on the pores (V_t). This illustrates that a significant amount of open porosity is formed in the PECVD film. From this result, the higher D^- intensity for the PECVD film in figure 2 can be ascribed to the larger surface area in comparison with the TO film, on which more D_2O adsorbs, possibly followed by the substitution of hydrogens at the pore surface by deuteriums.

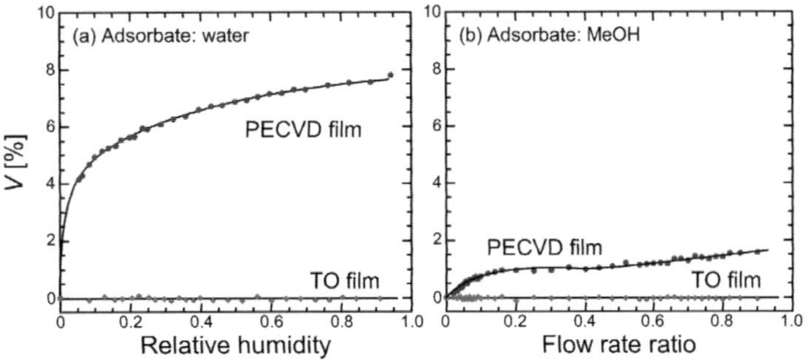

Figure 3. Adsorption isotherms at 26 °C for the present films. (a) H_2O and (b) MeOH.

The difference in V_t for water and MeOH adsorbed on the PECVD film suggests that the subnanoscaled pores, connected to the film surface, somewhat influences the diffusion of those adsorbates

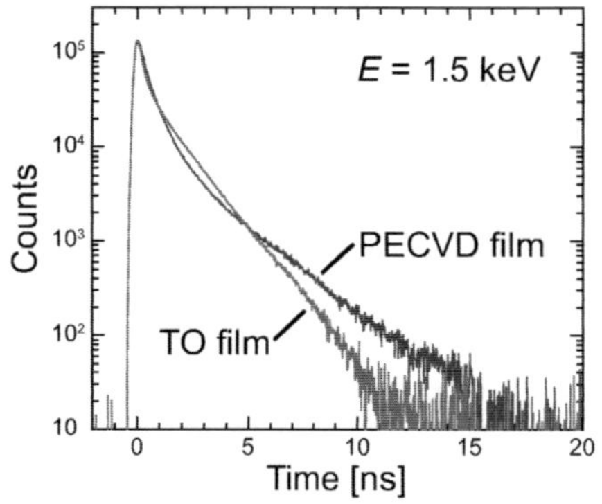

Figure 4. PALS data for the present films (corrected for the background).

Table 2. o-Ps lifetime τ and pore radius r_p estimated from equation 2 for the present films.

	τ [ns]	r_p [nm]
PECVD film	2.35	0.32
TO film	1.48	0.23

in the film. In order to evaluate pore sizes for the films, PALS spectra were observed at an incident positron energy E of 1.5 keV (figure 4), in which the long-lived o-Ps component is observable for both the films. The mean pore radius r_p for the PECVD film estimated from the o-Ps lifetimes τ with equation (2) is 0.32 nm (table 2). Based on the respective van der Waals radii for water (r_{water} = 0.14 nm) and MeOH (r_{MeOH} = 0.21 nm) [7], the mean radius is larger in the order of $r_p > r_{MeOH} > r_{water}$. Thus we can reasonably expect that the diffusion of MeOH molecules is inhibited more than that of water due to the bottlenecking pores, belonging to the smaller region of the pore size distribution, which leads to the reduced V_t for the MeOH adsorption in comparison with the water adsorption.

4. Summary

Subnanoscopic pore structure relevant to the molecular permeation for two kinds of the silica thin films was examined by means of D-SIMS, EP, and PALS. The D^- secondary ion intensity, originating from D_2O, for the PECVD film was much higher than for the TO film, indicative of the higher permeance of D_2O molecules in the former silica film. In light of the results from EP and PALS, it was demonstrated that significant amounts of open porosity in the subnanoscopic range are formed in the PECVD film and that bottlenecking pores with a radius smaller than ~0.3 nm may partly prevent MeOH from diffusing into interconnected pores in the film.

Acknowledgments

Part of this work was supported by JSPS KAKENHI Grant Number 15H02313. The authors thank T. Mino, A. Suzuki, Y. Tanahashi, and H. Seki (Toray Research Center) for their assistance in this study, and also thank Dr. Brian O'Rourke for careful proofreading of this manuscript. SY thanks Prof. Toshinori Tsuru for his encouragement to conduct this study.

References

[1] Davis K M and Tomozawa M 1996 *J. Non-Cryst. Solids* **201** 177
[2] Yoshimoto S, Ito K, Hosomi H and Takai Y 2014 *Jpn. J. Appl. Phys. Conf. Proc.* **2** 011205
[3] Tompkins H G and McGahan W A 1999 "Spectroscopic Ellipsometry and Reflectometry: A User's Guide", John Wiley & Sons: New York
[4] Tao S J 1972 *J. Chem. Phys.* **56** 5499
[5] Eldrup M, Lightbody D and Sherwood J N 1981 *Chem. Phys.* **63** 51
[6] Viserswaran B, Mandlik P, Mohan S H, Silvernail J A, Ma R, Sturm J C and Wagner S 2015 *J. Vac. Sci. Technol. A* **33** 031513
[7] Israelachvill J N 1992 "*Intermolecular and surface forces second edition*", Academic Press Limited: London

IOP Conf. Series: Journal of Physics: Conf. Series **791** (2017) 012028 doi:10.1088/1742-6596/791/1/012028

High-dose boron and silver ion implantation into PMMA probed by slow positrons: Effects of carbonization and formation of metal nanoparticles

T Kavetskyy[1,2], K Iida[3], Y Nagashima[3], A Kuczumow[2], O Šauša[4], V Nuzhdin[5], V Valeev[5] and A L Stepanov[5,6,7]

[1] Drohobych Ivan Franko State Pedagogical University, Drohobych, Ukraine
[2] The John Paul II Catholic University of Lublin, Lublin, Poland
[3] Department of Physics, Tokyo University of Science, Shinjuku, Tokyo, Japan
[4] Institute of Physics, Slovak Academy of Sciences, Bratislava, Slovak Republic
[5] Kazan Physical-Technical Institute, Russian Academy of Sciences, Kazan, Russia
[6] Kazan National Research Technological University, Kazan, Russia
[7] Kazan Federal University, Kazan, Russia

E-mail: kavetskyy@yahoo.com

Abstract. The Doppler broadening slow positron beam spectroscopy (SPBS) data for the previously observed effect of carbonization in high-dose (>10^{16} ion/cm^2) 40 keV boron-ion-implanted polymethylmethacrylate (B:PMMA) and another one obtained for the effect of formation of metal nanoparticles in high-dose 30 keV silver-ion-implanted polymer (Ag:PMMA) are compared. Following to the Doppler broadening SPBS results, a difference in the high-dose ion-irradiation-induced processes in B:PMMA and Ag:PMMA is detected.

1. Introduction

Ion implantation is well-known as a powerful experimental tool for altering and significant modification of surface structure of materials. Two main features, important for practical use, such as carbonization and formation of metal nanoparticles are expected for polymer materials subjected to high-dose (>10^{16} ion/cm^2) low-to-medium energy ion implantation (for reviews, see [1,2]).

High-dose ion-irradiation-induced processes in polymers are accompanied by substantial improvements in hardness, wear resistance, oxidation resistance, resistance to chemicals, and electrical conductivity due to carbonization [3-7].

Recently, carbonization or formation of carbon nanostructures has experimentally been verified [8] for 40 keV boron-ion-implanted polymethylmethacrylate (B:PMMA) by using slow positron beam spectroscopy (SPBS) based on Doppler broadening of positron annihilation γ-rays as a function of incident positron energy and positron annihilation lifetime at an positron energy of 2.15 keV. The SPBS results have also been found in consistent with Raman spectroscopy results for same structures B:PMMA [9,10].

The present work is aimed to compare the Doppler broadening SPBS data for the previously observed effect of carbonization in high-dose B:PMMA [8] with the Doppler broadening SPBS data for high-dose 30 keV silver-ion-implanted in same polymer (Ag:PMMA) with metal nanoparticles (NPs).

2. Experimental

Low-energy ion implantation of the polymer was performed under a pressure of 10^{-5} Torr at room temperature with an ILU-3 ion accelerator (KPTI, Russia) [2,11,12]. As substrates, 1.2-mm-thick PMMA plates with a high optical quality and transparency in a wide spectral range 400-1000 nm were used. Ag:PMMA samples were fabricated with an energy of 30 keV, an ion current density of 1 μA/cm^2 and doses of 2.5×10^{16} and 1.0×10^{17} ion/cm^2 (both doses are enough for the formation of Ag NPs in the surface region of polymer [2,11-16]) and were tested using Doppler broadening SPBS technique (TUS, Japan). The experimental system used was a magnetically guided slow positron beam apparatus with a trochoidal $\boldsymbol{E} \times \boldsymbol{B}$ filter [17]. The beam diameter was about 6 mm and the energy of the incident positrons ranged from 0.1 to 30 keV. The 511 keV γ-rays from the samples were detected using a Ge detector in coincidence with a NaI(Tl) detector.

3. Results and discussion

Figure 1 shows positron annihilation Doppler broadening *S*-parameter dependences for the virgin (non-irradiated) PMMA and B:PMMA [8] and Ag:PMMA samples formed with a dose of 2.5×10^{16} ion/cm^2 as a function of incident positron energy in the range 0-30 keV. Apart from the surface effect below 0.4 keV, the *S*-parameter for the pristine sample is practically constant, while that for the irradiated sample showed remarkable reduction below an incident positron energy $E_p \cong 4$ keV for B:PMMA and $E_p \cong 2$ keV for Ag:PMMA. Similar behavior of the *S*-E_p dependence for ion-irradiated polymeric materials has been observed for 150 keV He^+-implanted poly-L-lactic acid (PLLA) polymer [18,19]. The decrease in the *S*-parameter shows that the low momentum component of the Doppler broadened photon peak is decreased for implanted sample. This fact could be interpreted by a reduction of the positronium formation fraction as it has been mentioned for He:PLLA [19] and experimentally verified using positron annihilation lifetime measurements at incident positron energy of 2.15 keV for B:PMMA [8]. According to the analysis of *S*-E_p dependence performed in the work of Saito *et al.* [19] and recovery of the *S*-parameter for E_p larger than 8 keV for B:PMMA and 4 keV for Ag:PMMA as seen in Fig. 1, it is suggested that $E_p \cong 4$ keV for B:PMMA and $E_p \cong 2$ keV for Ag:PMMA, which correspond to the middle value of the *S*-parameter, could be explained as that the positrons implanted with such energy annihilate around the Bragg peak of the ion implantation.

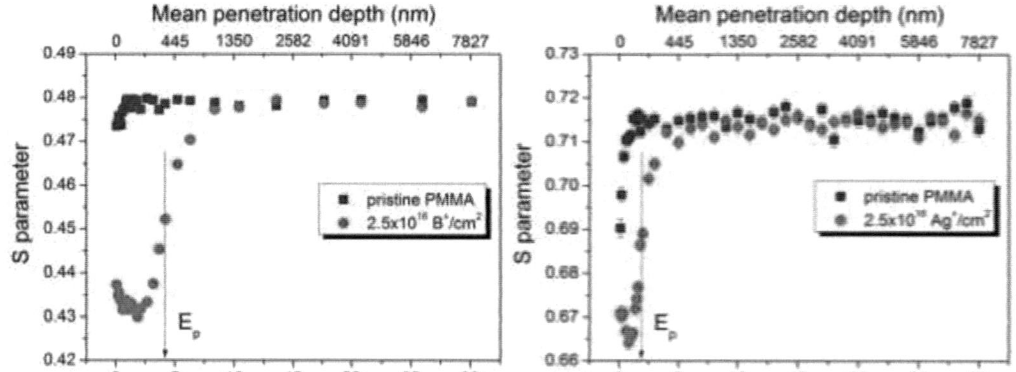

Figure 1. Positron annihilation Doppler broadening *S*-parameter for the pristine PMMA and irradiated B:PMMA (left, the error bars are within the size of the symbol) [8] and Ag:PMMA (right) formed with a dose of 2.5×10^{16} ion/cm^2 as a function of incident positron energy in the range 0-30 keV. Vertical arrows indicate incident positron energy, E_p, for the middle point of the *S*-parameter. The *S*-parameters for B:PMMA and Ag:PMMA are different from each other because the experiments were conducted with different measuring systems.

Figure 2 shows positron annihilation Doppler broadening *S*-parameter for the irradiated B:PMMA samples with doses of 2.5×10^{16}, 3.75×10^{16}, and 5.0×10^{16} ion/cm^2 [8] and Ag:PMMA samples with doses of 2.5×10^{16} and 1.0×10^{17} ion/cm^2 as a function of incident positron energy in the range 0-5 keV. It is seen that a rather stable behavior of $S(E_p)$ parameter in the near-surface range 0-1 keV for the high-dose B:PMMA samples is observed. At the same time, in the case of the higher dose Ag:PMMA the increase of $S(E_p)$ parameter in the near-surface range 0-1 keV is seen. It is suggested that the detected difference in the S-E_p dependences is probably due to formation of Ag nanoparticles for high-dose Ag:PMMA [2,16] and changes of density in the near-surface region of polymer. Besides, the surface sputtering effect for Ag:PMMA, which was first observed and published in the work [20], should be also taken into account as additional surface modification factor as shown in Fig. 3. In order to make more reliable conclusions the further Doppler broadening SPBS study of Ag:PMMA with higher ion doses and positron annihilation lifetime measurements are needed.

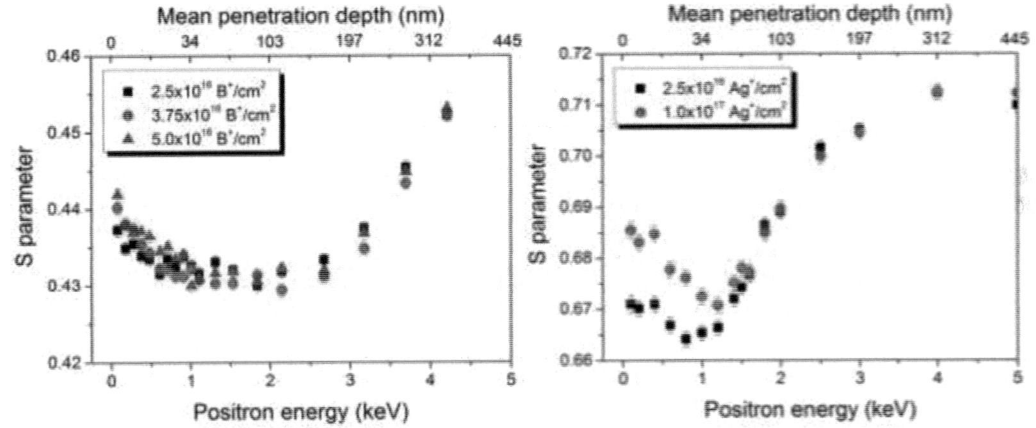

Figure 2. Positron annihilation Doppler broadening *S*-parameter for the irradiated B:PMMA samples with doses of 2.5×10^{16}, 3.75×10^{16}, and 5.0×10^{16} ion/cm^2 (left) [8] and Ag:PMMA samples with doses of 2.5×10^{16} and 1.0×10^{17} ion/cm^2 (right) as a function of incident positron energy in the range 0-5 keV. The *S*-parameters for B:PMMA and Ag:PMMA are different from each other because the experiments were conducted with different measuring systems.

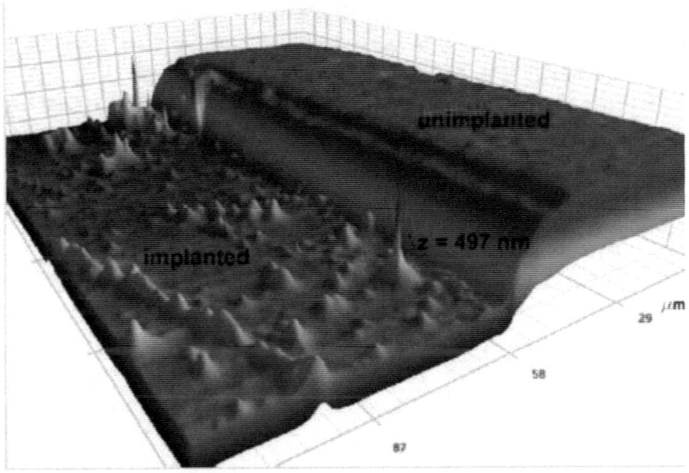

Figure 3. 3D image of surface sputtering for 30 keV Ag:PMMA with a dose of 2.5×10^{16} ion/cm^2 [20].

4. Conclusions

The Doppler broadening SPBS data for the previously observed effect of carbonization in 40 keV B:PMMA (2.5×10^{16}, 3.75×10^{16}, and 5.0×10^{16} ion/cm^2) [8] and for 30 keV Ag:PMMA (2.5×10^{16} and 1.0×10^{17} ion/cm^2) with metal nanoparticles have been compared. It has been established that a rather stable behavior of $S(E_p)$ parameter in the near-surface range 0-1 keV for the high-dose B:PMMA samples is seen; whereas in the case of the higher dose Ag:PMMA the increase of $S(E_p)$ parameter in the near-surface range 0-1 keV is detected.

Acknowledgments

The authors are grateful to A. Kinomura and Y. Kobayashi for encouragement in this work and for going through the manuscript. This work is supported in part by the MES of Ukraine (project No. 0116U004737). T. Kavetskyy acknowledges the SAIA for scholarships in the IPSAS within the National Scholarship Program of the Slovak Republic. O. Šauša acknowledges the Slovak Grant Agency VEGA (projects Nos. 2/0099/10, 2/0155/14 and 2/0164/ 14). A.L. Stepanov, who was responsible for ion implantation procedure, thanks for the financial support the RSF (project No. 14-13-00758) in Russia.

References

[1] Sviridov D V 2002 *Russ. Chem. Rev.* **71** 315
[2] Stepanov A L 2004 *Tech. Phys.* **49** 143
[3] Wasserman B, Braunstein G, Dresselhaus M S and Wnek G E 1983 *Mater. Res. Soc. Symp. Proc.* **27** 423
[4] Lee E H, Rao G R and Mansur L K 1992 *J. Mater. Res.* **7** 1900
[5] Bridwell L B 1992 *Solid State Phenomena* **27** 163
[6] Lee E H, Rao G R, Lewis M B and Mansur L K 1993 *Nucl. Instr. Meth. Phys Res.* B **74** 326
[7] Rao G R, Lee E H and Mansur L K 1993 *Wear* **162-164** 739
[8] Kavetskyy T, Tsmots V, Kinomura A, Kobayashi Y, Suzuki R, Mohamed H F M, Šauša O, Nuzhdin V, Valeev V and Stepanov A L 2014 *J. Phys. Chem.* B **118** 4194
[9] Kavetskyy T, Nowak J, Borc J, Rusnák J, Šauša O and Stepanov A L 2016 *Spectr. Lett.* **49** 5
[10] Kavetskyy T, Tsmots V, Nowak J, Kuczumow A, Kinomura A, Kobayashi Y, Suzuki R, Mohamed H F M, Šauša O, Nuzhdin V, Valeev V and Stepanov A L 2016 *Mater. Sci. Forum (Proc. "ICPA-17 IC Positron Annihilation 2015")* in press
[11] Stepanov A L, Abdullin S N, Petukhov V Yu, Osin Yu N, Khaibullin R I and Khaibullin I B 2000 *Phil. Mag.* B **80** 23
[12] Stepanov A L 2010 *Rev. Adv. Mater. Sci.* **26** 1
[13] Maier S A, Kik P G, Sweatlock L A, Atwater H A, Penninkhof J J, Polman A, Meltzer S, Harel E, Requicha A A G and Koel B E 2003 *Mater. Res. Soc. Symp. Proc.* **777** T7.1.1
[14] Boldyryeva H, Kishimoto N, Umeda N, Kono K, Plaksin O A and Takeda Y 2004 *Nucl. Instr. Meth. Phys. Res.* B **219-220** 953
[15] Boldyryeva H, Umeda N, Plaksin O A, Takeda Y and Kishimoto N 2005 *Surf. Coat. Technol.* **196** 373
[16] Galyautdinov M F, Nuzhdin V I, Fattakhov Ya V, Farrakhov B F, Valeev V F, Osin Yu N and Stepanov A L 2016 *Tech. Phys. Lett.* **42** 182
[17] Nagashima Y, Hakodate T, Miyamoto A, Michishio K and Terabe H 2009 *J. Phys.: Conf. Ser.* **194** 012039
[18] Saito F, Yotoriyama T, Nagashima Y, Suzuki Y, Itoh Y, Goto A, Iwaki M, Nishiyama I and Hyodo T 2004 *Mater. Sci. Forum* **445-446** 340
[19] Saito F, Yotoriyama T, Nishiyama I, Suzuki Y, Goto A, Nagashima Y and Hyodo T 2014 *Phys. Chem. Chem. Phys.* **16** 26991
[20] Stepanov A L, Nuzhdin V I, Valeev V F, Galyautdinov M F, Nuzhdin E V, Safina I R and Kavetskyy T S 2015 *Intern. J. Appl. Fund. Res.* **8** 676

Positron annihilation in the near surface of room temperature ionic liquids

T Hirade[1,3], B E O'Rourke[2] and Y Kobayashi[2]
[1]Nuclear Science and Engineering Center, Japan Atomic Energy Agency, Tokai, 319-1195 Japan
[2]National Institute of Advanced Industrial Science and Technology, Tsukuba, Ibaraki 305-8568, Japan
[3]Graduate School of Science and Engineering, Ibaraki University, 2-1-1, Bunkyo, Mito, Ibaraki 310-8512 Japan

E-mail:t.hirade@kurenai.waseda.jp

Abstract. Positronium (Ps; a bound state of an electron and a positron) formation in insulating materials is explained by the spur reaction model. According to the model, electron and/or positron mobility affects the yield of Ps formation. A vertical slow positron beam was used to investigate the surface of a room temperature ionic liquid, N,N,N-trimethyl-N-propylammonium bis(trifluoromethanesulfonyl)imide (TMPA-TFSI). Measurement of positron lifetimes indicated a higher Ps formation probability in near surface of TMPA-TFSI than in the bulk. This result suggests that the electron (and positron) mobility in the near surface may be larger than that in the bulk. Moreover, the longest annihilation lifetime of triplet Ps (ortho-Ps) in the near surface was found to be shorter than that measured in the bulk liquid. Ortho-Ps lifetimes in liquids are well correlated with the surface tension with a shorter lifetime corresponding to higher surface tension. The higher Ps formation yield and the shorter ortho-Ps annihilation lifetime were probably caused by the layered structure in near surface of TMPA-TFSI. A vertical slow positron beam is a strong tool to investigate the surface of room temperature ionic liquids.

1. Introduction

The structure of the liquid surface had been believed to be simple for a long time. However surface layering was observed by X-ray measurement for liquid metals [1]. Moreover, surface layering was also observed in molecular liquids [2]. It is very interesting to study liquid surface structure. However, the vapour pressure of molecular liquids makes studies in vacuum by many experimental methods difficult. Room temperature ionic liquids (IL's) have negligible vapor pressure and hence many experimental methods can be applied.

Slow positron beams can be applied to investigate the near surface structure. Previously, the surface of an IL was investigated using a horizontal ion beam [3]. However, it is more practical to use a vertical beam to study the stable horizontal surface of liquids. Recently a vertical slow positron beamline was developed at AIST in Tsukuba, Japan [4]. Therefore we decided to study the surface of IL's by using the AIST slow positron beam which can perform positron annihilation lifetime spectroscopy measurements. In particular the intensity and lifetime of the longest lifetime component from the annihilation of the triplet state of positronium (Ps, a bond state of an electron and a positron) are useful to understand the surface structure of IL's.

Ps formation in insulating materials, such as liquids or solids, is well explained by the spur reaction model [5]. According to the spur reaction model, a positron can form Ps with one of the excess electrons at the terminal spur, i.e. the end part of the positron track. Ps formation time is about 1ps, because the Ps precursor is free or quasi-free electrons and positrons. Hence, Ps formation is usually difficult after localization or solvation of electrons and positrons. Therefore, the electron and positron properties in the material, such as diffusion length or mobility, affect the Ps formation intensity, i.e. the yield of Ps formation.

Measurement of the Ps formation intensity is a strong tool to investigate the electron mobility especially in very small volumes. Carbon disulphide, CS_2, is an electron scavenger. However, it was previously found that further addition of CS_2 caused enhancement of the Ps formation intensity at higher concentration of CS_2 [6]. A possible explanation for this strange phenomenon was the higher electron (and positron) mobility in CS_2 clusters, even though a small electron mobility was reported by radiation chemists. After the report of the high Ps formation intensity in CS_2 as mentioned above, the electron mobility was carefully re-measured, and the high electron mobility was successfully observed [7].

Ps has a negative work function in many materials so that Ps is energetically more stable in vacuum than in the material. Therefore, if there is a vacant space or pore in the material, Ps prefers to be inside the vacant space. The triplet Ps (ortho-Ps) has the longest annihilation lifetime in a positron annihilation lifetime spectrum. In the case of liquids, Ps creates a sub-nano meter bubble by itself and localizes in it. This bubble is called a Ps bubble. The size of the Ps bubble is controlled by the balance between the zero point energy of Ps and the surface energy of the bubble. A beautiful relation between the surface tension and ortho-Ps annihilation rate (the reciprocal lifetime) in many liquids could be seen [8]. Osada et al. reported that surface tensions of TMPA-TFSI and PP13-TFSI were about 35mN/m and 34 mN/m at 30°C [9]. The o-Ps annihilation rates for TMPA-TFSI and PP13-TFSI at 30°C were 0.27 ns^{-1} and 0.29 ns^{-1} [10]. The values for TMPA-TFSI and PP13-TFSI are out of the correlation between ortho-Ps annihilation rates and surface tensions for many other liquids. In the case of IL's, there must be a local structure at small scales. The ortho-Ps annihilation rate tells us about the surface energy in microscopic region, i.e. at the nanometer scale, especially for IL's. Recently, GHz oscillation of Ps bubble in IL's were reported [11]. The measurements of Ps bubble oscillation must be a strong tool to investigate the viscoelasticity of IL's. The surface tension estimated by ortho-Ps annihilation lifetime must be also a strong tool to investigate the nano structure of IL's. Here, depth profiling of Ps formation intensity and ortho-Ps lifetime was used to investigate the surface structure.

2. Experimental

The IL sample used was N,N,N-trimethyl-N-propylammonium bis (trifluoromethanesulfonyl) imide (TMPA-TFSI) (Kanto Kagaku). The melting point of TMPA-TFSI is 19°C. TMPA-TFSI was used from a new bottle without any purification.

Positron annihilation lifetime measurements were performed using the vertical slow positron beamline installed at AIST [4]. The diameter of the positron beams was about 10mm which is small enough that almost all positrons can be injected into the IL surface. The IL was poured into a metal cup (see Figure 1(a)) which was placed in the vacuum chamber (see Figure 1(b)).

Measurements were performed at room temperature with positron energies of 1keV, 2keV and 15keV and every spectrum had 0.8-1.5 million counts. The time resolution was about 300ps (fwhm) which was enough to obtain lifetimes and intensities for the longest lifetime component from the annihilation of ortho-Ps.

3. Results and Discussion

Positron annihilation lifetime spectra obtained with positron energies of 1keV and 15keV are shown in Figure 2. We performed a free 3 component fit with the PALSfit program. [12] The results are shown in Table1. The lifetime and the intensity of the longest lifetime component for TMPA-TFSI measured by a radioisotope based lifetime apparatus were 3.858 ns and 13.6% respectively [13]. Two interesting phenomena were found. One was the higher Ps formation probability and the other was shorter annihilation lifetimes in the near surface region.

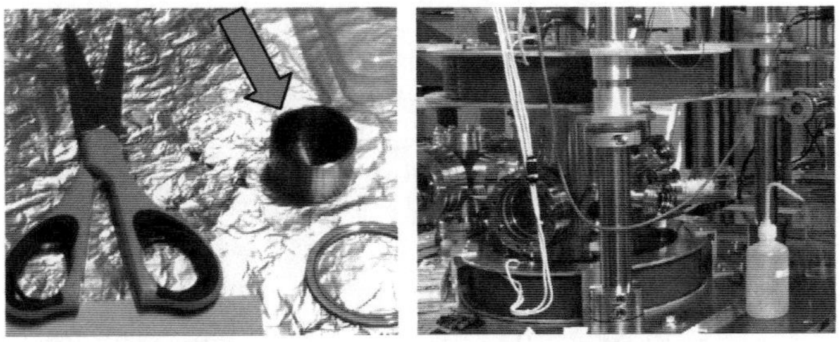

(a) (b)

Figure 1. A metal cup indicated by an arrow in **(a)** was used for the sample container of IL measurements. This cup was placed in the vacuum chamber (see **(b)**) of the vertical slow positron beamline.

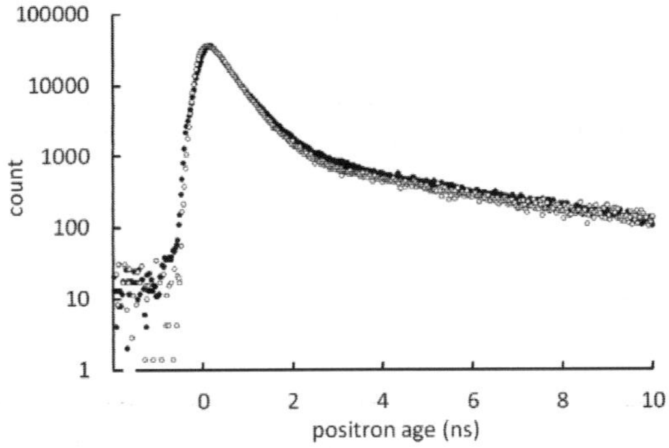

Figure 2. Positron annihilation lifetime spectra measured with positron energies of 1keV (filled circle) and 15keV (open circle). The 15keV spectrum was normalized to the 1 keV spectrum for easier comparison.

Table 1. Lifetime and intensity of the longest lifetime component with free 3 component fit. Incident depth means the average depth of positron implantation at respective energies

Positron Energy (keV)	Incident Depth (nm)	τ_3 (ns)	I_3 (%)
1	25	3.75 ± 0.07	17.2 ± 0.5
2	100	3.74 ± 0.05	15.9 ± 0.1
15	3000	3.97 ± 0.12	13.7 ± 0.2

The Ps formation intensity in the near surface of several ten nm depth was higher than the value observed in bulk of TMPA-TFSI by the conventional radioisotope based lifetime apparatus. This is probably caused by the layered structure. The most top layer consists of negative anions as indicated by Nakajima et al. [3]. Neutron reflectometry measurements suggest that the inhomogeneity of the surface region extends to ~4nm [14]. These things indicate that the structure in the near surface of IL's is similar to the crystalline state. Indeed, it was reported that there could be a crystalline structure on the surface of IL's even at temperature about 37°C higher than the melting point [15].

A crystalline structure in near surface of TMPA-TFSI is quite possible at room temperature (about 25°C) which is just 6°C higher than the melting temperature. In crystalline structures, light particles, such as electron, positron and Ps, are usually in a delocalized state. Therefore, the mobility of electrons (and positrons) in the near surface are probably higher than in the homogeneous structure found in the bulk liquid. This is the reason why Ps intensity is higher in the near surface.

Although a higher electron density at the surface was reported [16], the layer with higher electron density was only 2-3nm which is much smaller than the distribution of the terminal spur of the positron track. Therefore the higher electron density at the surface may not be the reason of the higher Ps formation intensity.

For the measurement at 2keV a time resolution of around 200ps was obtained which is sufficient to extract a GHz oscillation signal. Analysis of the 2keV spectrum found an oscillation of the annihilation rate which had a frequency of about 4.6GHz. It is known that this oscillation is caused by the oscillation of Ps bubble and the frequency corresponds to the oscillation at about 28°C according to the temperature dependence measurements performed by the conventional radioisotope based positron annihilation lifetime measurement apparatus [9]. We performed the experiments at the room temperature but did not know the precise temperature of the sample in this experiment. In the future we intend to study the temperature dependence of ortho-Ps lifetime and the oscillation of Ps bubble to understand the surface of IL's in detail.

4. Conclusion

We successfully observed the effect of the surface of TMPA-TFSI on the ortho-Ps lifetime and intensity. The lifetimes in near surface were shorter than that in bulk, indicating that the surface energy in near surface was larger, probably. The intensity of the long lifetime component was higher in near surface region. This increase was probably caused by the higher electron (and positron) mobility in crystalline like structure. Moreover, the oscillation of Ps bubble was also obtained successfully. Further investigation will be needed to understand more in detail. The vertical positron beamline installed at AIST will be a strong tool to investigate surface of IL's.

Acknowledgement

This research was partially supported by a Ministry of Education, Culture, Sports, Science and Technology Grant-in-Aid for Scientific Research (C), 23600011, 2011-2014 and 16K05026, 2016-2018.

References

[1] Magnussen O M, Ocko B M, Regan M J, Penanen K, Pershan P S, and Deutsch M 1995 *Phys. Rev. Lett.* **74** 4444

[2] Mo H, Evmenenko G, Kewalramani S, Kim K, Ehrich S N and Dutta P 2006 *Phys. Rev. Lett.* **96** 096107

[3] Nakajima K, Ohno A, Suzuki M and Kimura K 2009 *Nuclear Instruments and Methods in Physics Research B* **267** 605

[4] O'Rourke B E, Oshima N, Kinomura A and Suzuki R 2014 *Jpn. J. Appl. Phys. Conf. Proc.* **2** 011304

[5] Mogensen O E 1974 *J. Chem. Phys.* **60** 998

[6] Jansen, P and Mogensen, O E, 1977. *Chem. Phys.* **25** 75

[7] Gee N and Freeman G R 1989 *J. Chem. Phys.* **90** 5399

[8] Tao S J 1972 *J. Chem. Phys.* **56** 5499

[9] Hirade T, "Temperature dependence of positronium bubble oscillation in room temperature ionic liquids" 17th international conference on positron annihilation, Wuhan, China 20-25 September, 2015

[10] Osada R, Hoshino T, Okada K, Ohmasa Y and Yao M. 2009 J. Chem. Phys. **130** 184705

[11] Hirade T 2014 *JJAP Conf. Proc.* **2** 011003
Hirade T 2015 *Journal of Physics: Conference Series* **618** 012004

[12] Kirkegaard P, Olsen J V, Eldrup M and Pedersen N J 2009 PALSfit Risø-R-1652(EN)

[13] Hirade T 2009 *Materials Science Forum* **607** 232

[14] Bowers J and Vergara-Gutierrez M C 2004 *Langmuir* **20** 309

[15] Jeon Y, Vaknin D, Bu W, Sung J, Ouchi Y, Sung W and Kim D 2012 *Phys. Rev. Lett.* **108** 055502

[16] Sloutskin E, Ocko B M, Tamam L, Kuzmenko I, Gog T and Deutsch M 2005 *J. Am. Chem. Soc.* **127** 7796

IOP Conf. Series: Journal of Physics: Conf. Series **791** (2017) 012030 doi:10.1088/1742-6596/791/1/012030

Modelling the line shape of very low energy peaks of positron beam induced secondary electrons measured using a time of flight spectrometer

A J Fairchild[1], V A Chirayath[1], R W Gladen[1], M D Chrysler[1], A R Koymen[1] and A H Weiss[1]

[1]Deptartment of Physics, The University of Texas at Arlington, Arlington, Texas 76019-0059, USA

Email: alexander.fairchild@uta.edu

Abstract. In this paper, we present results of numerical modelling of the University of Texas at Arlington's time of flight positron annihilation induced Auger electron spectrometer (UTA TOF-PAES) using SIMION® 8.1 Ion and Electron Optics Simulator. The time of flight (TOF) spectrometer measures the energy of electrons emitted from the surface of a sample as a result of the interaction of low energy positrons with the sample surface. We have used SIMION® 8.1 to calculate the times of flight spectra of electrons leaving the sample surface with energies and angles dispersed according to distribution functions chosen to model the positron induced electron emission process and have thus obtained an estimate of the true electron energy distribution. The simulated TOF distribution was convolved with a Gaussian timing resolution function and compared to the experimental distribution. The broadening observed in the simulated TOF spectra was found to be consistent with that observed in the experimental secondary electron spectra of Cu generated as a result of positrons incident with energy 1.5 eV to 901 eV, when a timing resolution of 2.3 ns was assumed.

1. Introduction

The UTA TOF-PAES is capable of generating and transporting a very low, ~1 eV, positron beam and detecting the subsequent electrons emitted from the sample with energy as low as ~0 eV. The spectrometer consists of: (1) a permanent neodymium magnet behind the sample used to parallelize the ejected electron momentum, (2) a field free TOF tube, (3) a set of **E**×**B** plates that deflects electrons onto a micro-channel plate (MCP), and (4) detector electronics used in determining the TOF. The energy of the incident positron beam can be adjusted by applying a voltage bias to the sample, resulting in incident positron beam energies of ~1.5 eV to ~901 eV [1].

The work in this paper was motivated by the desire to better understand the nature of the spectrometer's instrument response function in order to obtain a quantitative estimate of the true energy spectrum of low energy electrons emitted from a polycrystalline Cu sample as a result of Auger mediated positron sticking (AMPS) and related positron induced electron emission processes [2].

2. TOF-PAES Spectrometer Model

The measured TOF distribution of electrons emitted from the sample surface is broadened by the instrument response function of the TOF-PAES spectrometer. The parameters that determine this broadening include: (1) the magnetic field gradient produced by the permanent magnet behind the sample, (2) the electric fields along the flight path of the electrons, and (3) the timing resolution of the detection electronics [3]. SIMION® 8.1 simulations, using the experimental parameters listed in table 1, were used to model the TOF broadening due to the combined effect of the magnetic gradient and the perpendicular electric and magnetic fields between the **E**×**B** plates. A more detailed description of the experimental system modelled can be found in reference [1].

Single energy electrons generated randomly on a 2.6 mm diameter circle and emitted with isotropic angular distribution, were flown in the SIMION® 8.1. The TOF of the electrons reaching the MCP was calculated using reverse timing to be consistent with the timing scheme adopted in the experiment and was counted to generate a histogram. That way low time of flights correspond to lower energies and high time of flights correspond to higher energies as is the case in the experiment. The modelled TOF spectrum was then convolved with a Gaussian function, meant to represent the timing resolution of the TOF system, and was converted to the energy domain using an experimentally derived conversion formula. The output of the simulated spectrometer, to the input delta function energy distributions, was used to obtain the broadening of trial functions, representing the true secondary electron distribution, using the principle of superposition. The timing resolution function used was a ~2.35 ns (FWHM) Gaussian which was found by convolving the simulated secondary electron TOF data, for electrons generated as a result of positrons incident with ~901 eV, with a series of Gaussians of different FWHMs until a match with experiment was found.

Table 1. Experimental parameters used in the SIMION® 8.1 model of the TOF-PAES spectrometer. The axial magnetic field is the magnetic field along the beam axis (south to north), the sample magnetic field is the magnetic field measured at the sample surface, and the east and west **E**×**B** plates refer to the plates east and west of the beam axis. See reference [1] for further details regarding the experimental apparatus.

Experimental Parameters	Value
Axial Magnetic Field	0.004 T
Sample Magnetic Field	0.046 T
East **E**×**B** Plate Voltage	-2.79 Volts
West **E**×**B** Plate Voltage	+3.19 Volts
E×**B** Plate Length	0.267 m
E×**B** Plate Spacing	0.025 m
Sample to MCP Distance	1 m

3. Results and Discussion

The choice of trial energy distribution functions used as inputs in our simulations was guided by the relevant physics of the positron induced electron processes measured. Following previous studies, we started with a parameterized trial function which has been used to describe the spectra of true secondary electrons produced by an electron beam [2, 4].

$$\mathrm{N}(E) = A\,\frac{E}{(E+E_0)}(E+\varphi)^m \tag{1}$$

The terms E_0, φ, and m are fitting parameters and were optimized to match the experimental data. Two regimes of parameters were found to match experiment best: (1) $E_0 = .35$ eV, $\varphi = 4.6$ eV, m = -2.5 and (2) $E_0 = .35$ eV, $\varphi = 4.6$ eV, m = -1.6.

If a positron transitions from a positive energy scattering state to the bulk ground state, then the maximum kinetic energy of the emitted secondary electron is given as:

$$E_{kin} = E_{e^+} + \varphi^+_{samp} - \varphi^-_{samp} \tag{2}$$

where E_{e^+} is the maximum kinetic energy of the positron incident on the sample taking into account the maximum kinetic energy of the positron inside the TOF tube, the contact potential between the sample and sample chamber, and the sample bias, φ^+_{samp} is the positron work function, and φ^-_{samp} is the electron work function. However, if the incident positron becomes trapped in a surface state directly from the scattering state then the maximum kinetic energy of the emitted electron is given as:

$$E_{AMPS} = E_{e^+} + E_b - \varphi^-_{samp} \tag{3}$$

where E_b is the surface state binding energy. This AMPS mechanism is dominant at very low energies but does not make a significant contribution to the electron yield for incident positron energies above ~10 eV [2, 5]. The function in equation 1 cuts off at an energy determined by equation 2 or 3 depending on the maximum incident positron energy. Using equation 2, which represents the most probable process of electron emission at incident positron energies > 10 eV, we find the maximum kinetic energy for secondary electrons due to 901.25 eV positrons incident on a polycrystalline Cu to be 896.6 eV (taking $\varphi_{cu}^- = 4.65$ eV and $\varphi_{cu}^+ = 0$ eV) [6]. Using the equation relevant at low positron energies, equation 3, we find the maximum kinetic energy for secondary electrons due to 3.75 eV positrons incident on a polycrystalline Cu sample to be 1.74 eV (taking $E_b = 2.64$ eV) [1].

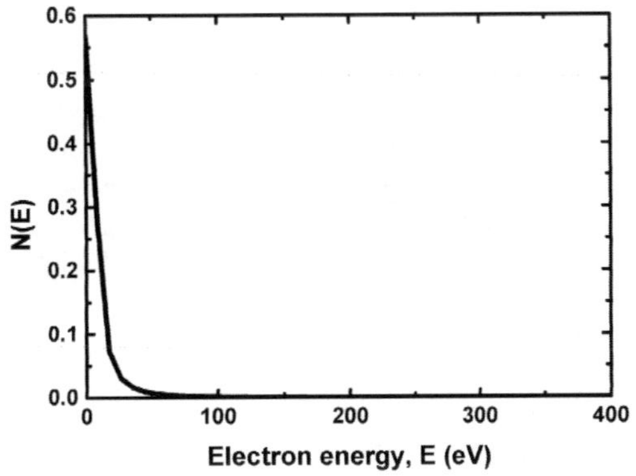

Figure 1. Electron energy distribution which was input to the simulated TOF spectrometer. This was chosen to represent the true energy distribution of secondary electrons from Cu for positrons incident at 901.25 eV. The distribution was generated using equation 1 with $E_0 = .35$ eV, $\varphi = 4.6$ eV, m = -2.5. The maximum secondary electron energy was calculated as 896.6 eV using equation 2.

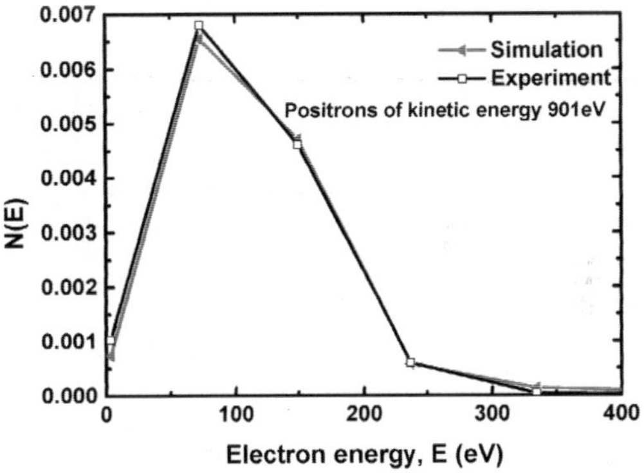

Figure 2. Comparison of the experimental and the simulated energy spectra of positron induced secondary electrons from the surface of a polycrystalline Cu at an incident positron energy of 901.25 eV. The legend indicating positron energy does not include the effect of contact potential. The simulated TOF spectrum was convoluted with a Gaussian of FWHM ~ 2.35 ns before conversion into energy.

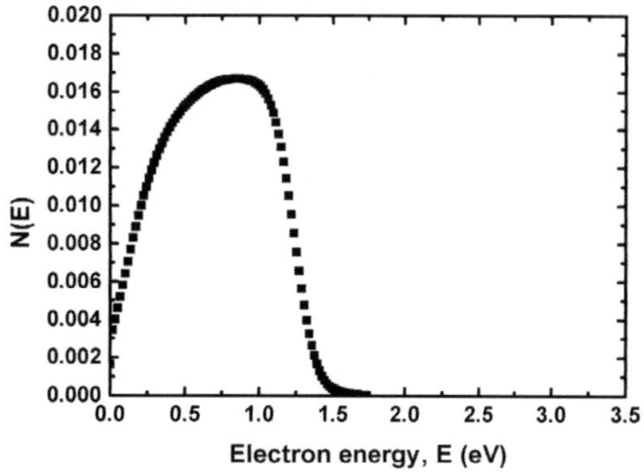

Figure 3. Electron energy distribution which was input to the simulated TOF spectrometer. This was chosen to represent the energy distribution of electrons emitted from Cu as a result of the sticking of 3.75 eV positrons on surface via AMPS process. The distribution was generated using equation 1 with E_0 = .35 eV, φ = 4.6 eV, m = -1.6. The maximum electron energy was calculated as 1.74 eV using equation 3.

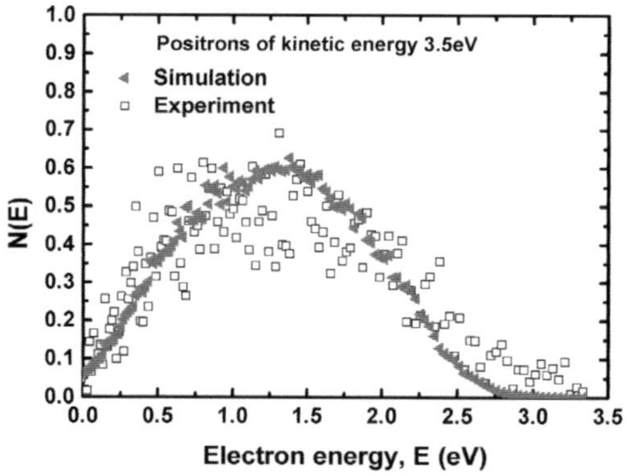

Figure 4. Comparison of experimental and simulated energy spectra of positron induced electrons from the surface of a polycrystalline Cu sample through the AMPS process. The legend indicating positron energy does not include the effect of contact potential. The simulated TOF spectrum was convoluted with a Gaussian of FWHM ~ 2.35 ns before conversion into energy.

Figure 1 shows the input electron energy distribution function, based on equation 1 with E_0 = .35 eV, φ = 4.6 eV, m = -2.5, for secondary electrons generated as a result of positrons incident with energy 901.25 eV. Figure 2 compares the respective simulated output with experimental Cu data. Figure 3 shows the input electron distribution function, based on equation 1 with E_0 = .35 eV, φ = 4.6 eV, m = -1.6 for electrons generated as a result of positrons incident with energy 3.75 eV. Figure 4 shows the simulated output compared with experimental Cu data. 50,500 particles were flown with the input electron energy distributions (figures 1 and 3) taking into account the distribution in the energy of positrons incident on the sample through a convolution of equation 1 with a Gaussian centred at ~0.5 eV and having a FWHM of 0.26 eV. The measured energy spectra can be seen to be significantly broadened in energy as compared to the model input spectra shown in figures 1 and 3. This broadening is well accounted for by the instrumental broadening found from our numerical modelling over a wide range of electron energies (~0.5 eV to ~900 eV) when a timing resolution of 2.35 ns was assumed.

4. Conclusion

In this paper, we have presented a model simulation of the University of Texas at Arlington's time of flight positron annihilation induced Auger electron spectrometer using SIMION® 8.1. Using this model we have determined the timing resolution of our system to be 2.3 ns. A parameterized function, equation 1, was used to model the initial kinetic energy of secondary electrons leaving the Cu sample as a result of positron annihilation inside the material. After comparison with experimental data, two regimes of parameters were found to match experiment best, one for incident positron energies above ~10 eV and one for incident positron energies below ~10 eV. We believe regime 1 to represent positrons which mostly transition from a scattering state to a bulk state before thermalization and annihilation in a surface state while regime 2 represents positrons which mostly transition from a scattering state directly to a surface state. However, more work needs to be done in order to verify this.

Acknowledgements
This work was supported by NSF grants DMR 1508719 and DMR 1338130.

References

[1] Mukherjee S, Shastry K, Anto C V, Joglekar P V, Nadesalingam M P, Xie S, Jiang N, and Weiss A H 2016 *Rev. Sci. Instrum.* **87** 035114
[2] Mukherjee S, Nadesalingam M P, Guagliardo P, Sergeant A D, Barbiellini B, Williams J F, Fazleev N G, and Weiss A H 2010 *Phys. Rev. Lett.* **104** 247403
[3] Lei C, Mehl D, Koymen A R, Gotwald F, Jibaly M, and Weiss A H 1989 *Rev. Sci. Instrum.* **60** 3656-60
[4] Ramaker D E, Murday J S, and Turner N H 1979 *J. Electron. Spectrosc. Relat. Phenom.* **17** 45
[5] Walker A B, Jensen K O, Szymanski J and Neilson D 1992 *Phys.Rev. B.* **46** 1687
[6] Farjam M and Shore H B 1987 *Phys. Rev. B* **36** 5089

IOP Conf. Series: Journal of Physics: Conf. Series **791** (2017) 012031 doi:10.1088/1742-6596/791/1/012031

Time of flight spectra of electrons emitted from graphite after positron annihilation

R W Gladen[1], V A Chirayath[1], M D Chrysler[1], A D Mcdonald[1], A J Fairchild[1], K Shastry[1], A R Koymen[1] and A H Weiss[1]

[1] Dept. of Physics, University of Texas at Arlington. Arlington, Texas 76019, USA

Email: randall.gladen@uta.edu

Abstract. Low energy (~2 eV) positrons were deposited onto the surface of highly oriented pyrolytic graphite (HOPG) using a positron beam equipped with a time of flight (TOF) spectrometer. The energy of the electrons emitted as a result of various secondary processes due to positron annihilation was measured using the University of Texas at Arlington's (UTA) TOF spectrometer. The positron annihilation-induced electron spectra show the presence of a carbon KLL Auger peak at ~263 eV. The use of a very low energy beam allowed us to observe a new feature not previously seen: a broad peak which reached to a maximum intensity at ~4 eV and extended up to a maximum energy of ~15 eV. The low energy nature of the peak was confirmed by the finding that the peak was eliminated when a tube in front of the sample was biased at -15 V. The determination that the electrons in the peak are leaving the surface with energies up to 7 times the incoming positron energy indicates that the electrons under the broad peak were emitted as a result of a positron annihilation related process.

1. Introduction

Experiments using positron beams with energies in the range of a few eV have resulted in the discovery of novel processes of electron emission [1], and have helped provide a quantitative estimate of the intensity under the low energy tail [2]. In this paper we present the TOF Positron-Induced Auger Electron Spectroscopy (PAES) spectra from HOPG with a 2 eV positron beam. The aforementioned sample, HOPG, is a technologically important material which is used as a substrate for a variety of nanostructures [3]. In addition, the surface of HOPG is a suitable model of the 2D allotrope of carbon, graphene, and could help to understand positron interactions with graphene. These experiments were made possible by the ability of the spectrometer to acquire positron stimulated electron data with positrons incident on the surface at energies as low as 1.5 eV. At such low incident energies, all impact induced secondary electron emission is precluded and any electron emission must be due to annihilation related processes [2]. The use of this low incident positron energy allowed us to obtain the annihilation induced electron spectrum from C free of background from impact induced secondaries over a wide range of emitted electron energies from ~500 eV down to ~1eV. Our measurement revealed that in addition to the previously observed annihilation induced C KVV Auger peak at ~263 eV, the spectra included a second, larger peak in the annihilation induced electron spectrum which reached its maximum around 4 eV and extended as high as 15 eV.

2. Experimental Apparatus

The data were acquired using a Time of Flight (TOF) PAES system at UTA described previously [4]. The positrons were obtained by the decay of a 10 mCi Na^{22} source and then moderated using a thin tungsten foil. The moderator is biased to a positive voltage ($V_{moderator}$) with respect to the chamber (source tube) which houses the source. A grid that is connected to the source tube and kept in front of the moderator extracts the slow positrons into the axial magnetic field. Following moderation, a tungsten barrier collimates these positrons to a diameter of ~8 mm. The slow positrons are filtered from the moderated and collimated beam using mutually perpendicular magnetic and electric fields which are generated by a set of electrically biased plates (**EXB** plates) and the axial magnetic field. The potential difference between the first set of **EXB** plates bends the slow positrons upward through the offset aperture in the tungsten barrier. The fast positrons that are not bent sufficiently, as well as the gammas from the source, are obstructed by this barrier. The second set of **EXB** plates bends the slow positrons back onto the axis of the beam to enter the TOF spectrometer after traveling through a graded accelerator.

The system is equipped with a TOF spectrometer which is used for the measurement of the energy of electrons emitted after the interaction of positrons with the sample and the associated electronics are shown in Figure 1. The low energy positrons are bent around the micro-channel plate (MCP) electron detector using two pairs of **EXB** plates and finally reach the sample after travelling through the TOF tube. The electrons emitted from the surface of the sample after positron implantation travel through the TOF tube and subsequently get deflected into the MCP. These electrons are detected in coincidence with the 511 keV gammas produced as a result of the annihilation of positrons with the electrons of the sample which are detected by the BaF_2 detector. A NaI (Tl) gamma detector kept on the opposite side of the sample also collects the annihilation gamma spectrum.

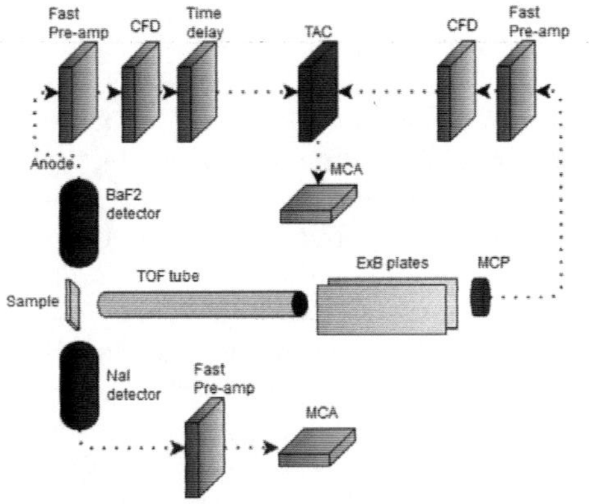

Figure 1. The TOF spectrometer and the associated nuclear electronics used to generate the PAES spectrum. The BaF_2 detector detects a gamma photon in coincidence with an electron detected by the MCP. The timing signal corresponding to the detection of gammas or electrons is produced by the corresponding constant fraction discriminators. The spectrometer is operated in reverse timing mode where the signal from BaF_2 detector is delayed and provided as the stop signal to the TAC. The start signal to the TAC is provided by the MCP detector.

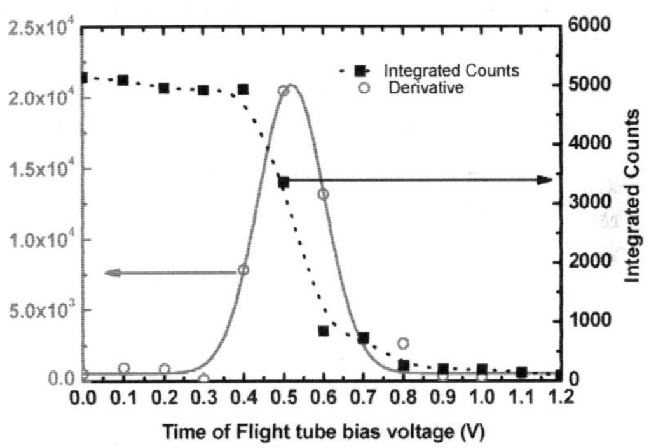

Figure 2. Positron beam energy profile Integrated coincidence counts vs. TOF tube bias (corresponding to positron energy), with the first derivative (red circles) providing the energy spectrum of the positron beam (Gaussian fit, x_c = 0.52, FWHM = 0.20). The sample was biased to -263 V and the total counts under the secondary electron peak are plotted as a function of the positive bias on the TOF tube. It can be seen that beyond ~ 1V, the coincidence count rate goes to near zero indicating that almost all positrons are reflected by the tube and do not reach the sample. The dotted line through the integrated counts is a guide to the eye. The solid line through the first derivative is a Gaussian fit.

2.1. Positron Beam Energy

The kinetic energy of the positrons hitting the sample was determined by measuring the positron energy distribution entering the TOF tube ($E_{ke,TOF}$) and then accounting for the potential difference between the TOF tube and the sample according to equation 1:

$$E_{KE} = -e \cdot V_{SB} + E_{KE,TOF} + \varphi^{-}{}_{sample} - \varphi^{-}{}_{chamber} \quad (1)$$

where e is charge of the positron, V_{SB} is the voltage on the sample, $E_{KE,TOF}$ is the kinetic energy of the positron entering the TOF tube, $\varphi^{-}{}_{sample}$ and $\varphi^{-}{}_{chamber}$ are the electron work functions of the sample and the chamber, respectively. The positron energy distribution entering the TOF tube was measured by using the TOF tube as a retarding field analyzer. For this experiment the sample was biased at -263 V (V_{SB}), which resulted in a strong secondary electron signal in the coincidence spectra when positrons were allowed to hit the sample. Applying a positive potential to TOF tube causes positrons to be reflected, resulting in fewer positrons reaching the sample. This reduces the intensity of secondary electrons detected at the MCP. Figure 2 displays the integrated coincidence counts under the secondary peak as a function of TOF tube bias, in addition to the first derivative of the integrated counts. It can be seen that nearly all of the positrons are reflected back with a positive bias of 1 V. Therefore, it can be concluded that the maximum kinetic energy of positrons travelling through the TOF tube is ~ 1 eV. The positron energy distribution given by the first derivative has a maximum at ~ 0.5 eV and a FWHM of 0.2 eV. In the data shown in Figure 3, the sample was biased at -1V, resulting in a maximum kinetic energy of positrons hitting the sample at ~2 eV.

3. Results and Discussion

The TOF-PAES spectra which were collected with a sample bias of -1 V, and with a positron energy distribution as shown in Figure 2, are given in Figure 3(a) – (d). Figure 3a shows the PAES spectra with 0 V bias on the TOF tube; allowing all of the electrons down to 0 eV to be detected at the MCP. The spectra in Figure 3(a) shows a broad low energy peak with a maximum around 4 eV and extending up to ~15 eV in addition to a peak corresponding to the annihilation induced KVV transition in carbon (~263 eV). This carbon sample was prepared by mechanically exfoliating HOPG in air and placing it in the sample chamber which was pumped down to a pressure of ~1×10^{-10} Torr. In order to confirm that this peak was really due to low energy electrons and to find the maximum energy of the electrons under the low energy peak, we again employed the TOF tube as a retarding field analyser this time to measure the energy of electrons leaving the sample. The TOF tube was biased at a series of negative voltages, thereby reflecting electrons with energies lower than $-\boldsymbol{e} \cdot \boldsymbol{V_{TOF}}$, and preventing them from reaching the MCP. The results are shown in Figures 3(b) – (d). It can be clearly seen that most of the signal in the low energy peak is gone when -7 volts is applied to the TOF tube (Figure 3(b)) and that almost all of the signal is gone when -15 volts is applied (Figure 3(c)). These results show that the broad peak seen in Figure 3(a) was due to electrons with energies less than 15eV.

4. Conclusion

The energy spectra of positron induced electrons leaving a HOPG surface were measured for an incident positron kinetic energy < 2eV. A broad peak was observed with a maximum at ~4 eV and extending up to ~15 eV in addition to the annihilation induced KVV Auger peak at ~263 eV. The integrated peak intensity of the low energy peak was an order of magnitude larger than the KVV Auger. Considering that the incident positron energy is < 2 eV, energy conservation rules out non-annihilation processes from producing electrons in the energy range of the low energy peak, unless additional energy is made available through an annihilation process. We postulate that these electrons may be due to an Auger process initiated by the annihilation of an electron deep in the valence band of

IOP Conf. Series: Journal of Physics: Conf. Series **791** (2017) 012031 doi:10.1088/1742-6596/791/1/012031

graphite. We are currently investigating the origin of this low energy peak, in addition to searching for its presence in other carbon allotropes, such as graphene.

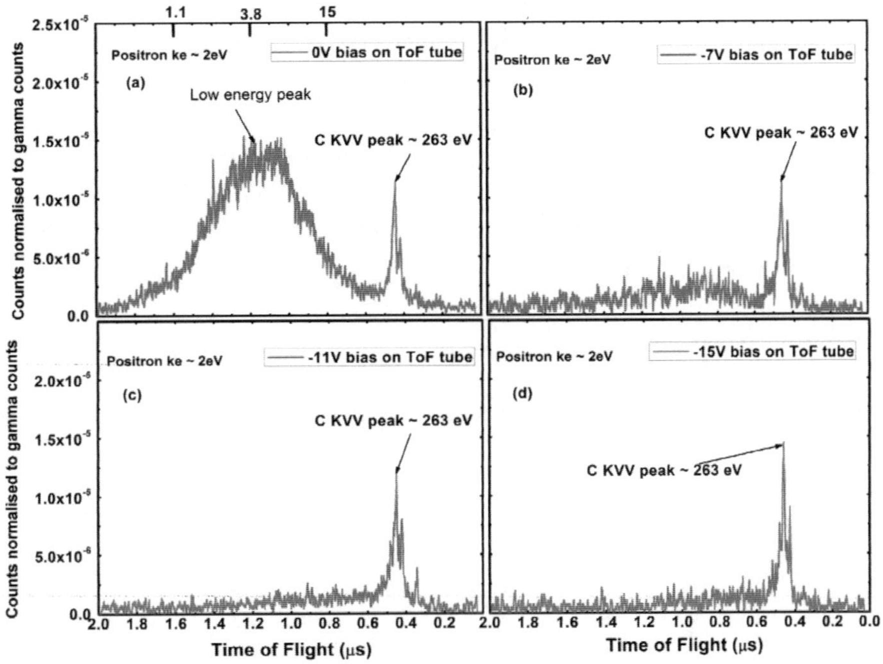

Figure 3. TOF PAES spectra from HOPG using positrons of kinetic energy < 2 eV. The indicated kinetic energy of the positrons does not correct for the additional energy due to the contact potential between the sample and the chamber. The TAC is set to observe a 2 μs time window. Pane (a) shows the PAES spectra with 0 V bias on the TOF tube. All of the spectra show the Auger peak corresponding to KVV transition in carbon at 263 eV. In the corresponding panes where the TOF tube is biased to different negative voltages, (b) -7 V (c) -11 V and (d) -15 V, it can be seen that the low energy peak is completely removed with a bias of -15 V. This indicates that the electrons under the low energy peak are < 15 eV.

Acknowledgments
This work was supported by NSF grant DMR 1508719.

References

[1] Mukherjee S, Nadesalingam M P, Guagliardo P, Sergeant A D, Barbiellini B, Williams J F, Fazleev N G, Weiss A H 2010 *Phys. Rev. Lett.* **104** 247403

[2] Mukherjee S F, Shastry K and Weiss A H 2011 *Phys. Rev. B* **84** 155109

[3] Scheier P, Marsen B, Lonfat M, Schneider W D and Sattler K 2000 *Surf. Sci.* **458** 113

[4] Mukherjee S, Shastry K, Anto C V, Joglekar P V, Nadesalingam M P, Xie S, Jiang N, and Weiss A H 2016 *Rev. Sci. Instrum.* **87** 035114

IOP Conf. Series: Journal of Physics: Conf. Series **791** (2017) 012032 doi:10.1088/1742-6596/791/1/012032

Investigation of graphene using low energy positron annihilation induced Doppler broadening spectroscopy

V A Chirayath[1], M D Chrysler[1], A D McDonald[1], R W Gladen[1], A J Fairchild[1], A R Koymen[1] and A H Weiss[1]

[1]Department of Physics, University of Texas at Arlington, Arlington Texas 76019-0059, USA

Email: koymen@uta.edu

Abstract. We report the first measurements on 6-8 layers graphene using a low energy positron beam employing a high efficiency rare gas moderator. Doppler broadening of the annihilation gamma was measured from graphene over layers at positron kinetic energies as low as 2 eV. The efficient trapping of positrons on the surface state of graphene at these low energies has been utilised to get Doppler broadened gamma spectra predominantly from graphene over layer on Cu. The ratio of the annihilation gammas measured at low positron energies (a few eV) to that measured at 20 keV show features corresponding to carbon. W-parameter calculated from gamma spectra at different positron energies point towards trapping of positrons in defected graphene as well as at the interface of graphene over layer and Cu. This is the first ever measurements of graphene film of 2-3 nm thickness on a substrate with a depth resolved Doppler broadening spectroscopy at positron kinetic energies below 10 eV.

1. Introduction

The excellent ballistic conduction, mechanical strength, optical properties etc. of graphene is related to the linear band dispersion at the Fermi level and due to these favourable properties graphene is touted to be widely used in future electronic applications like flexible electronics, organic LED, high frequency transistors etc [1]. However, in real world scenario there can be modification to its properties at Fermi level due to the mismatch and subsequent strain at the interface or due to impurities at the interface of graphene with other materials. Thus it is important to investigate the defects or chemical impurities at the interface of the multilayer or single layer graphene with technologically important materials. Further research into deposited graphene layers has been motivated by the discovery that the choice of substrate on which graphene is deposited plays an important role in the electronic, magnetic and chemical properties of the film [2, 3]. Hence it is important to have a non-destructive probe which can investigate graphene over layers on various substrates without extensive sample preparation or probe modification of its chemical or physical properties. Variable energy positron beams have been used to non destructively investigate over layers on substrate by analyzing the Doppler broadening of the 511 keV annihilation gammas after implanting the positrons at various depths [4]. However all these studies have used positrons of kinetic energy greater than 100 eV which is good for over layers or thin films which are a few tens of nanometer thick. In order to investigate two dimensional materials like graphene it is important to trap the positrons predominantly in the surface state, so that the Doppler broadened annihilation gamma carry information from graphene or other two dimensional materials. If positrons are implanted at energies less than 10 eV then it possible to get information from the top atomic layers of the sample as there is greater probability for trapping the positrons into the surface state as evidenced from positron annihilation induced Auger spectroscopy (PAES) [5].

We describe here first measurements of multilayer (6-8) graphene on polycrystalline Cu sample with an advanced positron beam attached with a high efficiency rare gas moderator. This positron beam system is under construction and when fully operational will allow for coincident PAES and coincidence Doppler broadening spectroscopy. The variable positron energy coincidence Doppler broadening system of the beam has been completed and was utilized for the present investigation. The spectra were taken with incident positron kinetic energies as low as 2 eV which allowed us to probe the surface state of graphene.

2. Experimental apparatus and sample

The advanced positron beam system consists of a high efficiency rare gas (neon) moderator system purchased from First Point Scientific Inc. and a positron transport system equipped with time of flight spectrometer. In the rare gas moderator system, crystalline neon is grown on a Cu cylinder kept at 6.8 K. The Cu cylinder also houses Na^{22}, the source of positrons that has an activity of 0.5 mCi. The Cu cylinder (moderator) is biased to a positive potential with respect to the ground. The slow positrons from the solid neon moderator are filtered from unmoderated fast positrons by bending the slow positrons through an aperture (offset from beam axis) on a tungsten barrier. The slow positrons are transported from the moderator to the sample using an axial magnetic field of ~60 Gauss. The positron beam system has a time of flight tube (~3 m) which will be utilized in the PAES experiments. This time of flight tube was utilized as a retarding field analyzer to find the kinetic energy of the positrons entering the sample chamber in this experiment. It was found that when the moderator is biased to 32 V, the kinetic energy of the positrons has a distribution which peaks at 33 eV and has an FWHM of ~2 eV [6]. The beam energy of the positron is defined as the peak value of the positron kinetic energy distribution i.e. in the case when moderator is biased to 32 V, the beam energy is defined as 33 eV. Lower energies are attained by biasing the sample to positive potentials. Lower energies can also be obtained by biasing the moderator to lower positive potentials with respect to the grounded sample. It was determined that both methods yield equal incident kinetic energies [6].

The two collinear 511 keV annihilation gammas are detected by a NaI (Tl) detector and a high purity germanium (HPGe) detector kept at diametrically opposite positions with respect to the sample. The HPGe detector has a resolution of 0.98 keV for the 356 keV line of Ba^{133}. The HPGe detector is gated with the output from NaI(Tl) detector for coincidence measurements. The measurements were carried out from positron kinetic energies of 2 eV to 20 keV in both coincidence and non coincidence mode. In coincidence mode, the lowest count rate obtained was ~ 8 cps under the 511 keV peak as detected by the HPGe detector. The annihilation gamma line shape was parameterized using the W parameter [7] which is defined as the ratio of counts in the wing region of 511 keV gamma peak to the total counts under the peak (defined within ±10 keV of 511 keV). The wing region in these experiments has been defined at two regions. The first wing region, corresponding to the W1 parameter was defined from 512.5 keV to 513.5 keV on the high energy side of the 511 keV peak and from 508.5 keV to 509.5 keV on the low energy side of the 511 keV peak. The second wing region, corresponding to the W2 parameter was defined from 513.5 keV to 518.5 keV on the high energy side of the 511 keV peak and from 503.5 keV to 508.5 keV on the low energy side of the 511 keV peak.

2.1. Sample

The 6-8 layers graphene sample, purchased from ACS materials, was grown using the CVD technique on polycrystalline Cu. The sample was characterized by Raman spectroscopy using 532 nm laser light. The resulting Raman spectrum is shown in Figure 1. The presence of multilayer graphene was confirmed from the shape of the 2D band at ~ 2700 cm^{-1} in comparison to that obtained from highly oriented pyrolytic graphite [8]. The Raman spectrum also shows the presence of defect peak at ~1320

cm^{-1} and the graphitic peak at ~1580 cm^{-1}. AFM images of the sample shows that it was covered uniformly with flakes of multilayer graphene (Figure 2). The polycrystalline Cu substrate was heated to 1100 K before graphene growth and hence no interfacial contamination is expected. PAES experiments conducted independently also gave evidence of no significant interfacial contamination. The analysis chamber was kept at a vacuum of 6.7×10^{-8} Pa during measurements.

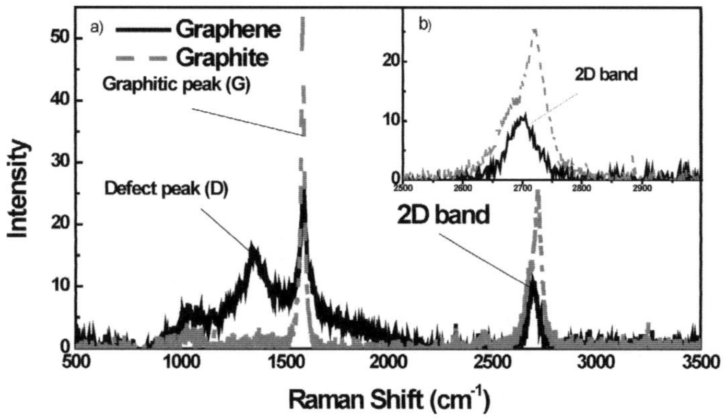

Figure 1. Raman spectra collected from multilayer graphene and highly oriented pyrolytic graphite using a 532 nm laser. A graphitic peak corresponding to sp^2 hybridisation is seen in both samples at ~1582 cm^{-1}. The 2D band of graphite (~2700 cm^{-1}) consists of two peaks which is absent in graphene and is an indication of the presence of graphene layers. Graphene also has a defect peak (~1320 cm^{-1}) which shows the presence of defects in the multilayer graphene sample. (color online)

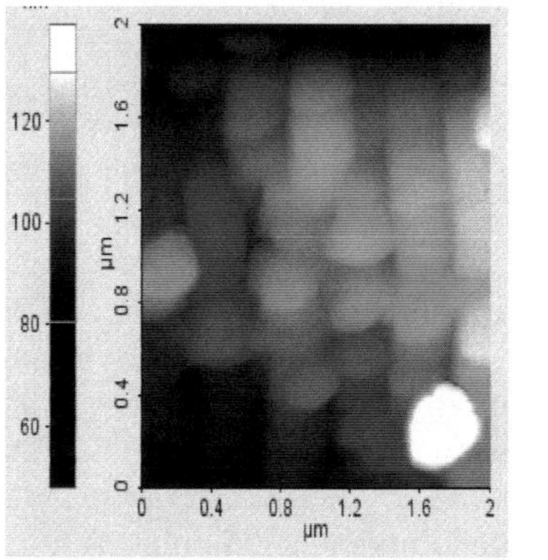

Figure 2. AFM image of multilayer graphene sample taken in contact mode. The scan was repeated in different areas of the sample which shows that flakes of graphene multilayer cover the sample almost uniformly. (color online)

3. Results

The gamma spectra collected in coincidence mode at positron kinetic energies of 5 eV and 20 keV is shown in Figure 3. The background has been subtracted from the data using the algorithm suggested by Mogensen *et al.* [9]. The ability of variable energy positron beam based Doppler broadening spectroscopy to distinguish between bulk (20 keV) and surface states (5 eV) is brought out from the difference in the peak and the wing regions of the annihilation gamma at two different positron energies. It shows that at the surface, positron has a greater probability to annihilate with the low momentum valence electron resulting in a sharper 511 keV annihilation gamma peak in comparison to the broader annihilation gamma peak from the bulk. The large changes observed in the Doppler broadening suggest that the spectra can be utilized in obtaining chemical information from the surface.

The chemical information in the gamma line shape is brought out by the ratio of the annihilation gamma line measured at various energies to that taken at 20 keV (Figure 4). The 511 keV gamma spectrum at an incident positron energy of 20 keV is approximated to be coming entirely from positron annihilation in the bulk of the Cu substrate. The x-axis of the plot is expressed in p_L representing the longitudinal component of the momentum of the annihilating electron positron pair [10]. A straight line at a ratio equal to one would represent bulk Cu. It can be seen that the gamma spectra at lower positron kinetic energies (energies are given in legends) have higher intensity at low momentum and a large dip at around ~15×10^{-3} m_0c. Analyzing the ratio curves given by Brusa *et al* for bulk carbon and bulk Cu [11], it can be seen that the dip at 15×10^{-3} m_0c corresponds to a highly reduced probability of positron annihilation with 3d electrons in Cu. As the positron kinetic energy is increased, the depth of the feature at 15×10^{-3} m_0c reduces and by 15 keV, the ratio curve corresponds to annihilations from bulk Cu. PAES spectra collected from the same sample using positrons of low kinetic energy (less than 10 eV) with the time of flight PAES spectrometer connected to a second positron beam system at University of Texas at Arlington [12] show an Auger electron peak from the KVV transition in carbon at 263 eV and no signature of the Cu Auger peak at 60eV due to the $M_{2,3}VV$ transition. Hence, the dip at 15×10^{-3} m_0c is due to enhanced annihilation of positrons with electrons of carbon atoms and reduced annihilation with 3d electrons of Cu atoms.

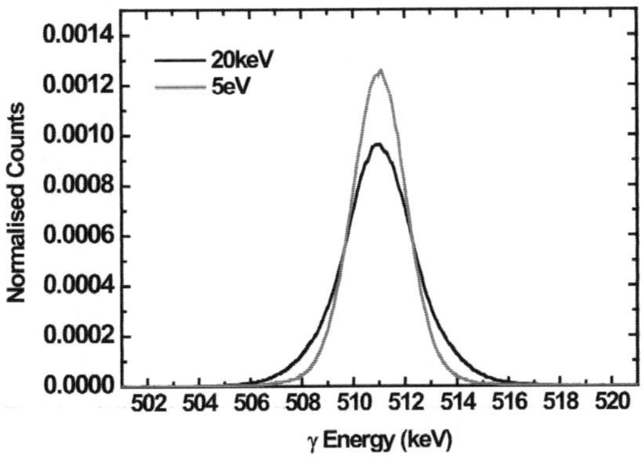

Figure 3. Gamma-ray energy spectra collected from multilayer graphene at two different positron energies (20 keV and 5 eV). Each spectrum is area normalized. The difference in spectral intensity distribution at the peak and the wings is indicative of the fact that spectral contributions from annihilations with core electrons are greatly reduced at the surface. (color online)

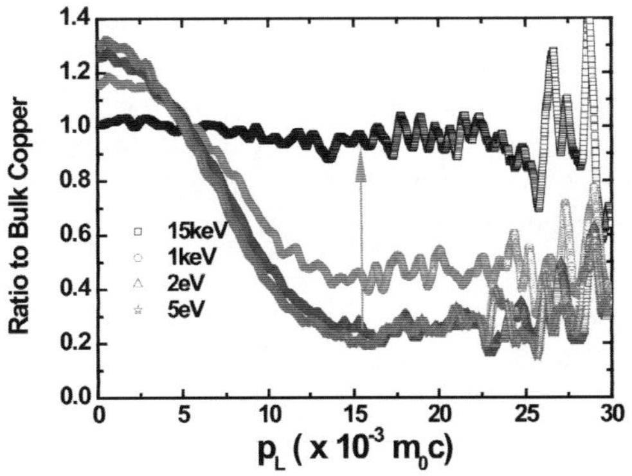

Figure 4. Ratio of the annihilation momentum distribution from multilayer graphene collected at different positron kinetic energies normalised to that taken at a positron energy of 20 keV (bulk Cu). The high intensity in the low momentum region of the ratio plots is due to annihilation with valence electrons (2p) of C and the dip at 15×10^{-3} m_0c is due to the reduced probability of annihilation with Cu 3d electrons (color online).

The gamma-ray spectra collected at different positron energies were parameterized by the W (wing)-parameter [7]. The parameter is reflective of the fraction of positrons annihilating with high momentum electrons. A decrease in the W-parameter is indicative of the presence of defects or surfaces in the material as the density of higher momentum electrons attached to the nuclear core are reduced in these regions. The variation in the W-parameter can also be due to changes in cthe hemical environment at the site of positron annihilation. The variation of the W-parameter calculated from gamma-ray spectra collected at various positron energies (2 eV-20 keV) for multilayer graphene is shown in Figure 5. Two different W parameters (W1 and W2) were defined. The parameter W1 was defined as the ratio of the integrated counts in the regions 1.5 keV to 2.5 keV above and below the central peak at 511 keV to the total peak count and is reflective of annihilation with outer most 2p electrons in C. Whereas W2 was defined as the ratio of the intensity in the region 2.5 keV to 7.5 keV from the central 511 peak. This region is reflective of annihilation with 3d electrons in Cu [11]. The calculated W-parameters (W1 and W2) have been normalized to their respective values at 20 keV representing bulk Cu. It can be seen that both W-parameters increase from a low value at the surface to a higher value in the bulk as the annihilation probability increases with higher momentum electrons in the bulk Cu. However, the W1 and W2 parameters show a difference in their variation with beam energy. The steady increase of W2 with beam energy is consistent with the variation seen in the ratio curve and is reflective of the increase in the fraction of positron annihilation with the Cu 3d electrons as the positrons are implanted more deeply and fewer of them return to C/Cu interface. On the other hand, W1 rises sharply and then levels off beyond 7.5 keV pointing to the fact that W1 is more reflective of annihilation with the C 2p electrons. This can explain why the major change in this parameter takes place over the incident positron energy range in which the fraction of positrons trapping in the C over layer is changing. The sharp dip in W1 at ~ 5 eV as seen in Figure 6 is proposed to be due to positrons annihilating in 6-8 layers graphene. We hypothesize that the fact that W1 has a lower value for incident positron energy of 5 eV than for 2 eV is due to the differences in the gamma-ray spectra caused by the trapping of positrons in defects within 6-8 layers of graphene. As the positron energy is increased more positrons are implanted near the graphene-Cu interface resulting in the increase of W1. It can be seen that the W1 values from 15 eV to 25 eV are equal within the experimental error pointing to annihilation of positrons at the same site which in this case can be graphene-Cu interface. This hypothesis is being currently tested by comparing the results to that obtained from measurements in polycrystalline Cu substrate after sputter cleaning the multilayer graphene. The influence of positronium formation on the surface of 6-8 layers graphene as a function of incident positron energy also needs to be investigated to further clarify on the trapping of positrons in graphene defects.

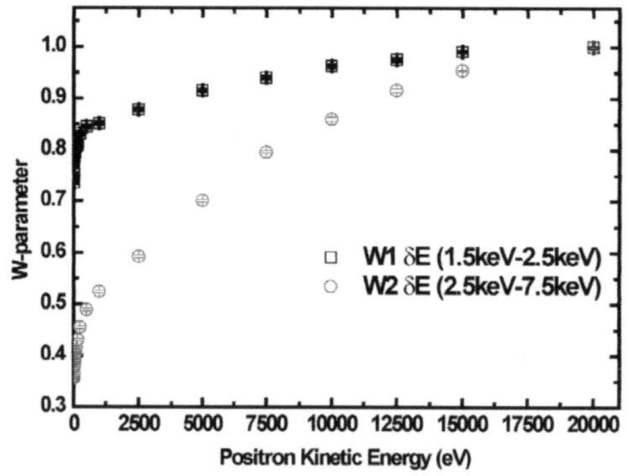

Figure 5. Variation of W-parameters W1 and W2 (defined in the text) as a function of incident positron kinetic energy which is varied from 2 eV to 20 keV. It can be seen that the variation of W2-parameter follows the typical trend for metals and reflects the increasing annihilation probability of positrons with 3d electrons. The variation of W1 is reflective of changes in the annihilation probability with 2p electrons of C. (color online)

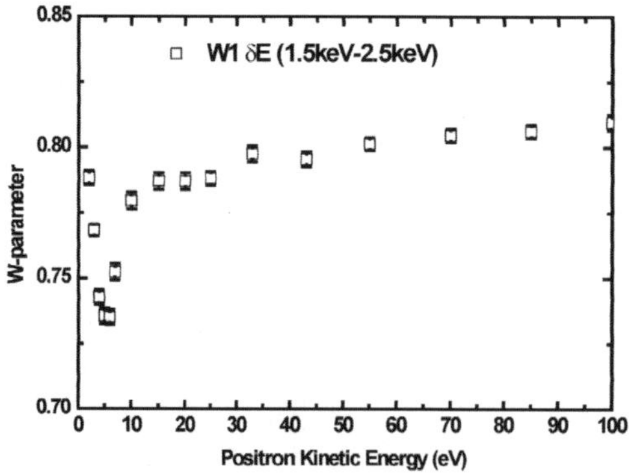

Figure 6. Variation of W1-parameter (defined in text) as a function of incident positron kinetic energy shown here from 2 eV to 100 eV. It is posited that the sharp dip in W1 at 5 eV is due to positrons annihilating in the 6-8 layers graphene over layer film on Cu.

IOP Conf. Series: Journal of Physics: Conf. Series **791** (2017) 012032 doi:10.1088/1742-6596/791/1/012032

4. Conclusion

Doppler broadening of the annihilation gamma-rays produced after implanting positrons at different energies (from 2 eV to 20 keV) in a multilayer graphene film on Cu substrate has been measured using an advanced positron beam system equipped with a high efficiency rare gas moderator. The measurements at low positron energies show the chemical sensitivity of the technique to a 2-3 nm thick graphene over layer on Cu substrate. The results also show the capability of the technique to deduce depth resolved information about positron trapping centres in graphene and at the interface of graphene and Cu substrate. These results suggest that it should be possible to utilize positron beam spectroscopy to characterize defects and chemical structure in nanometer thick 2D materials.

Acknowledgements
This work was supported by NSF grants DMR 1508719 and DMR 1338130.

References

[1] Novoselov K S, Geim A K, Morozov S V, Jiang D, Zhang Y, Dubonos S V, Grigorieva I V, Firsov A A 2004 *Science* **306** 666
[2] Sundaram R S, Gomez-Navarro C, Lee E J H, Burghard M and Kern K 2009 *Appl. Phys. Lett.* **95** 223507
[3] Zaniewski A M, Trimble C J and Nemanich R J *et al.*2015 *Appl. Phys. Lett.* **106** 123104
[4] Van Veen A, Schut A and Mijnarends P E 2000 *Positron beams and their applications* ed Coleman P G (Singapore: World Scientific Publishing Company) pp 191-225
[5] Mukherjee S, Nadesalingam M P, Guagliardo P, Segeant A D, Barbiellini B, Williams J F, Fazleev N G and Weiss A H 2010 *Phys. Rev. Lett.* **104** 247403
[6] Chrysler M D, Chirayath V A, Mcdonald A D, Gladen R W, Fairchild A J, Koymen A R and Weiss A H 2017 J. Phys.: Conf. Series in this volume
[7] Krause-rehberg R and Leipner H S 1999 *Positron Annihilation in Semiconductors:Defect Studies* (Berlin: Springer Verlag)
[8] Ferrari A C 2007 Solid State Communications **143** 47
[9] Mogensen O E and Hirade T 1993 *J. Phys. IV* **03** C4-17
[10] Asoka-Kumar P, Alatalo M, Ghosh V J, Kruseman A C, Nielsen B and Lynn K G 1996 *Phys. Rev. Lett.* **77**, 2097
[11] Brusa R S, Deng W, Karwasz G P and Zecca A 2002 Nuclear Instruments and Methods in Physics Research B **194** 519
[12] Mukherjee S, Shastrty K, Anto C V, Joglekar P V, Nadesalingam M P, Xie S, Jiang N and Weiss A H 2016 Review of Scientific Instruments **87** 035114

Positron induced electron emission from graphene

V A Chirayath[1], R W Gladen[1], K Shastry[1], M D Chrysler[1], A J Fairchild[1], A D McDonald[1], A R Koymen[1], B Barbiellini[2] and A H Weiss[1]

[1]Dept. of Physics, University of Texas at Arlington, Arlington 76019-0059, USA
[2]Dept. of Physics, Northeastern University, Boston, MA 02115, USA

email: chirayat@uta.edu

Abstract. We report here the energy spectra of positron-induced electrons emitted from single- and multilayer graphene films. One of UT Arlington's two Time of Flight positron beam systems was used to deposit very low energy (KE < 1.5 eV) positrons at the surface of graphene samples consisting of a) 1 layer of graphene on polycrystalline Cu and b) 6-8 layers of graphene deposited on polycrystalline Cu. A time of flight spectrometer was used to measure the energy of positron-induced electrons. A peak in the electron energy spectrum was observed at ~263 eV corresponding to annihilation induced Auger electrons as a result of the KVV transition from the two graphene samples, which were comparable in intensity to that observed from a bulk graphite sample. In addition, we have observed a low energy peak in the annihilation induced electron energy spectra from graphene which extends up to ~15 eV. Our observation that the positron annihilation induced KVV Auger signal from graphene has a significant intensity indicates that the low energy positrons can be efficiently trapped on the surface of graphene layer on Cu down to a single-layer.

1. Introduction

Low energy positrons incident on a surface induce electron emission through several mechanisms. Secondary electron emission resulting from positron impact was observed in the first measurements with a low energy positron beam [1-3]. Weiss *et al.* [4] identified a second process in which electron emission occurred as a result of an Auger transition involving the filling of core hole created by electron-positron annihilation. This effect resulted in a new surface characterization technique called positron annihilation induced Auger electron spectroscopy (PAES). Since the starting core hole is created through an annihilation process, a PAES signal can be obtained with arbitrarily low positron beam energies. The secondary and redistributed primary background, which makes quantitative analysis difficult in electron induced Auger electron spectroscopy (EAES), can be completely eliminated as a result of this merit. Moreover, PAES has superior surface sensitivity due to the fact that almost all of the signal stems from the annihilation of positrons trapped in a surface state localized at an image potential well just outside the topmost layer of surface atoms [5]. Recently Mukherjee *et al.* [6] identified a third process by which positrons can induce electron emission. They observed a secondary electron peak for positron beam energies well below the electron work function. In this process, the energy released during the transition of a positron from a positive energy scattering state directly into a bound surface state is taken up by an electron near the Fermi level, giving it enough energy to escape. Because of the similarity with an Auger transition, this process was named Auger mediated positron sticking (AMPS).

Application of conservation of energy shows that the AMPS process can lead to electron emission when the incident positron beam energy is above a threshold: $E_{th,AMPS} = \phi^- - E_{SS}$, where ϕ^- is the electron work function, E_{SS} is the positron surface state binding energy and $E_{th,AMPS}$ is the threshold for the AMPS process. Mukherjee *et al.* [6] used this relation and an experimental determination of the AMPS threshold energy to obtain an independent estimate of the binding energy of the positrons in the surface state of copper and gold which was in close agreement with previous estimates based on positronium thermal desorption measurements. A similar argument shows that the threshold energy, $E_{th,sec}$, for electron emission in the case where the positron first makes a transition to a bulk state can be written as $E_{th,sec} = \phi^- - \phi^+$ where ϕ^+ is the positron work function (binding energy of positrons in the bulk ground state) and $E_{th,sec}$ is the threshold for secondary electron emission. Typically $E_{SS} > \phi^+$ and consequently $E_{th,AMPS} < E_{th,sec}$. AMPS is the only significant non-annihilation induced electron emission process for positrons incident in the energy range above $E_{th,AMPS}$ and below $E_{th,sec}$. Once the incident positron energy is greater than $E_{th,sec}$ by a few eV, non-annihilation induced electron emission is dominated by processes in which the positron loses energy to an electron while making a transition into a bulk state.

It should be noted that the yield of secondary electrons due to the photo-absorption or Compton scattering of annihilation gamma rays is estimated to be smaller than 10^{-6} electrons per incident positron by taking the ratio of the secondary electron escape depth to the gamma adsorption length. This estimate is consistent with the experimental observation that the gamma induced electron background is negligible compared to yield of electrons resulting from secondary emission, annihilation induced Auger emission or the AMPS process.

In this paper, we report the energy spectra of electrons emitted from single and multilayer (6-8 layers) graphene on a polycrystalline Cu substrate at different positron kinetic energies. The spectra, taken with an incident positron kinetic energy of 1.5 eV, allowed us to observe a new low energy electron emission peak from single and multilayer graphene layers deposited on Cu, which is not present on the clean Cu surface. It is posited that this emission feature is associated with Auger like processes initiated by the creation by positron annihilation of a deep hole in the valence band.

2. Experimental apparatus and sample

The experimental apparatus, consisting of a low energy positron beam equipped with a time of flight spectrometer, has been described in detail in a previous publication [7]. However, for completeness, the key features of the system are described here. Slow positrons are obtained using a 10 mCi Na^{22} source mount with a tungsten moderator in transmission geometry. The slow positrons are filtered from the fast positrons using mutually perpendicular magnetic and electric fields generated using a set of electrically biased plates (**E**×**B** plates) and an axial magnetic field (~ 100 Gauss). This is followed by a nine stage graded accelerator /decelerator which enables us to accelerate or decelerate the positrons by appropriate biasing.

The time of flight (TOF) spectrometer and sample were biased so as to allow us to measure the energy of electrons emitted from the sample with kinetic energies from 0 eV to 500 eV. The micro-channel plate (MCP) detector used for the detection of electrons emitted from the sample is in the path of the positron beam. An additional set of **E**×**B** plates (C and D in figure 1) is used to bend the positrons around the MCP. The positrons then fly through a grounded TOF tube (~ 600 cm) before reaching the sample. The magnetic field at the TOF region and the MCP is kept at 40 Gauss. A permanent magnet kept behind the sample produces a large magnetic field gradient near the sample. The field on the surface of the sample because of this permanent magnet is ~450 Gauss. The magnetic field gradient from the sample to the entrance of TOF tube reduces the transverse velocity component of the emitted electrons as a result of positron implantation and thus reduces the width of the distribution of TOF of electrons emitted from the sample [7, 8]. The electrons, after entering the TOF tube, travel through a field free region and are deflected by the **E**×**B** plates, D, onto the MCP.

The two 511 keV gammas produced after positron annihilation with electrons from the sample are detected by a BaF_2 scintillation detector and a NaI(Tl) detector placed on opposite sides of the sample. The TOF of the electrons are determined by the time difference between the signal generated by the BaF_2 scintillation detector after the detection of 511keV gamma and the signal generated by the MCP after electron detection. The experiment was performed using "reverse timing" in which the gamma signal was delayed and used as the stop signal to the time to amplitude converter (TAC). The MCP signal is used as the start signal. A histogram of the time of flight is generated from the TAC output using a multichannel analyzer. The TOF histogram is converted to an energy histogram by using an appropriate conversion function determined experimentally using the secondary electron peak generated at different incident positron energies. The relation between TOF and the inverse of the square root of energy is expected to be linear, though higher order corrections are required to correct for the time electrons spend between the **E**×**B** plates.

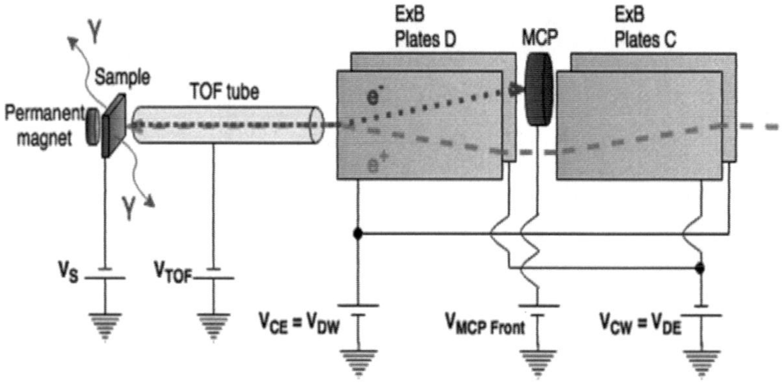

Figure 1. Shows the schematic of the TOF spectrometer attached to the positron beam system. **E**×**B** plates C and D bends the positrons (blue dash) around the MCP. The electrons (red dots) emitted from the sample after annihilation of positrons are bend by **E**×**B** plates D into the MCP. The TOF tube is connected to ground for the present experiment and the sample is applied a negative potential with respect to the TOF tube.

2.1. Sample

Single-layer graphene and 6-8 layers graphene grown on polycrystalline Cu using chemical vapor deposition was purchased from ACS materials. The samples were characterized with Raman spectroscopy and the presence of single-layer and multi-layer graphene were confirmed from the shape of the 2D band at ~ 2700 cm^{-1} [9]. The analysis chamber was kept at a vacuum of 1.33×10^{-8} Pa during measurements. The single-layer graphene on polycrystalline Cu was sputter cleaned in Argon atmosphere at a vacuum of 8×10^{-3} Pa and at a sputter current of <1 μA to obtain a clean sample of polycrystalline Cu. It was observed that the Auger peak of Cu could be obtained within minutes of sputtering which is indicative of the presence of clean Cu on which graphene was grown. The data on polycrystalline Cu shown here were collected after Auger peaks of carbon or oxygen were completely absent after sputtering out the single-layer graphene over-layer. In order to avoid contamination by surface adsorption, the polycrystalline Cu was sputtered every 14 hours during data acquisition.

3. TOF-PAES results

In order to obtain the Auger spectrum free from secondary electron background, it was important to set the incident kinetic energy of the positron at the sample below the threshold required for secondary electron emission in Cu and graphene. This threshold was found to be ~2 eV [7]. We have achieved beam settings that allow both positron and electron transport at these very low energies. For these settings, measurements of the positron kinetic energy showed that almost all positrons had a kinetic energy less than 1 eV [10] while traveling through the grounded TOF tube. The kinetic energy of positrons hitting the sample surface was determined by adding a term:

$$\Delta KE = e(-V_s) + (\phi^-_s - \phi^-_{TOF\ tube}), \quad (1)$$

to the kinetic energy of the positrons measured at the TOF tube. Here e, V_s, ϕ^-_s, and $\phi^-_{TOF\ tube}$ are the charge of the electron, the sample bias, the electron work functions of the sample and TOF tube respectively. $(\phi^-_s - \phi^-_{TOF\ tube})$ takes into account the contact potential developed between the sample and the TOF tube because of the difference in their electron work functions. For the settings used in this work, the maximum KE of positrons hitting the sample, KE_s, was taken to be equal to ΔKE +1 eV.

3.1. Electron Energy Spectra Obtained with Positron Incident below the AMPS Threshold

The TOF-PAES spectra obtained with a positron kinetic energy of 1.5 eV are shown in figure 2. The spectra of polycrystalline Cu, single-layer graphene and multilayer graphene are shown together for comparison of various spectral features. The TOF's are sampled at an interval of ~0.00195 μs from 0 to 4μs. The peaks are identified after converting the TOF to energy using the non linear conversion function and have been labeled in the figure. The energy per TOF increases as TOF decreases which causes the low energy peaks to be broader and the high energy Auger peaks to be sharper in the TOF spectrum. At a TOF of 1.3 μs which corresponds to 15 eV, energy increases by 0.15 eV per TOF step where as at a TOF of 1 μs which corresponds to 350 eV, energy increases by 16 eV per TOF step. All data are normalized to the counts under the 511 keV peak as detected by NaI (Tl) detector. The PAES spectrum from Cu shows the $M_{2,3}VV$ peak and M_1VV peak at 60 eV and 108 eV respectively. Spectral intensity for TOF flights above 1.2 μs in the electron spectra from Cu is mostly due to the inelastic loss of the Auger peaks. Within the detection limit of the experiment neither carbon nor oxygen are present on the surface of polycrystalline Cu. The spectral weight under the inelastic tail (defined up to ~15 eV) is approximately three times the intensity in $M_{2,3}VV$ peak at 60 eV.

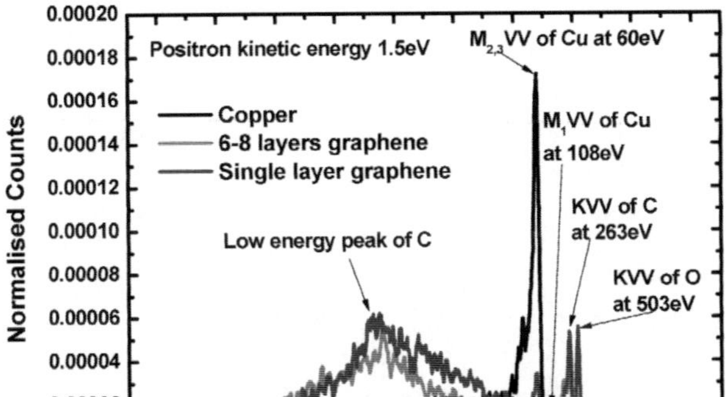

Figure 2. TOF spectra of electrons resulting from positrons incident at 1.5 eV on single and multilayer graphene (deposited on Cu) and clean Cu. The sample was biased at -0.5 V in each case. Major peaks, corresponding to Auger transitions in Cu and C, are labeled along with their corresponding energies. The counts in the TOF spectra have been normalized to annihilation gamma counts. (color online)

The spectra of single-layer and multilayer graphene show Auger peaks corresponding to KVV transition in Carbon at 263 eV. It was found that the intensity of KVV peak is comparable to that obtained from graphite [10]. Therefore, low energy positrons can be trapped efficiently on the surface state of graphene down to a single-layer of graphene. Both single-layer graphene and multilayer graphene show presence of adsorbed oxygen on the surface through the Auger peak corresponding to KVV transition in oxygen. However, the intensity of oxygen peak is higher in single-layer graphene than in multilayer graphene, which is indicative of the surface chemical reactivity of single-layer graphene compared to multilayer graphene. The data from single-layer graphene shows a peak corresponding to the $M_{2,3}VV$ Auger transition in Cu which has ~ 0.2 times the intensity under the $M_{2,3}VV$ peak observed for the clean Cu surface. This indicates that there is significant overlap of the positron surface state wave function with Cu atoms even though the graphene film is expected to be continuous. A possible explanation for this overlap is that the positron surface state on single-layer graphene has significant penetration into the interface between the graphene layer and Cu substrate [9]. In contrast, it may be seen in figure 2 that the Cu $M_{2,3}VV$ peak is barely visible in the spectra obtained from multilayer graphene on Cu.

A key feature in the Auger spectra of graphene (both single-layer and multilayer) is a broad peak at ~1.75 μs. Note that this peak is not seen in the spectra of clean Cu. The higher intensity in the broad peak for single-layer graphene when compared to multilayer graphene could be due to the additional contributions of the inelastic tail of the $M_{2,3}VV$ peak of Cu present in the single-layer graphene spectra. The broad peak cannot be accounted for in terms of the inelastic tail of the KVV peaks of carbon or oxygen as its shape, and more importantly, its intensity (~ 16 times the KVV peak at 263 eV) does not follow the expected trend of an Auger inelastic tail (of which the tail seen in the Cu spectra is typical). A similar broad peak was also seen for highly oriented pyrolytic graphite [10]. Since the incident positron energy was below both secondary electron and AMPS threshold, energy considerations rule out non-annihilation induced processes for the production of the peak between 2 μs to 1.3 μs (corresponding to an energy range of 2 eV to 15 eV). Considering that the intensity of the low energy peak is more than 6 times larger than the C KVV PAES peak (which is associated with the annihilation of C 1s electrons) we posit that the low energy peak is due to the annihilation of the positrons with valence electrons. It is energetically possible to account for the observed upper bound of ~15 eV on the low energy peak in terms of known width and depth of the valence band of graphene [11]. A hole created deep enough in the valence band can result in an intra-band Auger process releasing enough energy for a third electron from the band to make transition to states outside vacuum. Such transitions are extremely difficult to observe in photoemission spectroscopy or in conventional electron Auger spectroscopy because of the background of photoelectrons or secondary electrons [12]. To our knowledge ours is the first direct observation of this low energy spectral feature in graphene.

3.2. Electron Energy Spectra Obtained with Positron Incident above the AMPS and Secondary emission threshold

Panels (a)-(d) in figure 3 show a series of TOF positron induced electron spectra taken with increasingly more negative sample biases. An increase in the magnitude of the (negative) sample bias results in an increase of ΔKE (see eq. 1) in the kinetic energy of positrons hitting the surface. The negative bias also results in a corresponding acceleration and increase in the energies of electrons as they travel from the sample and enter into the TOF tube. This increase results in a shift of the entire TOF spectrum to the right (lower TOF), as the lowest energy of the electrons starting from the sample varies from 0 eV to a value equal to ΔKE. Referring to figure 3, as the incident positron kinetic energy increases above the AMPS threshold, a low energy peak becomes visible in both Cu and graphene TOF spectra as shown in figure 3 (a) and (b). In graphene, the AMPS feature is merged with the low energy peak and hence it appears as a shoulder on the low energy peak. The presence of an AMPS peak at the incident positron energy of 2 eV and 2.5 eV demonstrates the efficient trapping of positrons on the surface state of a single-layer graphene. When the positron energy reaches a value above the threshold for secondary emission, the spectra at low electron energies are dominated by secondary electrons emitted from the bulk. By the time the positron kinetic energy reaches 6 eV, the spectral contributions of positron induced secondary electron emission merge with the low energy C peak seen in panel (a) and they mask it. The difficulty of separating the low energy C peak from a secondary background helps explain the failure of conventional surface spectroscopic methods in detecting this peak previously. Note that the distribution of the positron impact induced secondary electrons from Cu is different from that obtained from a single-layer graphene on Cu (see figure 3 (c) and (d)). Changes observed in the electron induced secondary electron spectrum at low energies caused by the adsorption of graphene on a metal have been ascribed to changes in the surface density of states [13]. The large changes seen in our data in going from clean Cu to single layer graphene indicate that positron induced secondary electron spectra are even more sensitive to changes in the surface density of states.

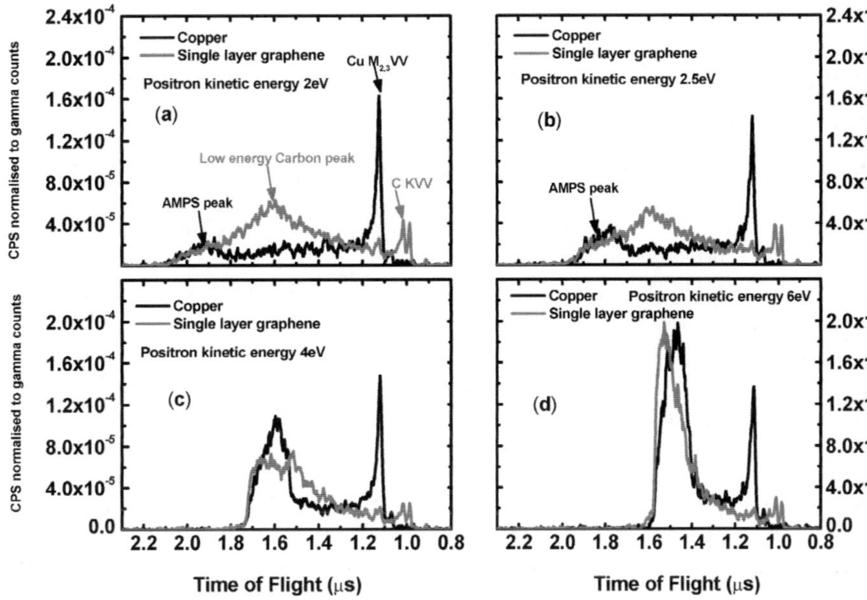

Figure 3. TOF spectra of positron induced electrons emitted from a single-layer of graphene on Cu and from clean Cu taken for sample biases: (a) -1.5V (b) -2.V (c) -3.5 V and (d) -5.5V (corresponding to incident energies of 2 eV, 2.5 eV,4 eV, and 6 eV respectively). Note: an increase in the magnitude of the (neg.) sample bias results in an increase in the kinetic energy (and a decrease in the time of flight) of electrons emitted from the sample as they traverse the TOF analyzer.(color online)

4. Conclusion

A very low energy positron beam was used to obtain the spectrum of positron annihilation induced electrons from surfaces consisting of a single and multilayer graphene on polycrystalline Cu and clean polycrystalline Cu. The backgrounds from non-annihilation related electron emission processes were eliminated by maintaining the incident positron kinetic energy below the threshold for electron emission through either impact induced secondary emission or the AMPS process. A strong positron annihilation induced Auger peak corresponding to KVV transition in carbon was observed for both single-layer and multilayer graphene indicating that positrons are trapped efficiently in surface state on top of graphene. The very low background obtained with 1.5 eV incident positron beam allowed us to observe a low energy feature in the Auger spectra of graphene that is ~ 16 times the intensity in the carbon KVV peak. We posit that these emitted electrons are a result of the filling of deep holes in the graphene valence band that were created by the annihilation of positrons in a bound surfaces state.

Acknowledgment
This work was supported by NSF grant DMR 1508719.

References

[1] Cherry W H 1958 PhD Thesis Princeton University

[2] Mayer R and Weiss A 1988 *Phys.Rev. B* **38** 11927; Mayer R, Gramsch E and Weiss A 1989 *Phys. Rev. B* **40** 11287

[3] Overton N and Coleman P G 1997 *Phys. Rev. Lett.* **79** 305

[4] Weiss A, Mayer R, Jibaly M, Lei C, Mehl D and Lynn K G 1988 *Phys. Rev. Lett.* **61** 2245

[5] Fazleev N G, Fry J L and Weiss A H 2004 *Phys. Rev. B.* **57** 12506

[6] Mukherjee S, Nadesalingam M P, Guagliardo P, Segeant A D, Barbiellini B, Williams J F, Fazleev N G and Weiss A H 2010 *Phys. Rev. Lett.* **104** 247403

[7] Mukherjee S, Shastrty K, Anto C V, Joglekar P V, Nadesalingam M P, Xie S, Jiang N and Weiss A H 2016 *Rev. Sci. Instrum.* **87** 035114

[8] Fairchild A J, Chirayath V A, Gladen R W, Chrysler M D, Koymen A R and Weiss A H 2017 *Journal of Physics. Conference Series*. in this volume

[9] Chirayath V A, Chrysler M D, Mcdonald A D, Gladen R W, Fairchild A J, Koymen A R and Weiss A H 2017 *Journal of Physics. Conference Series*. in this volume

[10] Gladen R W, Chirayath V A, Chrysler M D, Mcdonald A D, Fairchild A J, Shastry K, Koymen A R and Weiss A H 2017 in this volume

[11] Ooi N, Rairkar A and Adams J B 2006 *Carbon* **44** 231

[12] Braicovich L 1981 Comparing Valence Auger Lineshapes and Photoemission Spectra: The Cd and CdO Cases *Emission and Scattering Techniques - Studies of Inorganic Molecules, Solids, and Surfaces* (Nato Science Series C, vol 73) ed P Day (Netherlands: Springer) chapter 15 pp 375-384

[13] Riccardi P, Cupolillo A, Pisarra M, Sindona A and Caputi L S 2010 A*ppl. Phys. Lett.* **97** 221909; Riccardi P, Cupolillo A, Pisarra M, Sindona A and Caputi L S 2012 A*ppl. Phys. Lett.* **101** 183102

IOP Conf. Series: Journal of Physics: Conf. Series **791** (2017) 012034 doi:10.1088/1742-6596/791/1/012034

Iterative fitting method for the evaluation and quantification of PAES spectra

Samantha Zimnik, Mathias Hackenberg and Christoph Hugenschmidt

Heinz Maier-Leibnitz Zentrum (MLZ) and Physik Department E21, Technische Universität München, Lichtenbergstraße 1, 85748 Garching, Germany

E-mail: `samantha.zimnik@frm2.tum.de`

Abstract. The elemental composition of surfaces is of great importance for the understanding of many surface processes such as catalysis. For a reliable analysis and a comparison of results, the quantification of the measured data is indispensable. Positron annihilation induced Auger Electron Spectroscopy (PAES) is a spectroscopic technique that measures the elemental composition with outstanding surface sensitivity, but up to now, no standardized evaluation procedure for PAES spectra is available. In this paper we present a new approach for the evaluation of PAES spectra of compounds, using the spectra obtained for the pure elements as reference. The measured spectrum is then fitted by a linear combination of the reference spectra by varying their intensities. The comparison of the results of the fitting routine with a calculation of the full parameter range shows an excellent agreement. We present the results of the new analysis method to evaluate the PAES spectra of sub-monolayers of Ni on a Pd substrate.

1. Introduction

Positron annihilation induced Auger Electron Spectroscopy (PAES) is known as an analysis method with outstanding surface sensitivity [1], [2]. Auger electrons are emitted with an element-specific energy from ionized atoms in a radiation-free transition. Low energy positrons from the positron source NEPOMUC with an energy of less than 20 eV are used to ionize an atom and induce the Auger process [3]. For this technique, the vast majority of the Auger electrons originate from the topmost atomic layer of the sample [4], allowing the investigation of the elemental composition of the surface layer. So far, there is no standard evaluation method for PAES spectra as implemented in e.g. CasaXPS for the spectra of X-ray induced photo electron spectroscopy. Hence, the evaluation of PAES spectra is of great importance for quantification and the interpretation of the results.

2. Evaluation of PAES spectra

Before going into detail of the evaluation, the main components of the PAES spectrum will be explained. A typical PAES spectrum is shown in Figure 1. In addition to the outstanding surface sensitivity of PAES using low energy positrons of some eV, a second advantage is that the Auger transitions are free from secondary electron background. Low energy Auger transitions have a FWHM of several eV, which results from the broad energy distribution of the valence band and the lifetime of the states involved. Besides, the Auger peak of the according transition shows a low energy tail of inelastically scattered Auger electrons. A very low electron background

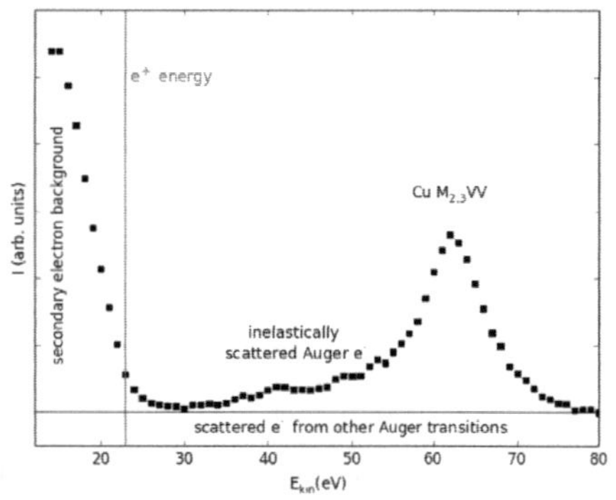

Figure 1. PAES survey spectrum of polycrystalline Cu showing the main contribution to a PAES spectrum: Due to the low energy of the incoming positrons of only ~20 eV, the Cu $M_{2,3}$VV-Auger transition is free from secondary electron background. The Auger transition itself shows a tail of inelastically scattered Auger electrons with lower energy. A very low electron background results from scattered Auger electrons from transitions at higher energies and appears as a constant contribution to the spectrum.

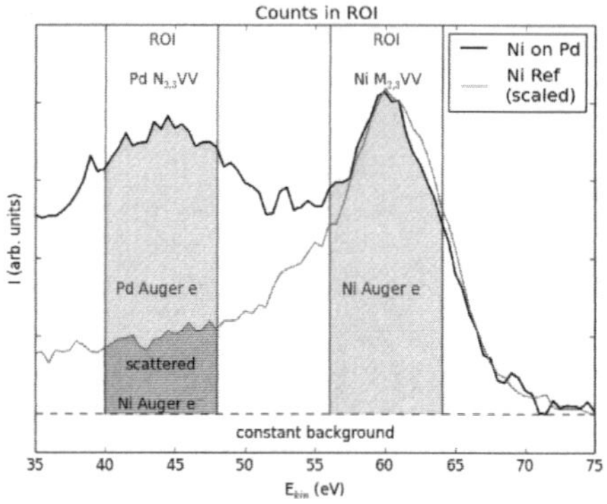

Figure 2. First approach for the quantification of PAES spectra: Energy intervals (ROIs) around the Auger peak positions are defined wherein the events are counted. For this, the spectra of the pure elements are necessary for calibration. The scattered Ni Auger electrons in the Pd ROI have to be taken into account. Therefore, this weighted fraction of scattered Ni Auger electrons is subtracted from events in the Pd ROI to not overestimate the Pd contribution in the spectra.

originates from scattered Auger electrons from transitions at higher energies and appears as a constant contribution to the spectrum. Conventional AES uses electrons or X-rays with some keV to induce the Auger process. These high energies lead to a large secondary electron background, especially in the range of the low energy Auger transitions, which are most surface sensitive in conventional AES. For this reason, usually the derivative of an AES spectrum is shown, to make the peaks visible but a quantitative analysis of the spectra is almost impossible. The initialization of the Auger process using positrons overcomes all these disadvantages, the spectra don't need to be differentiated and a quantification becomes possible.

2.1. Counts in ROI

In a first approach to quantify the contribution of each element to the PAES spectra, the events in fixed energy Region of Interest (ROI) around the Auger peak positions were summed up. Each ROI was set to a width of 8 eV corresponding to the FWHM of a Gaussian fit of the Auger peaks. For this purpose, PAES spectra of the pure elements were recorded and used as reference spectra for calibration. For several reasons, the Auger transitions do not appear as sharp lines in the recorded spectra. Especially, the low energy tail, which stems from inelastically scattered Auger electrons, can extend to another Auger peak with a lower energy spuriously pretending a higher intensity of this transition. In the measurements discussed here, scattered electrons from the Pd N_1VV transition at 78 eV disturb the Ni $M_{2,3}$VV and the Pd $N_{2,3}$VV transitions at 61 eV and 43 eV, respectively. However, this is a rather small effect due to the low core annihilation probability for the Pd 4s electrons, so the contribution to the background is negligibly small. Further, the Ni $M_{2,3}$VV transition leads to a significantly increased Pd $N_{2,3}$VV peak. For the data analysis, the amount of the scattered electrons in the respective ROIs of the lower energy

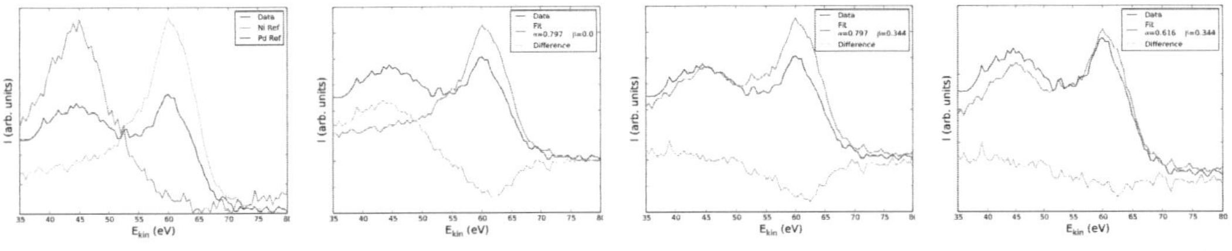

Figure 3. Illustration of the first steps of the minimization algorithm using the example of 0.5ML Ni on Pd. Starting from the left, the plots show the reference spectra of pure Ni (yellow) and pure Pd (green) and the measured spectrum of the compound (black). In the first iteration step, the scaling factor α of Ni is varied to minimize the difference R (cyan) between the measured spectrum and the fit (red). In the second step, the Pd scaling factor β is varied, while α is kept constant. Subsequently, Ni is varied again, while Pd is fixed.

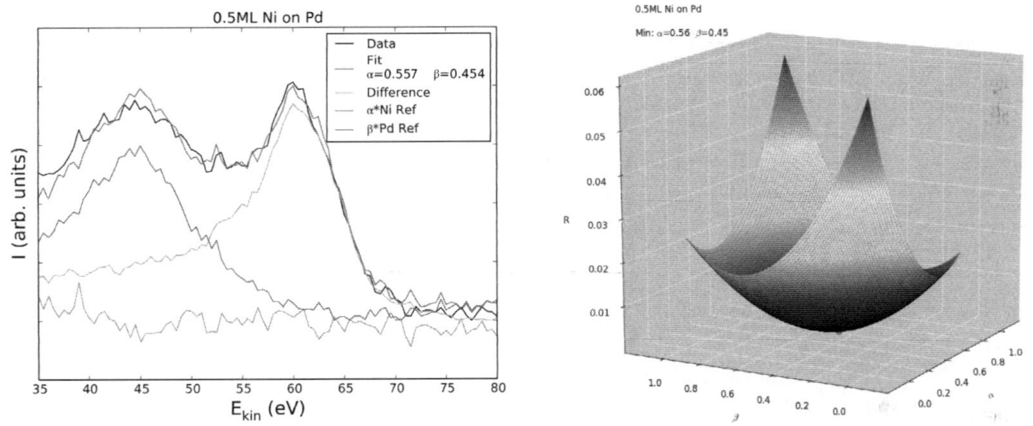

Figure 4. Result of the approximation of the measured data by reference spectra.
Left: The measured spectrum (black) is fitted (red) by independent scaling of the spectra of pure Ni (yellow) and pure Pd (green) in order to minimize the difference R (cyan) between the measured spectrum and the fit. Right: By plotting the R value for all combinations of the scaling factors α and β for Ni and Pd respectively with a step size of 0.01, it can be shown that the result of the new fitting procedure corresponds to the global minimum for the full parameter range. The minimum of R is indicated by the red dot.

peaks was determined from the reference materials and subtracted as shown for Pd in Figure 2.

2.2. Approximation by reference spectra

To overcome the special handling of scattered Auger electrons, a new approach to quantify the spectra was used. The recorded data M of the compound is reproduced by the function F, which is a linear combination of independently scaled fractions of the reference spectra R of the pure elements X and Y [5]:

$$F(i) = \alpha \cdot R_X(i) + \beta \cdot R_Y(i) \tag{1}$$

The reference spectra R_j were recorded in advance with the same measurement parameters. As a figure of merit, the fit is characterized by the value R which represents the difference between the recorded data and the fit function. This value is minimized in the fitting procedure.

$$\mathrm{R} = \sum_{i}(\mathrm{M}(i) - \mathrm{F}(i))^2 \tag{2}$$

It is not a requirement that the sum of the scaling factors has to be one. This condition would presuppose that the line shape of an Auger transition is not influenced by the presence of another element, or by chemical bonding. For this reason, the scaling factors are varied independently. For the fitting procedure, a constant background represented by the minimum value is subtracted from each spectrum and all spectra are normalized to one. Figure 3 illustrates the first few steps of the iteration on a 0.5 ML cover layer of Ni on a Pd substrate. Initially, the measured data and the reference spectra of pure Ni and Pd are shown. Then, the algorithm starts to vary the scaling factor α for the Ni reference to minimize the difference R between data and fit. In the next step, the scaling factor β for Pd is varied to minimize R, while the scaling factor α for Ni is kept constant. After this, again the Ni fraction is varied and Pd is kept constant. Up to now, the iteration is executed 20 times, though no reduction of the R value is found any more after about ten iterations for most spectra. For a faster data evaluation, an abort criterion will be implemented, to stop the fit as soon as the variation of R between two iterations is less than e.g. 1‰. The result of the minimization is shown in Figure 4. The measured data can be excellently described by a linear combination of the two reference materials over the whole energy range. By plotting the full parameter range of the scaling factors α and β, it can be shown that the result of the fitting procedure corresponds to the global minimum. The fit algorithm was also applied to Ni on Pt, Fe on Pt and Fe on Pd. Fitted and measured spectra showed a very good agreement and the resulting fractions match with the amount of cover layers deposited on the substrate. It can hence be assumed that the reproduction of the measured spectrum by a linear combination of references is possible for these systems. None of the element combinations mentioned above exhibits island growth. PAES is sensitive only to the topmost layer, and due to the three dimensional structure in case of island growth, the surface appears to have a lower coverage than the nominal value. In this case the fit might show a good agreement with the spectra, but the scaling factors differ from the amount of cover layer deposited.

3. Conclusion

In this paper we presented a new approach for quantifying PAES spectra of binary compounds by their replication with PAES reference spectra of the pure elements. The fit function is a linear combination of the reference spectra of the pure elements, each scaled with an independent factor. The quality of the fit can be quantified by the R value, representing the difference between the measured spectrum of the compound and the fit. For several combination of elements, an agreement between the measured and the fitted spectrum was achieved and the scaling factors were in good agreement with the fractions expected from the sample preparation. A comparison of the results of the algorithm with a calculation of all combinations of the scaling factor of each element proved, that the algorithm identified the best accordance between the measured spectrum and the fit.

4. Acknowledgment

Financial support by the German Federal Ministry of Education and Research within the project no. BMBF-05K13WO1 is gratefully acknowledged.

References

[1] Mehl D, Köymen A R, Jensen K O, Gotwald F and Weiss A 1990 *Phys. Rev. B* **41** 799
[2] Koymen A R, Lee K H, Mehl D, Weiss A and Jensen K O 1992 *Phys. Rev. Lett.* **68** 2378

[3] Hugenschmidt C, Dollinger G, Egger W, Kögel G, Löwe B, Mayer J, Pikart P, Piochacz C, Repper R, Schreckenbach K, Sperr P and Stadlbauer M 2008 *Appl. Surf. Sci.* **255** 29
[4] Fazleev N G, Fry J L and Weiss A H 2004 *Phys. Rev. B* **70** 165309
[5] Koymen A R, Lee K H, Yang G, Jensen K O and Weiss A H 1993 *Phys. Rev. B* **48**(3) 2020

IOP Conf. Series: Journal of Physics: Conf. Series **791** (2017) 012035 doi:10.1088/1742-6596/791/1/012035

Detection of re-emission positrons on metal surfaces during slow positron measurements

A Kinomura[1], R Suzuki[2], H Ogawa[2], N Oshima[2], B E O'Rourke[2] and A Yabuuchi[1]

[1] Research Reactor Institute, Kyoto University, Kumatori-cho, Osaka 590-0494, Japan
[2] National Metrology Institute of Japan (NMIJ), National Institute of Advanced Industrial Science and Technology (AIST), Tsukuba, Ibaraki 305-8568, Japan

E-mail: akinomura@rri.kyoto-u.ac.jp

Abstract. Detection methods for re-emitted positrons from Ni samples were investigated for in-situ positron annihilation lifetime spectroscopy during ion beam irradiation. A conventional mesh electrode and a ring electrode were examined to detect re-emitted positrons. Results of positron trajectory calculation were in good agreement with the experimental results on the mesh electrode. Simulation and experiments indicated that the ring electrode can form almost uniform electric fields in front of the sample and thus it can be used instead of the mesh electrode.

1. Introduction

It is well known that positrons have negative work functions for many metals [1]. When the positron work function is negative, some of slow positrons implanted into metals show diffusion to the surface and re-emission from the surface with kinetic energies corresponding to the absolute values of the work functions. Such unique characteristics of positrons are used in (1) positron moderators to obtain slow positron beams from high energy positron sources [1,2], (2) re-moderators for brightness enhancement to reduce positron beam sizes [3,4], and (3) re-emission microscopy [5,6]. In the re-emission microscopy including both transmission and reflection geometries, two dimensional defect distributions can be visualized due to the difference in re-emission yields between the defective and defect-free regions. Positrons in the defective region are preferentially trapped in defects and less positrons can diffuse to the surface, leading to a reduction in re-emission yield.

The method used in re-emission microscopy can be extended to in-situ positron annihilation spectroscopy under simultaneous ion beam irradiation. A dual-beam analysis system has been developed in AIST for in-situ positron annihilation lifetime spectroscopy (PALS) of ion-irradiation induced defects [7,8]. In this system, irradiation by pulsed ion and slow-positron beams can be controlled to investigate transient changes of irradiation-induced defects. Transient changes of defects were found to be detectable by re-emission yields as well as positron annihilation lifetimes. Thus, it is necessary to efficiently detect re-emitted positrons in PALS measurements during ion beam irradiation. In this study, detection methods for re-emitted positrons were investigated based on experiments as well as trajectory calculations and an optimum design for the dual-beam analysis system was proposed.

2. Experimental procedure

Positron experiments were performed at the AIST dual-beam analysis system for in-situ PALS during simultaneous ion beam irradiation [7,8]. This system consists of an electron linear accelerator (LINAC) based slow-positron beamline (positron source) combined with a small ion-accelerator (ion source). The electron beam from the AIST electron LINAC was used with pulse repetition rates of 25-50 pps. A slow positron beam was generated by a Ta converter and a W moderator and magnetically guided to the sample chamber. The positron beamline has pulsing electrodes (i.e., chopper, prebuncher and buncher electrodes) for lifetime spectroscopy. Annihilation gamma-rays were detected by a BaF_2 scintillation detector. Lifetime spectra were obtained by a fast waveform digitizer controlled by a personal computer [9]. An inert-gas ion beam (e.g., Ar^+ and He^+) was generated by the ion accelerator connected to the positron beamline with an acceleration voltage up to 150 kV. The ion-beam current was measured by a Faraday cup installed on the sample holder. The incident angle of the ion beam was 45° off the positron beam direction. Spot sizes of the positron and ion beams were approximately 1 cm and 1.5 cm in diameter, respectively. Mirror-polished pure Ni samples after annealing at 900 °C for 1h in vacuum were used in this study. The samples were heated to 500 °C to observe re-emitted positrons during the PALS measurements. Positron trajectory calculations were performed using the SIMION code [10].

Figure 1 shows the configurations of suppression electrodes around the sample holder used in this study. After the pulsing electrodes at the right-hand side (not shown in this figure), the positron beam is incident on the sample after passing through a drift tube and a cup electrode. The ion beam is also directed to the sample through a hole on the side of the cup electrode. Two type of suppression electrodes were installed in front of the sample to examine the effect of re-emitted positrons on lifetime spectra. The electrodes in figure 1 are cylindrically symmetric except for small parts including the ion-beam entrance hole of the cup electrode. The exit of the drift tube and positron-beam entrance hole of the cup electrode have meshes to ensure a uniform electric field for acceleration between two electrodes. High voltage (HV) was applied to the sample (sample holder) and the cup electrode to determine the final positron acceleration energy. A bias voltage to control the suppression electric fields near the sample was applied to the mesh- or ring-electrode.

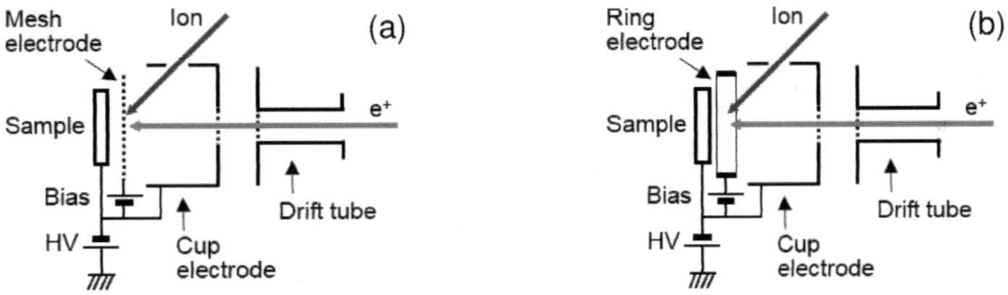

Figure 1. Configurations of suppression electrodes around the sample holder.

3. Results and discussion

Positrons re-emitted from the surface can be detected by using suppression electrodes that push re-emitted positrons back to the sample, leading to formation of satellite peaks in lifetime spectra. Suzuki *et al.* evaluated re-emission efficiencies for several remoderator materials and indicated that the positions of satellite peaks originating from re-emitted positrons depend on the bias voltage applied to the suppression electrode [2]. They used a mesh electrode as the suppression electrode to detect re-emitted positrons. When appropriate bias voltages are applied to the suppression electrode, incident positrons with an acceleration energy of several keV can go through the electrode but re-emitted positrons with energies of several eV are pushed back to the sample. The same type of the suppression electrode was used during the initial stage of this study as shown in figure 1(a).

The bias voltage applied to the mesh electrode of figure 1(a) was changed during PALS measurements of Ni with a positron energy of 2 keV. Figure 2 shows the PALS spectra of Ni as a function of the bias voltage, where E_p is the positron energy. When the bias voltage was 0 V, no appreciable peak was observed [figure 2(a)]. However, with a bias voltage of 2 V, a satellite peak was observed around 17 ns. With increasing bias voltage from 2 V to 20 V, the time difference between the primary peak (annihilation peak at 0 ns) and the satellite peak decreased [figure 2(b)]. This trend was exactly the same as that observed in the previous study [2]. Positrons re-emitted from the sample have kinetic energies comparable to the negative work function of the sample. Initially the re-emitted positrons are emitted in the opposite direction to the incident positron beam. However, such positrons are pushed back to the sample by the electric field formed by the suppression electrode, giving rise to another annihilation peak corresponding to flight times (i.e., satellite peak). Trajectory simulation for the re-emitted positrons were performed using the SIMION code. Structures of the sample holder and surrounding electrodes were modelled as closely as possible in the simulation. The electric field formed between the sample and the suppression mesh electrode was found to be uniform (not shown). Flight times of re-emitted positrons were plotted as a function of the bias voltage for three re-emission energies (0.5, 1.0 and 1.5 eV) in figure 3, together with experimental values from figure 2. The time of flight (vertical axis) decreased with increasing the bias voltage. The experimental values were in good agreement with the curve calculated for the re-emission energy of 1.0 eV. This value (1.0 eV) is comparable with the reported work function of Ni [1], although more precise measurements are required to determine the work function. At least, the results in figure 3 indicate that the SIMION simulation is effective to evaluate the suppression behavior of re-emitted positrons.

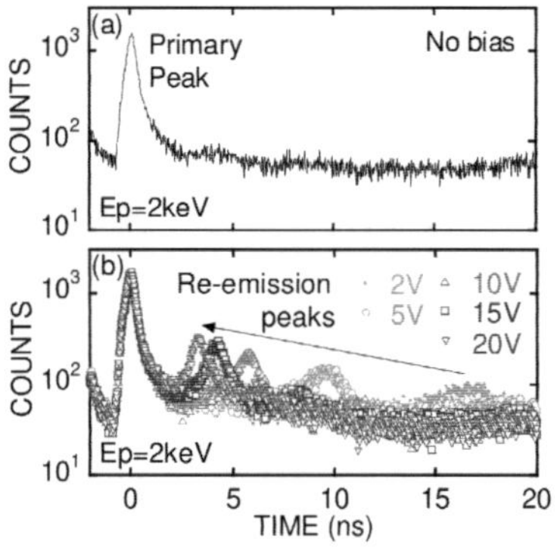

Figure 2. Positron lifetime spectra indicating re-emission peaks as a function of bias voltage to the mesh electrode.

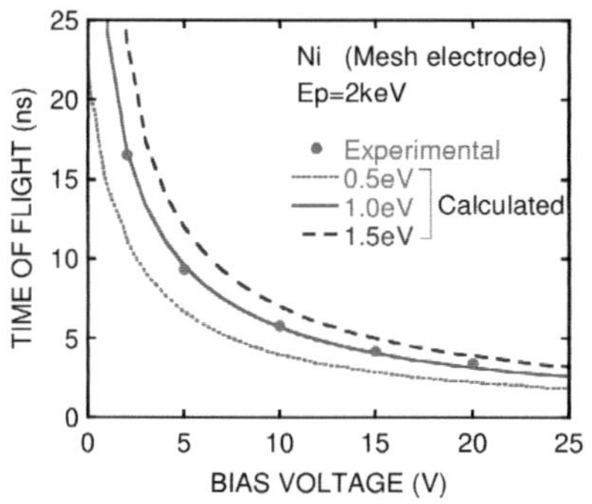

Figure 3. Calculated time of flight as a function of bias voltage to the mesh electrode for different re-emission energies.

The mesh electrodes were effectively used to observe re-emitted positrons. However, in the case of the dual-beam system, an ion beam is incident on the sample to introduce defects simultaneously with the positron beam for PALS measurements. If the ion beam hits the mesh electrode, it induces sputtering particles and secondary electrons from the mesh. Such secondary particles may negatively influence in-situ PALS measurements during ion beam irradiation. Thus, we introduced an alternative suppression electrode as shown in figure 1(b). This electrode with a ring shape allows us to introduce the ion beam without hitting a suppression electrode. It is easy to understand that the mesh electrode forms the uniform electric field for the suppression of re-emitted positrons, as it is a typical parallel-plate electrode structure. However, it is necessary to confirm the electric field formed by the ring electrode.

In the same way as the mesh electrode, the simulation for the ring electrode was performed to confirm the electric field and flight times of re-emitted positrons. Figure 4 shows the equipotential lines calculated by the SIMION code in the range of -5000 V to -4980 V (2 V step) when the bias voltage of 10 V was applied to the ring electrode. The structures shown in figure 4 include a rod to support the sample holder (upper side), a Faraday cup to measure ion beam current (lower side) and other supporting plates. The cup electrode was installed to reduce the irregular change of the electric field between the sample and the drift tube. The combination of the cup and ring electrodes successfully achieved the almost uniform electric field in front of the sample, as shown in figure 4.

Figure 5 shows the positron lifetime spectrum of Ni at 500 °C with a bias voltage of 10 V. A positron energy of 5 keV was chosen to efficiently detect transient defect changes in in-situ PALS measurements. In addition to the primary peak at 0 ns, two satellite peaks were observed at 2.5 and 5.2 ns. The flight time of positrons re-emitted with an energy of 1.0 eV was calculated to be 4.6 ns, in reasonable agreement with the experimentally obtained value. The potential distribution formed by a ring electrode may be more sensitive to the small differences between the actual and the modelled electrode structures, compared with that of the mesh electrode. The simulation also showed that positrons backscattered at 2.5 keV form a peak around 2.5 ns. The satellite peaks observed in figure 5 can be attributed to the re-emitted and backscattered positrons. Thus, figures 4 and 5 showed that the ring electrode can be used instead of the mesh electrode to detect re-emitted positrons in lifetime spectra.

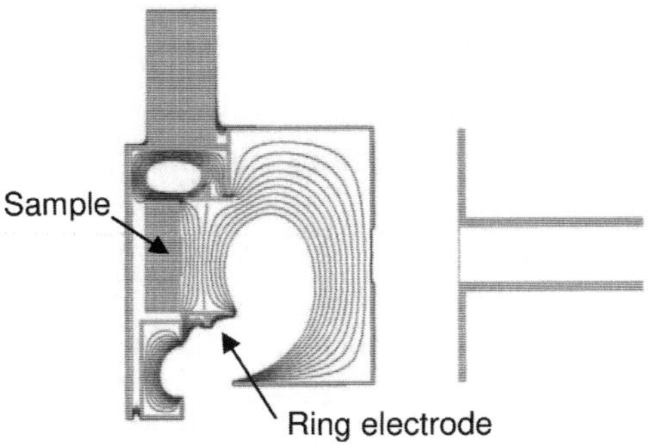

Figure 4. Equipotential lines (2 V step) calculated by the SIMION code for the ring electrode with a bias voltage of 10 V.

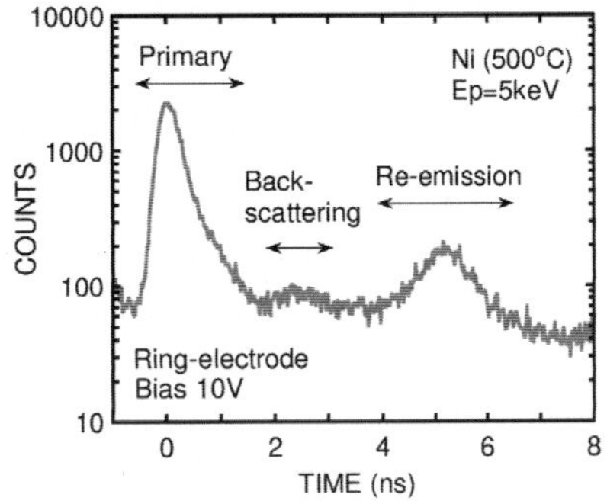

Figure 5. Positron lifetime spectrum of Ni at 500 °C with the ring electrode and bias voltage of 10 V.

4. Conclusion

Detection methods for re-emitted positrons were investigated for an in-situ PALS system during simultaneous ion beam irradiation. Mesh and ring electrodes were examined as suppression electrodes to detect re-emitted positrons from Ni samples. Positron trajectory simulation effectively explained positron flight times (i.e., satellite peak positions in lifetime spectra) in experiments with the mesh electrode. Both simulation and experiment showed that a ring electrode can be used instead of a mesh electrode to detect re-emitted positrons.

Acknowledgments

The authors would like to thank T. Nishijima and H. Tsuchida for their assistance on this study. This study was partially supported by JSPS KAKENHI Grant Number 2431008.

References

[1] Schultz P J and Lynn K G 1988 *Rev. Mod. Phys.* **60** 701
[2] Suzuki R *et al.* 1998 *Jpn. J. Appl. Phys.* **37** 4636
[3] Mills A P Jr 1980 *Appl. Phys.* **23** 189
[4] Oshima N *et al.* 2008 *J. Appl. Phys.* **103** 094916
[5] Brandes G R *et al.* 1988 *Phys. Rev. Lett.* **61** 492
[6] Van House J and Rich A 1988 *Phys. Rev. Lett.* **61** 488
[7] Kinomura A *et al.* 2012 *Phys. Procedia* **35** 151
[8] Kinomura A *et al.* 2013 *J. Phys. Conf. Ser.* **443** 012043
[9] Kinomura A *et al.* 2014 *Rev. Sci. Instrum.* **85** 123110
[10] See for example http://simion.com/

IOP Conf. Series: Journal of Physics: Conf. Series **791** (2017) 012036 doi:10.1088/1742-6596/791/1/012036

Surface states and positron annihilation spectroscopy: results and prospects from a first-principles approach

V Callewaert[1], R Saniz[1], B Barbiellini[2] and B Partoens[1]

[1] Departement Fysica, Universiteit Antwerpen, Groenenborgerlaan 171, B-2020 Antwerpen, Belgium

[2] Department of Physics, Northeastern University, Boston, Massachusetts 02115, USA

E-mail: vincent.callewaert@uantwerpen.be

Abstract. The trapping of positrons at the surface of a material can be exploited to study quite selectively the surface properties of the latter by means of positron annihilation spectroscopy techniques. To support these, it is desirable to be able to theoretically predict the existence of such positronic surface states and to describe their annihilation characteristics with core or valence surface electrons in a reliable way. Here, we build on the well-developed first-principles techniques for the study of positrons in bulk solids as well as on previous models for surfaces, and investigate two schemes that can improve the theoretical description of the interaction of positrons with surfaces. One is based on supplementing the local-density correlation potential with the corrugated image potential at the surface, and the other is based on the weighted-density approximation to correlation. We discuss our results for topological insulators, graphene layers, and quantum dots, with emphasis on the information that can be directly related to experiment. We also discuss some open theoretical problems that should be addressed by future research.

Surface phenomena play a fundamental role in determining the properties of materials of great current interest. Prominent examples are topological insulators and nanostructured systems, such as quantum dots and graphene layers. To study these systems it is preferable to use a technique capable of probing the surface of a material with as little interference from the bulk as possible. Indeed, if the information collected can be ascribed to the response of the surface alone, it can provide deeper and unambiguous insight into the properties of the material studied. Advanced positron annihilation spectroscopy (PAS) techniques are nearly ideally suited for this purpose. Positrons can get trapped at the surface of a material via, e.g., diffusion or Auger Mediated Positron Sticking (AMPS) [1, 2]. In such a case, the radiation following annihilation will contain quite direct information on the surface electronic properties, structure and/or composition [2–4].

To disentangle the information that experiments collect, it would be advantageous to count on theoretical information in various respects. First, it would be useful to know whether a surface can support positronic surface states, and what their binding energies would be. Furthermore, details about annihilation characteristics, such as lifetimes and annihilation probabilities with core and valence electrons, would be of great value. First-principles methods offer the most reliable tools for this purpose. To apply these, however, we face the challenge of describing

IOP Conf. Series: Journal of Physics: Conf. Series **791** (2017) 012036 doi:10.1088/1742-6596/791/1/012036

properly electron-positron correlations in the surface region. Indeed, while quite reliable approximations to electron-positron correlation for studies in the bulk exist [5], this is not the case for surfaces. The image form of the correlation potential in the vacuum away from the surface is not reproduced by the current approximations, and the appropriate form of electron-positron correlation in the low electron density region at the surface is difficult to model. Although this has been recognised early on [1, 5, 6], a completely satisfactory solution has not been found so far.

The present work is performed in the zero positron density limit of the well-known two-component electron-positron density functional theory [1, 5, 7]. In this limit the positron does not perturb the electron density. We thus calculate the latter following a standard ab initio method. For this we use the VASP ab initio package [8, 9], using the projector-augmented-wave (PAW) method [10] and the Perdew-Burke-Ernzerhof (PBE) functional [11]. We take into account spin-orbit coupling. The density thus obtained is used to calculate the electron-positron correlation potential in the Kohn-Sham hamiltonian for the positron. The Kohn-Sham equations corresponding to the positron are then solved using the Doppler/MIKA package [12].

To model electron-positron correlation at the surface we follow two different approaches, applying them to well-studied systems, as mentioned above. Our first approach consists in using the local-density (LDA) form of the correlation in the bulk region, and supplement it with the corrugated mirror, or image potential, model in the surface region [13]. Specifically, if z is the coordinate perpendicular to the surface, the correlation potential is

$$v_{\text{corr}}(\mathbf{r}) = \begin{cases} v_{e\text{-}p}^{\text{LDA}}(\mathbf{r}), & z \leq z_0 \\ \max(v_{\text{im}}(\mathbf{r}), v_{e\text{-}p}^{\text{LDA}}(\mathbf{r})), & z > z_0. \end{cases} \tag{1}$$

Here, the corrugated image potential is given by

$$v_{\text{im}}(\mathbf{r}) = -\frac{e}{4(z_{\text{eff}}(\mathbf{r}) - z_0)}, \tag{2}$$

with z_0 the image potential reference plane and

$$z_{\text{eff}}(\mathbf{r}) = \int_{z_0}^{\infty} dz\, z\, \delta(n_e(\mathbf{r}) - \langle n_e \rangle(z)), \tag{3}$$

with n_e the electronic density and the brackets indicating its average over the unit cell in planes parallel to the surface. We applied this model to Bi_2Te_2Se, a well-known topological insulator. A fuller account of our results will be published elsewhere [14]. Here we focus on the existence of a surface state and on some of the information it can yield. We model the system with a four quintuple-layer slab perpendicular to the [111] direction (rhombohedral unit cell), exposing its Te-terminated surface, as in experiment [15]. To give an idea of the potential seen by the positron near the surface, in Fig. 1(a) we plot the average of the potential, over planes parallel to the surface, within the unit cell. One can see that the correlation potential has an incorrect behaviour away from the surface, and deviates sharply from the image potential. The latter is also shown. The total potential shown is the image-potential corrected potential as defined in Eq. (1). The solution of the Kohn-Sham equations results in a positron groundstate residing in the potential well at the surface, with a binding energy $E_b = 2.69$ eV. The experimental binding energy can be deduced as follows. In AMPS measurements, the maximum kinetic energy of the emitted Auger electrons is related to the incident positron energy, E_p, and the positron binding energy by $E_{\text{max}} = E_p + E_b - \phi_e$, with ϕ_e the electron work function. Experiment shows that the incident positron threshold energy for Auger electron emission is 1.8 eV [14], i.e., $E_p + E_b - \phi_e = 0$. On the other hand, the positron binding energy is related to the

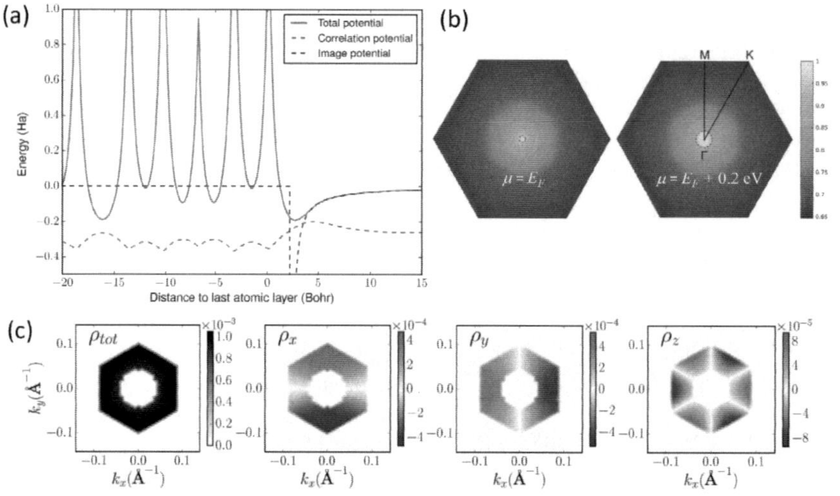

Figure 1. (a) Potential seen by the positron at the surface of Bi_2Te_2Se. The total potential results from replacing the LDA correlation potential away from the surface by the image potential, as in Eq. (1). (b) The LCW maps over the (2D) Brillouin zone of Bi_2Te_2Se, for the electronic chemical potentials indicated in the plots. The break a the Fermi momenta clearly reveals the small circular Fermi surfaces (dotted circles). (c) Magnetisation components for states around the Fermi surface (between E_F and $E_F + 0.2$ eV). The leftmost plot shows the total amplitude of the LCW map. The other plots show the components indicated. (Figs. 1(b) and 1(c) adapted with permission from Ref. 14. Copyrighted by the American Physical Society.)

positronium thermal activation energy, E_a, by $E_b = E_a - \phi_e + 6.8$ eV [16]. Given that E_a has been measured to be 0.4 eV [14,17], the experimental value of E_b is found to be 2.7 eV, showing excellent agreement between theory and experiment. We consider now the possible information that the annihilation of such a state can deliver. In Bi_2Te_2Se, the Fermi level is just above the Dirac point (and can actually be tuned by alloying with Sb), so the Fermi surface consists of a relatively small circle around the Γ point [15]. Since the positron is at the surface and the Dirac cone in Bi_2Te_2Se arises from electronic surface states, two-dimensional angular correlation of annihilation radiation (2D-ACAR) measurements can be expected to detect the Fermi surface. In Fig. 1(b) we show the calculated Lock-Crisp-West (LCW) maps [18] for two different values of the electronic chemical potential, namely at the undoped material Fermi energy value, and 0.2 eV above it. The Fermi surface is indicated by the dotted line. We note that the drop in intensity at the Fermi momenta is 5%-7%, so should be clearly observable in experiment. Indeed, the drop is much larger than the 1% in the case of the $Nd_{2-x}Ce_xCuO_{4-\delta}$, in which 2D-ACAR has been successfully applied to map the Fermi surface [19] and, moreover, the momentum resoltion of current 2D-ACAR spectrometers is sufficiently high [20]. A further feature of the Dirac cone that is important, but difficult to observe experimentally, is the spin texture associated with the hexagonal warping [21,22]. We can extract the signal from the Dirac cone in the vicinity of the Fermi surface by taking the difference of the LCW maps at $\mu = E_F + 0.2$ eV and $\mu = E_F$. The magnetisation components of the momentum density from the Dirac cone should be observable using spin-polarised positron beams. The calculated magnetisation components in the region delimited by the two chemical potentials mentioned are shown in Fig. 1(c). These clearly show (i) the hexagonal warping as momenta move away from the Fermi surface, (ii) that the in-plane (xOy) magnetisation rotates (clockwise) with momentum, and (iii) that there is an out-of-plane

(z direction) component that increases with warping. Thus, these important features can be experimentally studied using spin-polarised PAS techniques.

As mentioned in the introduction, quantum dots and nanocrystals are among the systems in which surface states are determinant. PAS techniques, such as 2D-ACAR and positron annihilation lifetime spectroscopy (PALS) were used to show that positrons can get trapped at the surface of CdSe and PbSe nanocrystals [3, 4]. We applied the preceding scheme to see if we could observe these states and calculate their binding energy. We modeled the surface of the nanocrystals with slabs. Our first results indicated that while for CdSe the potential well at the surface was deep enough to trap a positron, this was not the case in PbSe. Also, in the case of CdSe, the positron state seems to overlap more strongly with the surface than in experiment, resulting in a too short lifetime. At this point we must recognize that the approach based on Eq. (1) is ad hoc, and there is no guarantee that it is reliable in all cases. Our second approach to describing a positron at the surface of a solid is based on the the weighted-density approximation (WDA) [23], as modified by Rubaszek for positron surface states [24, 25]. It relies on a non-local electron-positron correlation functional that connects smoothly the LDA correlation inside the solid to the mirror-like correlation far outside from the surface. Rubaszek used it to describe positron states at the surface of metals, describing the latter with the jellium model. Briefly, the electron-positron correlation potential can be written

$$v_{e\text{-}p}(\mathbf{r}_p) = -\frac{1}{2}\int d\mathbf{r}_e \frac{\Delta n_c(\mathbf{r}_e|\mathbf{r}_p)}{|\mathbf{r}_e - \mathbf{r}_p|}, \tag{4}$$

with the screening cloud given by $\Delta n_c(\mathbf{r}_e|\mathbf{r}_p) = n_e(\mathbf{r}_e)[g_{n_e}(\mathbf{r}_e|\mathbf{r}_p) - 1]$, where g_{n_e} is the electron positron pair correlation function (a functional of the electron density and dependent on the electron and positron coordinates) [26]. In the LDA approximation,

$$\Delta n_c^{\mathrm{LDA}}(\mathbf{r}_e|\mathbf{r}_p) = n_e(\mathbf{r}_p)[g_{n_e(\mathbf{r}_p)}(|\mathbf{r}_p - \mathbf{r}_e|) - 1], \tag{5}$$

i.e., the pair correlation function is given the form it has for a positron in a homogeneous electron gas (n_e is evaluated at the position of the positron only and the pair correlation function depends on the distance between electron and positron). In the WDA, the screening cloud is written

$$\Delta n_c^{\mathrm{WDA}}(\mathbf{r}_e|\mathbf{r}_p) = n_e(\mathbf{r}_e)[g_{n^*(\mathbf{r}_p)}(|\mathbf{r}_p - \mathbf{r}_e|) - 1]. \tag{6}$$

Here, the n_e is evaluated at the position $\mathbf{r}_e$. In contrast with the LDA, the screening cloud depends non-locally on the electron density. The pair correlation has the same form as in the LDA, but it depends on an effective, or "weighted" density, obeying to the sum rule

$$\int d\mathbf{r}_e \Delta n_c^{\mathrm{WDA}}(\mathbf{r}_e|\mathbf{r}_p) = 1. \tag{7}$$

We first applied this approach to our Bi_2Te_2Se, and CdSe and PbSe surfaces. In the first case, the WDA and the LDA+corrugated image potential surface states are nearly the same. Our preliminary results in the case of CdSe and PbSe, on the other hand, show a stronger difference. In particular, the positron state in PbSe is clearly located on the vacuum side of the slab, instead of between the first and second layer of the material (which is the case in the LDA+corrugated image potential approach). In the case of CdSe, the positron surface state is similarly pushed further away from the surface. We note, however, that these systems are more complicated than layered systems like Bi_2Te_2Se in that surface relaxation is stronger and therefore more difficult to describe accurately and compare with experiment. In the case of Bi_2Te_2Se, experiments are carried out on nearly perfect surfaces, so comparison with calculations is more straightforward. In the case of the nanocrystals, it might be necessary to take into account the effect of the

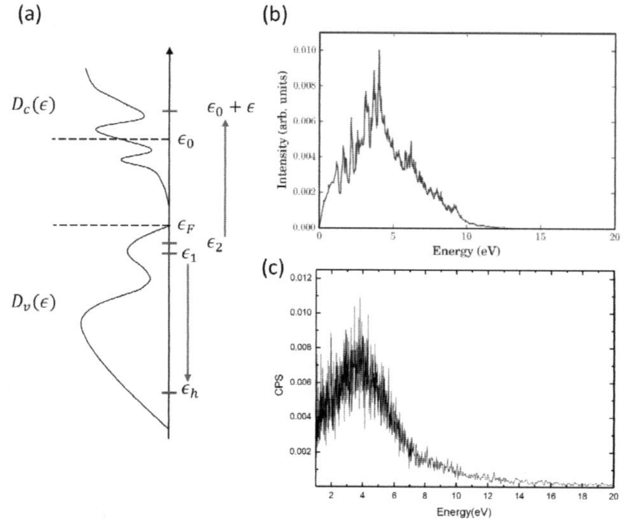

Figure 2. (a) Illustration of a VVV PAES process. A positron creates a hole in the valence band at energy ϵ_h. An electron with energy ϵ_1 recombines with the hole, transferring the energy difference to an electron of energy ϵ_2 that might escape if its final energy is above the vacuum level, ϵ_0. Plots (b) and (c) show the calculated and measured Auger spectra, respectively, showing remarkable overall agreement.

ligands [3, 4] in our calculations. Further research is necessary to verify our results and to obtain lifetimes and other annihilation characteristics.

We further applied this approach to determine whether a nanofilm of graphene sustains a positron surface state and, if so, to calculate its wavefunction. This is another layered system of great interest because recent experiments by Weiss and co-workers strongly suggest that in their positron-annihilation Auger-electron spectroscopy (PAES) measurements they have observed VVV processes, i.e., Auger processes involving solely valence states (in contrast to standard Auger spectroscopy, where the initial hole is at a core state) [27]. The process is illustrated in Fig. 2(a). Given the positron surface state wavefunction, the rate at which it annihilates with a valence electron with energy ϵ_h (thus creating an initial hole there) is readily given by

$$\lambda(\epsilon_h) = \frac{1}{\Omega_{\rm BZ}} \sum_i \int_{\Omega_{\rm BZ}} d\mathbf{k}\lambda_i(\mathbf{k}) f(\epsilon_i(\mathbf{k}))\, \delta(\epsilon_h - \epsilon_i(\mathbf{k})), \tag{8}$$

with f the Fermi distribution and the annihilation rate with an electron in band i and momentum $\mathbf{k}$ given as usual by $\lambda_i(\mathbf{k}) = \pi r_e^2 c \int d\mathbf{r} |\psi_{i,\mathbf{k}}(\mathbf{r})|^2 |\psi_p(\mathbf{r})|^2 \gamma(n_e(\mathbf{r}))$, where ψ_p denotes the positron wavefunction and γ the enhancement factor [1, 5]. Following the work of Hagstrum on Auger processes [28], the Auger spectrum in our case is given by

$$A(\epsilon) = \int d\epsilon_h\, d\epsilon_1\, d\epsilon_2 \lambda(\epsilon_h) D_v(\epsilon_1) \Theta(\epsilon_1 - \epsilon_h) D_v(\epsilon_2) D_c(\epsilon_0 + \epsilon) P(\epsilon_2, \epsilon) \delta((\epsilon_1 - \epsilon_h) + (\epsilon_2 - \epsilon_0 - \epsilon)). \tag{9}$$

Here D_v and D_c denotes the valence and conduction bands densities of states, ϵ_0 is the vacuum level energy, and $P(\epsilon_2, \epsilon) = [1 - \sqrt{(\epsilon_0 - \epsilon_2)/(\epsilon_0 + \epsilon - \epsilon_2)}]/2$ is the electron "escape function" [28]. The calculations were done for a graphene layer. Fig. 2(b) shows the calculated Auger spectrum, while Fig. 2(c) shows one of various preliminary experimental results. The common feature in experiment is a dominant peak around 4 eV, in striking coincidence with experiment. Further details will be presented in a forthcoming publication. The main purpose here is to show the usefulness of first-principles calculations and the power of PAS techniques.

Acknowledgments
We thank A. Weiss for very useful conversations and for his preliminary Auger results. We further benefited from clarifying discussions with S. Eijt on the positron annihilation characteristics of the nanocrystals investigated in his laboratory. We acknowledge financial support from FWO-Vlaanderen (projects G.0150.13 and G.0224.14N). This work was carried out using the HPC infrastructure of the University of Antwerp (CalcUA), a division of the Flemish Supercomputer Center (VSC), funded by the Hercules foundation and the Flemish Government (EWI Department). The work at Northeastern University was supported by the US Department of Energy (DOE), Office of Science, Basic Energy Sciences grant number DE-FG02-07ER46352 (core research), and benefited from Northeastern University's Advanced Scientific Computation Center (ASCC), the NERSC supercomputing center through DOE grant number DE-AC02-05CH11231, and support (applications to layered materials) from the DOE EFRC: Center for the Computational Design of Functional Layered Materials (CCDM) under DE-SC0012575.

References

[1] Puska M J and Nieminen R M 1994 *Rev. Mod. Phys.* **66** 841
[2] Mukherjee S, Nadesalingam M P, Guagliardo P, Sergeant A D, Barbiellini B, Williams J F, Fazleev N G and Weiss A H 2010 *Phys. Rev. Lett.* **104** 247403
[3] Eijt S W H, Van Veen A, Schut H, Mijnarends P E, Denison A B, Barbiellini B and Bansil A 2006 *Nat. Mater.* **5** 23
[4] Chai L *et al.* 2013 *APL Materials* **1** 022111
[5] Tuomisto F and Makkonen I 2013 *Rev. Mod. Phys.* **85** 1583
[6] Rubaszek A, Kiejna A and Daniuk S 1993 *J. Phys.: Condens. Matter* **5** 8195
[7] The zero density limit of the two-component density-fnctional theory remains valid at a perfect surface as the positron ground state will be delocalized in at least in the two directions parallel to the surface.
[8] Kresse G and Furthmüller J 1996 *Comput. Mat. Sci.* **6** 15
[9] Kresse G and Furthmüller 1996 *Phys. Rev. B* **54** 11169
[10] Kresse G and Joubert D 1999 *Phys. Rev. B* **59** 1758
[11] Perdew J P, Burke K and Ernzerhof M 1996 *Phys. Rev. Lett.* **77** 3865
[12] Makkonen I, Hakala M and Puska M J 2006 *Phys. Rev. B* **73** 035103
[13] Nieminen R M and Puska M J 1983 *Phys. Rev. Lett.* **50** 281
[14] Callewaert V, Shastry K, Saniz R, Makkonen I, Barbiellini B, Assaf B A, Heiman D, Moodera J S, Partoens B, Bansil A and Weiss A H 2016 *Phys. Rev. B* **94** 115411
[15] Neupane M *et al.* 2012 *Phys. Rev. B* **85** 235406
[16] Chu S, Mills A P and Murray C A 1981 *Phys. Rev. B* **23** 2060
[17] Shastry K, Weiss A H, Barbiellini B, Assaf B A, Lim Z H, Joglekar P V and Heiman D 2015 *J. Phys. Conf. Ser.* **618** 012006
[18] Lock D G, Crisp V H C and West R N 1973 *J. Phys. F: Metal Phys.* **3** 561
[19] Shukla A, Barbiellini B, Hoffmann L, Manuel A A, Sadowski W S, Walker E and Peter M 1996 *Phys. Rev. B* **53** 3613
[20] Dugdale S B, Laverock J, Utfeld C, Alam M A, Haynes T D, Billington D and Ernsting D 2013 *J. Phys. Conf. Ser.* **443** 012083
[21] Basak S, Lin H, Wray L A, Xu S Y, Hasan M Z and Bansil A 2011 *Phys. Rev. B* **84** 121401
[22] Fu L 2009 *Phys. Rev. Lett.* **103** 266801
[23] Gunnarsson O, Jonson M and Lundqvist B I 1979 *Phys. Rev. B* **20** 3136
[24] Rubaszek A 1989 *J. Phys.: Condens. Matter* **1** 2141
[25] Rubaszek A 1991 *Phys. Rev. B* **44** 10857
[26] Note that it can be shown that if the generalised-gradient approximation (Barbiellini B and Kuriplach J 2015 *Phys. Rev. Lett.* **114** 147401) were used instead of the LDA, the correlation potential outside the surface would have an unphysical faster-than-exponential decay.
[27] Chirayath V A *et al* 2016 arXiv:1610.09575v2 [cond-mat.mess-hall]
[28] Hagstrum H D 1954 *Phys. Rev.* **96** 336; 1961 *Phys. Rev.* **122** 83

IOP Conf. Series: Journal of Physics: Conf. Series **791** (2017) 012037 doi:10.1088/1742-6596/791/1/012037

Spin-polarized first-principles calculation of momentum densities of Fe

Itaru Sugita[1] and Mineo Saito[2]
[1]Graduate School of Natural Science and Technology, Kanazawa University, Kanazawa 920-1192, Japan
[2]Division of Mathematical and Physical Sciences, Kanazawa University, Kanazawa 920-1192, Japan
E-mail: m-saito@se.kanazawa-u.ac.jp

Abstract. We perform first-principles calculations on the momentum densities in Fe. The ridge and valley appear on the Σ and Δ lines, respectively, on the (001) plane. This anisotropic distribution of the momentum densities is found to originate from the difference between symmetries of the $\Sigma(C_{2v})$ and $\Delta(C_{4v})$ lines. The calculated one-dimensional momentum densities of the majority spin have a broader distribution than those of the minority spin, which is expected to be due to the fact that the distribution in the real space of the majority spin is narrower than that of the minority spin.

1. Introduction

Electron momentum densities are detected by experiments of positron annihilations and Compton scattering and provide useful information on electronic structures [1-2]. In positron annihilation experiments, two-dimensional angular correlation of annihilation radiation (2D-ACAR) and Doppler broadening annihilation radiation (DBAR) spectra detect two-dimensional and one-dimensional momentum densities, respectively. Recently, spin-polarized positron annihilation experiments attracted scientific interests since they can be applied to studies of spintronics and magnetic materials [3-4]. Spin-polarized DBAR spectra have been observed and thus the study of the momentum densities for the majority and minority spins becomes important [3-4].

In this paper, we perform calculations of momentum densities of the spin-polarized material, Fe. First, we find the prominent anisotropy of the momentum densities on the (001) plane; the momentum densities on the Δ line are smaller than those on the Σ line. We analyze the anisotropy based on the group theory [1,5] and find that the Δ and Σ lines have high (C_{4v}) and low (C_{2v}) symmetries, respectively. This difference in symmetry is the origin of the anisotropy. Next, we find that the momentum distribution of the majority spin is broader than that of the minority spin, which is consistent with experimental results of DBAR [3]. This difference is expected to originate from the fact that distribution of the majority spin densities in the real space is narrower than that of the minority spin densities [6].

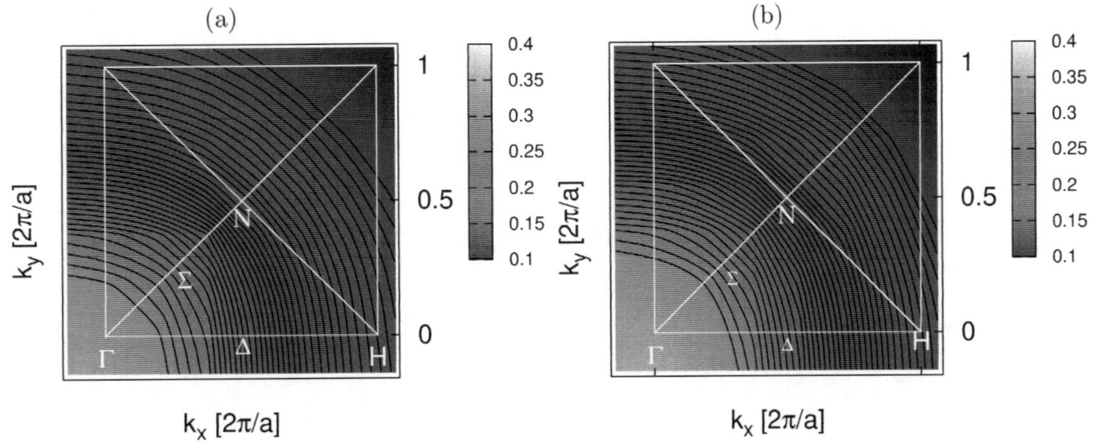

Figure 1. Momentum densities on the (001) plane of the majority spin (a) and minority spin (b).The values of the contour lines are in the arbitrary units (a.u.).

2. Method

The Bloch wave function, $\Psi_n^{\mathbf{k}}$, is given by

$$\Psi_n^{\mathbf{k}}(\mathbf{r}) = \psi_n^{\mathbf{k}}(\mathbf{r}) \exp(i\mathbf{k} \cdot \mathbf{r}), \tag{1}$$

where n and $\mathbf{k}$ is the band index and the wavenumber, respectively, and $\psi_n^{\mathbf{k}}(\mathbf{r})$ is the periodic function. The electron momentum density for the momentum $\mathbf{p}$ in the first Brillouin zone (FBZ) is given by [1]

$$\rho(\mathbf{p}) = \sum_n^{occ.} \left| \int_{cell} \psi_n^{\mathbf{p}}(\mathbf{r}) d\mathbf{r} \right|^2. \tag{2}$$

When the momentum is outside the FBZ ($\mathbf{p}' = \mathbf{p} + \mathbf{G}$), the momentum density is given by

$$\rho(\mathbf{p}') = \sum_n^{occ.} \left| \int_{cell} \psi_n^{\mathbf{p}}(\mathbf{r}) \exp(i\mathbf{G} \cdot \mathbf{r}) d\mathbf{r} \right|^2, \tag{3}$$

where $\mathbf{G}$ is a reciprocal lattice vector. The momentum densities deduced from positron annihilation experiments include the positron wavefunction effect but we do not consider this effect in Eqs.(2) and (3). The density is also affected by electron-positron correlation and electron-electron correlation which is imperfectly included in the density functional theory (DFT) calculations. Indeed a previous study suggests these effects on the 2D-ACAR by comparing experimental and calculational results [7]. However, since these effects are small, we neglect these effects in our study.

We perform DFT calculations to obtain the Block wave function $\Psi_n^{\mathbf{k}}(\mathbf{r})$. We use the spin-polarized generalized gradient approximation. The cutoff energies of the wave function and charge density are 25.0 Ry and 225.0 Ry, respectively. The number of the sampling k points is $20 \times 20 \times 20$. The lattice constant is deduced from a structure optimization calculation (2.845 Å).

3. Results and discussion

3.1. Three-dimensional momentum densities

We calculate momentum densities of the majority and minority spins on the (001) plane (Figure 1). We find that the ridge appears on the Γ-N (Σ) line and the valley appears on the Γ- H (Δ) line

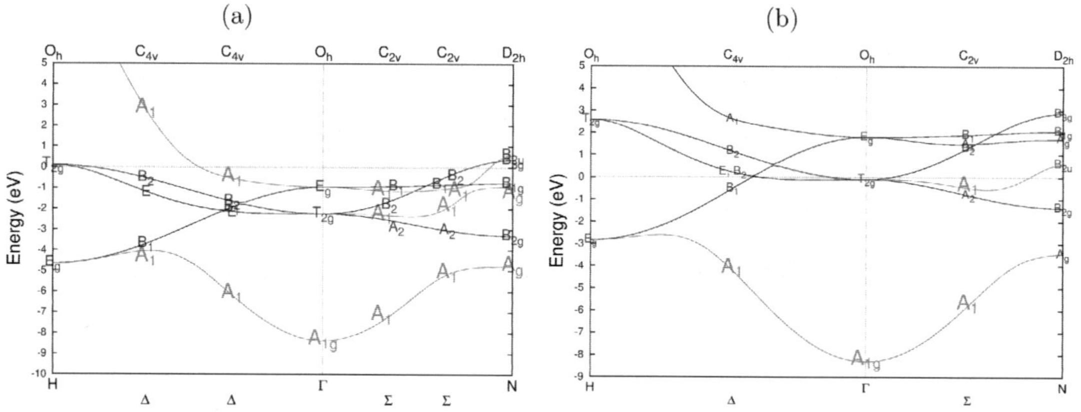

Figure 2. Band structure of the majority spin(a) and minority spin(b). The red lines represent totally symmetric occupied bands and black lines do non-totally-symmetric and/or unoccupied bands. The band energies are measured from the Fermi energy.

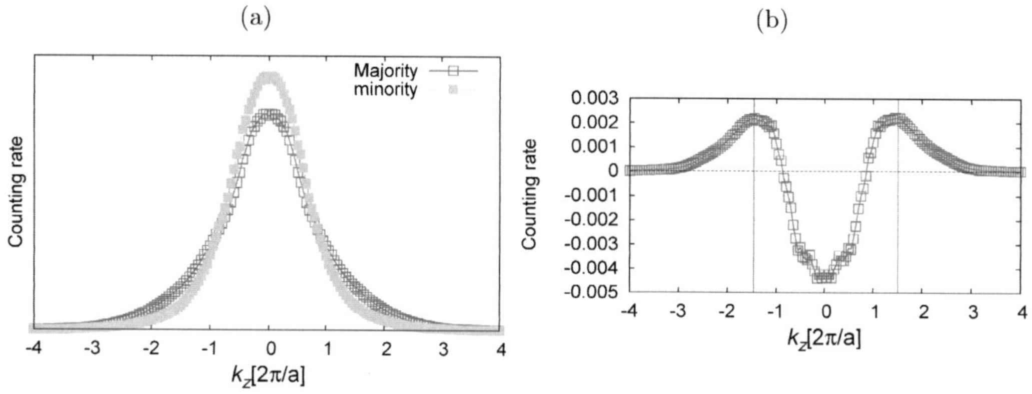

Figure 3. One dimensional momentum densities of the majority spin and the minority spin (a), and difference between majority spin and minority spin (b).

in both cases of the majority and minority spins.

We next analyze the anisotropy by using the group theory. The band structures with the irreducible representations are shown in Figure 2. In the case of the majority (minority) spin, we find three (two) totally symmetric occupied bands on the Σ line, and only one or two (one) totally symmetric occupied bands on the Δ line. Therefore, the Σ line has more totally symmetric occupied bands than the Δ line. This difference in the number of the irreducible representation originates from the fact that the Δ line (C_{4v}) has higher symmetry than the Σ line (C_{2v}). Since only the totally symmetric representation contributes to the momentum density, the momentum density on the Σ line is expected to be larger than that on the Δ line and this expectation is consistent with our calculational results mentioned above.

3.2. One-dimensional momentum densities

Next, we calculate one-dimensional momentum densities which are spatially averaged [3]. We find that the momentum distribution for the majority spin is broader than that for minority spin (Figure 3). This feature is consistent with the result of the experiment of DBAR [3]. When $p = 0$, the sign is minus and we find that the positive peak appears at $p = 1.4$ in the $2\pi/a$ unit in FIG.3 (b).

In a previous study, it was reported that the distribution of the majority spin in the real space is narrower than that of the minority spin [6]. This difference in the real-space distribution is expected to be the origin of the difference in the momentum density distribution.

4. Conclusion

In this study, we perform calculations of the spin-polarized momentum densities of Fe using the DFT. We first find the prominent anisotropy of the momentum densities on the (001) plane; the ridge and valley appear on the Σ and Δ lines, respectively. This anisotropy originates from the fact that the symmetry of the Σ line (C_{2v}) is lower than that of the Δ line (C_{4v}). We also find that the one-dimensional momentum densities of the majority spin is broader than that of the minority spin. The difference in the distributions of the momentum densities is expected to originate from the fact that the distribution in space of the majority spin density is narrower than that of the minority spin density.

References

[1] Saito M, Oshiyama A and Tanigawa S 1991 *Phys. Rev.* B **44** 10601
[2] Rao D V *et al* 2004 *J. Phys. Chem. Ref. Data* **33**, 627
[3] Kawasuso A *et al* 2011 *Phys. Rev.* B **83** 100406
[4] Kawasuso A *et al* 2012 *Phys. Rev.* B **85** 024417
[5] Mijnarends P E, 1973 *Physica* **63**, 235
[6] Lin J, Yamasaki T and Saito M 2014 *Jpn. J. Appl. Phys.* **53**, 053002
[7] Genoud P *et al* 1988 *J. Phys. F: Met. Phys.* **18**, 1933

IOP Conf. Series: Journal of Physics: Conf. Series **791** (2017) 012038 doi:10.1088/1742-6596/791/1/012038

Development of a commercial positron annihilation lifetime measurement system

M Yamawaki[1], **K Ito**[1], **K Hattori**[2] **and N Uesugi**[2]

[1]National Institute of Advanced Industrial Science and Technology, Higashi, Tsukuba, Ibaraki 305–8565, Japan
[2]TOYO SEIKO Co., Ltd., Yatomi, Aichi 490–1412, Japan

E-mail: yamawaki.masato@aist.go.jp

Abstract. In order to realize a commercial system with a user-friendly interface for positron annihilation lifetime (PAL) measurements, we have applied our previously developed anti-coincidence method to a compact system, controlled by dedicated software with a data analysis module. The functionality of the data-analysis code was confirmed by examining the reproducibility of input average lifetimes for calculated PAL histograms. A prototype for the commercial system was constructed and the validity of the analysis using the system was ensured by measuring a reference material.

1. Introduction

The positron annihilation lifetime (PAL) technique is known as a powerful method to detect open spaces, such as lattice defects and free-volume holes in materials. However, for non-experts of this technique there is some difficulty in adjusting to practical applications due to the complicated data recording and analysis process with a conventional system for bulk-sample measurements. In this work, in order to achieve a commercial digital PAL measurement system with a user-friendly interface, we have applied the anti-coincidence (AC) method [1–3] to a compact measurement system, controlled by dedicated software with a data analysis module.

2. Technical elements for the PAL measurement system

2.1. Anti-coincidence (AC) method

Previously our group developed the AC method for PAL measurements [1–3], in order to overcome the disadvantage of the conventional "sandwich" method [4, 5] using a Na-22 radioisotope as a positron emitter, which requires two specimens so that the positron source is sandwiched by these specimens. A block diagram of PAL measurement based on the AC method is shown in Figure. 1(a). In this technique, a Na-22 positron source with an activity of 1 MBq is placed on a plastic scintillator (Pilot-U) combined with a photomultiplier tube (PMT; R7400U, HAMAMATSU JPN) so as to construct a positron detector (Figure. 1(b)). The positron source, sealed with 7.5-μm-thick Kapton films, was provided by the Japan Radioisotope Association. The lifetime of a single positron is determined from the time interval ΔT between the start signal due to the emission of near-simultaneous γ-ray from the Na-22 nucleus and the corresponding stop signal due to its annihilation in the sample. A lifetime histogram for the sample is obtained by accumulating the annihilation events

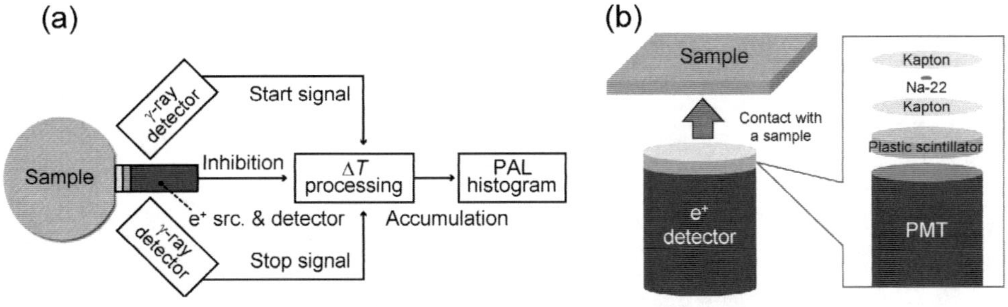

Figure 1. (a) Block diagram of the AC method. (b) Positron source and detector.

with adequate counting-statistics. Annihilation events in the positron detector, corresponding to events where the positron is not annihilated in the sample, are rejected through the AC method. During actual measurements the positron source side of the positron detector is brought into contact with the sample surface so that emitted positrons are introduced into the sample, meaning the AC method enables us to analyze a single specimen without to cut out or prepare a pair of samples from the object under study.

2.2. Dedicated software for the measurement system

Dedicated software was developed for a PAL measurement system equipped with a digital storage oscilloscope (DSO; HDO4024, Teledyne LeCroy). Up to 200 signals from the γ-ray and positron detectors are stored in the DSO memory with a sampling rate of 2 GS/s and those are processed to obtain ΔT, followed by the accumulation of a PAL histogram. To determine the detected times of the start and stop signals from the two γ-ray detectors, consisting of BaF_2 scintillator with truncated conical shape of ϕ30 mm × ϕ40 mm × h40 mm and PMT (H3378-51, HAMAMATSU Japan), the constant fraction [6] was fixed to 30 % of the signal height, and the energy windows for the signals were assigned to the corresponding photo peaks of 1.27 MeV and 0.511 MeV, respectively (Figure. 2). In order to set the upper- and lower-level discriminators (ULD and LLD) for both the windows, the pulse-height histograms of the signals are accumulated, followed by fixing the trigger level for LLD and the maximum level for ULD of the amplitude ranges of the DSO. Prior to

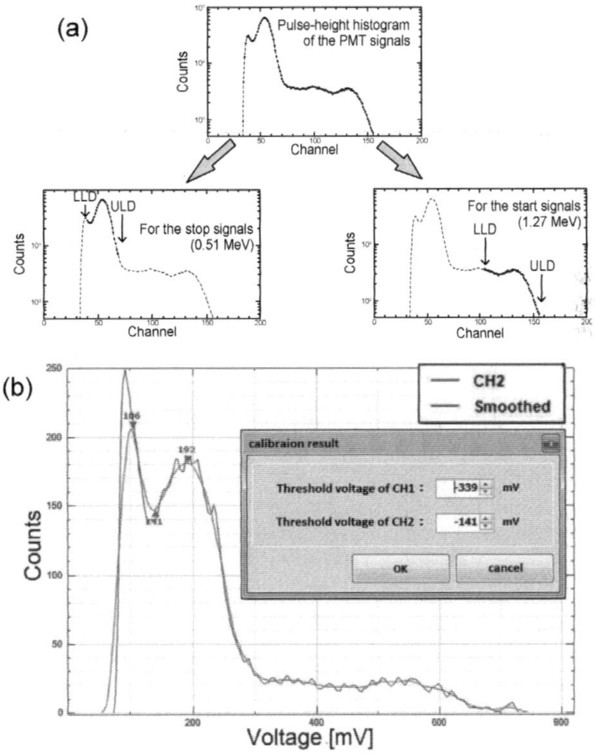

Figure 2. (a) Assignment of lower-level discriminator (LLD) and upper-level discriminator (ULD) for the detected γ-ray signals. (b) Software dialog for adjusting LLD and ULD.

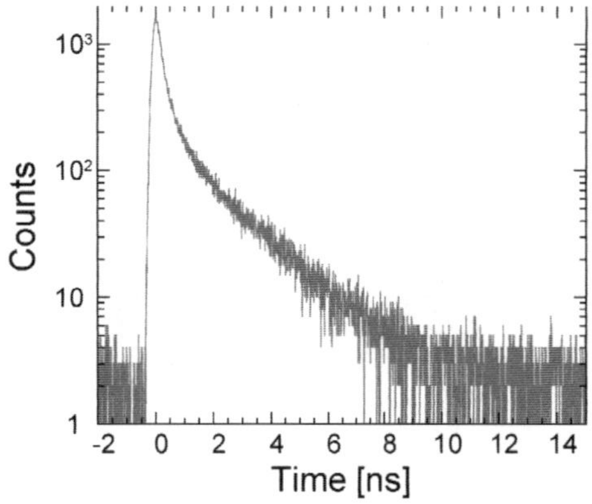

Figure 3. Typical calculated PAL histogram.

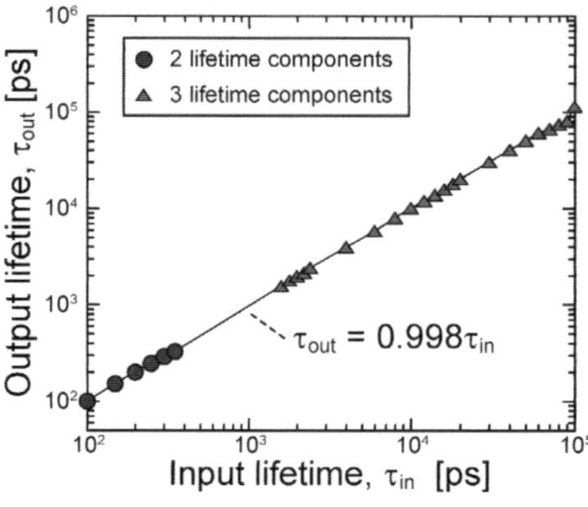

Figure 4. Relationship between the input average lifetime and the output average lifetime.

starting measurements, the software can automatically adjust these amplitude ranges.

The software also has a data analysis module, which can evaluate mean positron lifetimes through a least-square fit to the accumulated PAL histogram. In order to ensure the validity of the developed code for the fitting, the reproducibility of input positron lifetimes was examined by analyzing lifetime histograms, mathematically generated with a variation of average positron lifetimes, which simulate a data set for various materials such as metals, alloys, semiconductors, polymers and porous oxides. In the calculations for two kinds of the lifetime histograms with 2 and 3 lifetime components (see Table 1), the time resolution in FWHM, the background level with the Gaussian error function due to random γ-rays coincidence, and the total counting statistics were fixed to 150 ps, 2 counts/ch, and 100 k counts, respectively. Here, spectra with low counting statistics were examined in order to confirm that the analysis program is apposite for onsite short-time measurements. A typical calculated histogram is shown in Figure. 3. Figure 4 shows the relationship between the input average lifetime and the output one analyzed by the developed code. As seen in the figure, a good linear correlation is obtained throughout the whole time range of the input lifetimes from 100 ps to 100 ns, indicating that the analysis code can properly evaluate positron and positronium lifetimes for practical materials.

Table 1. Input average lifetime and intensity for calculated PAL histograms.

Number of components	τ_1 [ps]	I_1 [%]	τ_2 [ps]	I_2 [%]	τ_3 [ns]	I_3 [%]
2	100–350	85	380	15	N/A	N/A
3	150	30	400	30	1.5–100.0	40

3. Prototype for the commercial measurement system

Figure 5 shows the analysis chamber, equipped with the detectors and a high voltage bias, for the prototype of the commercial measurement system. The geometry of the detectors is comparable to that reported previously [1]. We confirmed that the system can measure PAL spectra with a counting rate of ~75 cps and a time resolution of ~220 ps. Figure 6 displays a PAL histogram obtained for fused silica reference material (NMIJ CRM 5601a) at room temperature. The obtained average lifetime for *ortho*-positronium (*o*-Ps) was (1.605 ± 0.007) ns, in agreement with the certified value of CRM, (1.62 ± 0.05) ns, indicating the validity of the measurement system. The geometry of the detectors is now under optimization so that the target values for counting rate (100 cps) and time resolution (~180 ps) are attained.

Figure 5. Analysis chamber of the prototype system (HWD: 430 mm x 400 mm x 400 mm). The chamber contains γ-ray and positron detectors with a Na-22 positron source of 1 MBq.

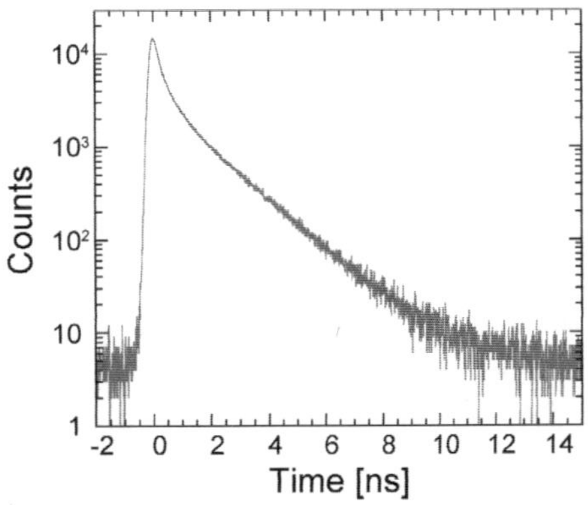

Figure 6. PAL data observed for fused silica reference material.

4. Summary

We applied the AC method to a compact PAL measurement system, controlled by newly developed software with a data analysis module, in order to achieve a commercial digital measurement system with a user-friendly interface. The software can automatically adjust the amplitude ranges of the equipped DSO to set properly the energy windows for the γ-ray detectors. Testing the calculated lifetime histograms using the developed program signified that the analysis code is able to evaluate properly positron and positronium lifetimes for practical materials. A prototype for the commercial system was constructed, and it was confirmed that the average *o*-Ps lifetime for fused silica reference material observed by the prototype system agrees well with the certified value of the reference material.

Acknowledgements

This research was supported in part by a grant from Chubu Bureau of Economy, Trade and Industry METI. We gratefully acknowledge helpful discussions with Dr. Yoshinori Kobayashi, Dr. Nagayasu Oshima and Dr. Brian O'Rourke of AIST.

References

[1] Yamawaki M, Kobayashi Y, Hattori K and Watanabe Y 2011 *Jpn. J. App. Phys.* **50** 086301
[2] Yamawaki M, Kobayashi Y, Ito K, Matsumoto M, Ishizu H, Umino A, Hattori K and Watanabe Y 2013 *Mater. Sci. Forum* **733** 310
[3] Patent No. 5843315(JP) and Patent No. 8785875(US).
[4] Shirai Y and Araki H 2005 Japan Patent No. 3671223.
[5] Shirai Y, Sakamura M, Shishido I and Yamaguchi M 1995 *J. Japan Inst. Metals* **59** 679
[6] Saito H, Nagashima Y, Kurihara T and Hyodo T 2002 *Nucl. Instrum. Methods Phys. Res., Sect. A* **487** 612